Flexible Incentives for the Adoption of Environmental Technologies in Agriculture

NATURAL RESOURCE MANAGEMENT AND POLICY

Editors:

Ariel Dinar
Agricultural and Natural Resources Dept.
The World Bank
1818 H Street, NW
Washington, DC 20433

David Zilberman
Dept. of Agricultural and
Resource Economics
Univ. of California, Berkeley
Berkeley, CA 94720

EDITORIAL STATEMENT

There is a growing awareness to the role that natural resources such as water, land, forests and environmental amenities play in our lives. There are many competing uses for natural resources, and society is challenged to manage them for improving social well being. Furthermore, there may be dire consequences to natural resources mismanagement. Renewable resources such as water, land and the environment are linked, and decisions made with regard to one may affect the others. Policy and management of natural resources now require interdisciplinary approach including natural and social sciences to correctly address our society preferences.

This series provides a collection of works containing most recent findings on economics, management and policy of renewable biological resources such as water, land, crop protection, sustainable agriculture, technology, and environmental health. It incorporates modern thinking and techniques of economics and management. Books in this series will incorporate knowledge and models of natural phenomena with economics and managerial decision frameworks to assess alternative options for managing natural resources and environment.

Environmental effects of agricultural activities have been in the center of public debate and government intervention for some time. Top-down regulations have proven to be less efficient than expected and it seems as environment and agriculture couldn't operate side by side. This book focuses on identifying flexible and sustainable regulations that provide economical incentives for improving environmental quality, while allowing agricultural development. Using both a conceptual framework and empirical-based examples, the book will attract a variety of readers interested in environmental regulation in agriculture

The Series Editors

Recently Published Books in the Series

Bauer, Carl J.:
Against the Current: Privatization, Water Markets, and the State in Chile
Easter, K. William, Rosegrant, Mark W., and Dinar, Ariel:
Markets for Water: Potential and Performance
Smale, Melinda:
Farmers, Gene Banks, and Crop Breeding: Economic Analyses of Diversity in Wheat, Maize, and Rice

Flexible Incentives for the Adoption of Environmental Technologies in Agriculture

EDITED BY

Frank Casey
Andrew Schmitz
Scott Swinton
David Zilberman

Kluwer Academic Publishers

Distributors for North, Central and South America:
Kluwer Academic Publishers
101 Philip Drive
Assinippi Park
Norwell, Massachusetts 02061 USA
Telephone (781) 871-6600
Fax (781) 871-6528
E-Mail <kluwer@wkap.com>

Distributors for all other countries:
Kluwer Academic Publishers Group
Distribution Centre
Post Office Box 322
3300 AH Dordrecht, THE NETHERLANDS
Telephone 31 78 6392 392
Fax 31 78 6546 474
E-Mail <services@wkap.nl>

 Electronic Services <http://www.wkap.nl>

Library of Congress Cataloging-in-Publication Data

Flexible incentives for the adoption of environmental technologies
  in agriculture / edited by Frank Casey ... [et al.].
     p. cm. -- (Natural resource management and policy)
  Papers derived from a symposium held at the University of Florida,
Gainesville, June 8-9, 1997.
  Includes bibliographical references and index.
  ISBN 0-7923-8559-4 (alk. paper)
  1. Agricultural innovations--Environmental aspects.
2. Agricultural innovations--Economic aspects. 3. Agriculture and
state. I. Casey, Frank. II. Series
S494.5.I5F59   1999
363.73'7--dc21   99-23739
                                                     CIP

Printed on acid-free paper.

Printed in the United States of America

Contents

Acknowledgements

The editors thank all who attended and participated in the symposium entitled "Flexible Incentives for the Adoption of Environmental Technologies in Agriculture," from which this book is derived. The symposium was held at the University of Florida, Gainesville, from June 8–9, 1997. Special thanks go to the authors of the chapters in this book who put forth a great effort in producing a thoughtful and innovative set of papers. We deeply appreciate the thoughtful contributions of the discussants whose ideas improved the papers as they matured into book chapters; the discussants included Terry Francl of the American Farm Bureau Federation, Jack Laurie, President of the Michigan Farm Bureau, and Gary Fairchild and Ken Tefertiller at the University of Florida.

A grateful acknowledgement is extended to the various agencies and institutions that provided the financial support to organize the symposium and to prepare this book for publication. Included as financial supporters are the Resource Economics Division of the Economic Research Service at the U.S. Department of Agriculture, The Ben Hill Griffin, Jr., Endowed Chair in the Department of Food and Resource Economics at the University of Florida, The Elton R. Smith Endowed Chair in Food and Agricultural Policy at Michigan State University, The Department of Food and Resource Economics at the University of Florida, the Department of Agricultural Economics at Michigan State University, the Department of Agricultural and Resource Economics at the University of California, Berkeley, the Florida Farm Bureau, Northwest Economic Associates, Henry A. Wallace Institute for Alternative Agriculture, and the Institute of Food and Agricultural Sciences at the University of Florida.

A special thank you is bestowed upon H. Carole Schmitz for her invaluable assistance in organizing the symposium, formatting the book and carrying out its copy-editing. This book would not have been possible without her participation. Tom Stevens provided invaluable computer and editing skills, and Kim Box contributed her consultation services regarding the presentation of the book materials.

About the Contributors

Sandra S. Batie, Elton R. Smith Professor in Food and Agricultural Policy, Department of Agricultural Economics, Michigan State University, East Lansing, Michigan.

Jeffrey R. Blend, Tax Policy and Research Economist, Montana Department of Revenue, Helena, Montana.

Darrell J. Bosch, Associate Professor, Department of Agricultural and Applied Economics, Virginia Polytechnic Institute and State University, Blacksburg, Virginia.

William P. Browne, Professor of Political Science, Central Michigan University, Mount Pleasant, Michigan.

C. Line Carpentier, Formerly with the International Food Policy Institute, now is a Policy Analyst, Henry A. Wallace Institute, Greenbelt, Maryland.

Frank Casey, Economist, Northwest Economic Associates, Vancouver, Washington.

Margriet F. Caswell, Chief, Production Management and Technology Branch, Economic Research Service, U.S. Department of Agriculture, Washington, DC.

Mei-chin Chu, Associate Researcher, Taiwan Institute of Economic Research, Taipei, Taiwan, Republic of China.

M.S. Deepak, Economic Analyst, Europe, Africa and Middle East Branch, Economic Research Service, U.S. Department of Agriculture, Washington, DC.

David E. Ervin, Director of the Policy Studies Program for the Henry A. Wallace Institute, Greenbelt, Maryland.

Ray Huffaker, Associate Professor, Department of Agricultural Economics, Washington State University, Pullman, Washington.

Madhu Khanna, Assistant Professor, Department of Agricultural and Consumer Economics, University of Illinois, Urbana, Illinois

Donna J. Lee, Assistant Professor, Department of Food and Resource Economics, University of Florida, Gainesville, Florida.

Stina Levin, Ph.D. Graduate, Department of Agricultural Economics, Washington State University, Pullman, Washington.

Stephen B. Lovejoy, Professor, Agricultural and Environmental Policy, Purdue University, Lafayette, Indiana.

Gary D. Lynne, Professor and Department Chair, Department of Agricultural Economics, University of Nebraska, Lincoln, Nebraska.

Martin B. Main, Assistant Professor of Wildlife Ecology and Conservation, Southwest Florida Research and Education Center, Institute of Agricultural Sciences, University of Florida, Immokalee, Florida.

Katti Millock, Post Doctoral Research Fellow, Institute of Economics, University of Copenhagen, Denmark.

J. Walter Milon, Professor, Department of Food and Resource Economics, University of Florida, Gainesville, Florida.

Patricia E. Norris, Associate Professor, Department of Agricultural Economics and the Department of Resource Development, Michigan State University, East Lansing, Michigan.

Douglas D. Parker, Assistant Professor and Extension Economist, Department of Agricultural and Resource Economics, University of Maryland, College Park, Maryland.

A. Clayton W. Ogg, U.S. Environmental Protection Agency, Washington, DC.

Nicole N. Owens, Economist, U.S. Environmental Protection Agency, Washington, DC.

Leo Polopolus, Professor Emeritus, Department of Food and Resource Economics, University of Florida, Gainesville, Florida.

Alan Randall, Professor and Department Chair, The Ohio State University, Department of Agricultural, Environmental and Development Economics, Columbus, Ohio.

Mark Ribaudo, Agricultural Economist, Resource and Environmental Policy Branch, Economic Research Service, U.S. Department of Agriculture.

Fritz M. Roka, Assistant Professor of Agricultural Economics, Southwest Florida Research and Education Center, Institute of Agricultural Sciences, University of Florida, Immokalee, Florida.

Andrew Schmitz, Eminent Scholar, Ben Hill Griffin, Jr., Endowed Chair, Department of Food and Resource Economics, University of Florida, Gainesville, Florida.

Kathleen Segerson, Professor, Department of Economics, University of Connecticut, Storrs, Connecticut.

Thomas H. Spreen, Professor, Department of Agricultural Economics, University of Florida, Gainesville, Florida.

Scott M. Swinton, Associate Professor, Agricultural Economics, Michigan State University, East Lansing, Michigan.

Amy P. Thurow, Assistant Professor, Department of Agricultural Economics, Texas A&M University, College Station, Texas.

Eileen O. van Ravenswaay, Professor, Agricultural Economics, Michigan State University, East Lansing, Michigan.

John J. VanSickle, Professor, Department of Agricultural Economics, University of Florida, Gainesville, Florida.

David Zilberman, Professor and Department Chair, Department of Agricultural and Resource Economics, University of California, Berkeley, California.

Preface

Ken Tefertiller and Frank Casey
University of Florida, Gainesville, FL
Northwest Economic Associates, Vancouver, WA

The environmental regulation of agriculture is a major issue in the United States. One only needs to read a local newspaper to find that there are several articles dealing with agricultural runoff, pesticide use, or compliance with the Endangered Species Act. For the agricultural sector, the regulatory environment has become more complex in the past few years, and has resulted in higher compliance costs and disputes with regard to property rights. Although environmental regulations that promote improved water quality, higher standards of food safety, and the protection of endangered species are necessary, implementation problems have occurred. These problems include the definition of appropriate indicators of compliance, the measurement of these indicators, the costs of compliance, and non-funded environmental mandates that are directed to state and local governments.

Environmental regulations are based on the perception and/or reality that negative social externalities result from agricultural production processes. Generally, these regulations have a greater impact on agriculture than they do on other industries because agricultural production processes are based on the extensive use of natural resources (water, soil and air). Of course, it is the combination of these same resources that create agricultural products.

The cost of regulation varies by production input and by commodity group, but in some cases the private cost of compliance may exceed its benefit to society. Furthermore, the impact of a specific regulation may not adequately reflect the cumulative impact of the total regulatory environment. Farmers must often work with a number of separate agencies to address the same problem, and often it can result in duplication of their time and effort. Generally, conflicts between agricultural producers and regulators have become increasingly serious in the more urban states in which land and water resources—both in terms of quantity and quality—have become increasingly scarce.

Steps are being taken to resolve these problems. Recently, federal and state governments have taken major efforts to decrease the number of new regulations and to make existing regulations more flexible. For example, in February of 1995 the President announced a program of "Reinventing Regulation." This program mandated federal agencies to cut obsolete regulations, to reward results instead of red tape, to create grassroots partnerships for solving local problems, and to negotiate solutions instead of dictating them.

The theme of this book is how to identify and structure more flexible economic incentives for the achievement of environmental goals in agriculture. It provides a conceptual framework and presents case studies that analyze how flexible incentives can address environmental problems that are caused by agricultural production. A group of papers prepared for the symposium "Flexible Incentives for the Adoption of Environmental Technologies in Agriculture" that was held in Gainesville, Florida, from June 8–10, 1997, became the chapters in this book. The symposium brought together economists, agency personnel and political economists for the purpose of exploring how new cutting-edge economic tools could be developed and applied to environmental problems. The goal of the symposium was to complement and to expand the economic theory of environmental regulation and technology adoption with new research findings.

The key theme of this book is the important role technology takes when addressing environmental problems. New technologies and technological development are broadly defined to include economic instruments (such as new contract designs), innovative ways to communicate environmental information, new economic institutions, and education. Throughout the book keep in mind the importance of the institutional structure in which incentives for technology adoption take place and the importance of property rights and laws that govern resource use.

We stand at the threshold of an era in which environmental legislation that affects agriculture is now being extended from point-source regulation to nonpoint-source regulation. We have the opportunity to benefit from the experience of other industries. We also benefit from a history of research and outreach in agriculture that provides us with more knowledge than we had in the first generation of environmental policy. The knowledge gained from the effects of past environmental policies, economics, agricultural sciences, and pioneering experiments (some of which are reported in this book) offer us the opportunity to improve the second generation of environmental policy. Based on the principle of flexibility, the new generation of environmental policy (as discussed in this book) aims to achieve social environmental goals with minimal disruption to agriculture.

This book is designed for public and private policymakers, government analysts, teachers, researchers and students who specialize in the fields of natural resources, agricultural economics and environmental regulation. It complements theoretical textbooks in these subject areas by presenting an array of case studies that examine different economic instruments for the promotion of the adoption of environmental technologies in agriculture. It provides a fresh perspective on what types of incentives may be used to lead us to the desired environmental outcomes and offers new ideas about the types of economic instruments that may achieve these outcomes.

1 INTRODUCTION

David Zilberman
University of California, Berkeley, CA

The evolution of agriculture has been strongly influenced by government policies, in spite of the popular view of agriculture as a classic case of free competition. Land and resource policies were dominant in shaping the expansion of U.S. agriculture in the 18th and 19th centuries (Cochrane, 1993). Publicly financed research and extension played a crucial role in the intensification and productivity growth of U.S. agriculture from the 1860s on. Since the Great Depression, commodity programs and support policies have enabled the farm sector to withstand its tendency to oversupply. Since the 1970s, the environmental side effects of agriculture have become a major focus of government policies, and that is likely to intensify in the new millennium.

This book provides a perspective on the design and implementation of environmental policies in agriculture. It suggests a transition from the command-and-control policy emphasis of the past toward policies that recognize heterogeneity among producers and ecosystems, and enables diversified behavioral responses to achieve environmental quality objectives.

While many of the chapters in this book are based on the premise that applied welfare economics provides a broad set of tools to design and apply policy reform in agriculture, the book as a whole takes an interdisciplinary approach. It emphasizes that policy changes are very closely linked to technological and attitudinal changes. It also emphasizes that policy modeling is futile without the understanding of preferences and technologies. It combines economic principles with basic findings and key features of models in agricultural and the environmental sciences. It emphasizes that environmental policymaking in agriculture is constrained by legal considerations and political feasibility. It has to be viewed within the context of the evolution of U.S. agricultural policy.

The new emphasis on the environmental side effects of agriculture is part of the more general trend toward environmentalism. The impacts of other major trends are woven throughout the text. For example, the notion of flexible policies (defined in Batie and Ervin and in Segerson) and the transfer of responsibilities from federal to state governments (Lovejoy; Ogg; Ribaudo and Caswell) are manifestations of the general trends of privatization and devolution (Khanna, Millock and Zilberman).[1] Globalization and the concern for global environmental issues lead to regulations of chemicals, such as methyl bromide, that affect competitiveness (Deepak, Spreen, and Van Sickle). Consumerism and the growing effect of consumer beliefs of product design and production practices may improve the profitability of "greener" agricultural products and will become an important component of flexible policies (van Ravenswaay and Blend; Swinton, Chu, and Batie). The major themes in this book, and the chapters that address these themes are presented below.

Agricultural production systems are heterogeneous in terms of productivity and environmental side effects. Heterogeneity suggests that optimal technological solutions vary across locations. Thus, incentives for environmental quality enhancement should likewise vary across locations because the value of environmental preservation may be more significant in some areas than in others. This theme is emphasized in the general theoretical discussion (Batie and Ervin; Segerson; and Khanna et al.), the dairy case study (Carpentier and Bosch), the livestock case study (Norris and Purvis), and in the discussion of current environmental policy approaches (Ribaudo and Caswell; Ogg; Lovejoy).

Existing or nearly developed technologies may improve productivity while significantly reducing environmental side effects. There are some technologies that can (1) adjust input applications to variations in ecological conditions, (2) have high accuracy in input application, and (3) reduce residues that may harm the environment. Khanna et al. refer to these technologies generically as precision technologies. Other chapters present specific examples. Casey and Lynne analyze the adoption of water conservation technologies, and Parker and Caswell present the economics of the use of an additive, polyacrylamide polymers, which slows chemical transport on soil particles. Carpentier and Bosch provide several examples from dairy management, and Norris and Thurow give examples of other livestock technologies.

Some consumers are willing to support environmentally friendly technologies. Evidence of such behavior is presented in a discussion on ecolabeling in van Ravenswaay and Blend. This support may be expressed in a willingness to pay more for products produced with environmentally friendly technologies and willingness to contribute efforts to encourage such technologies. The reasons for the support of environmentally friendly products may include concern for personal and environmental health, concern for animal rights, and preferences for natural products.

Farmers and other producers are motivated to adopt environmentally friendly technologies for a multitude of reasons. These reasons may include profitability, concern about personal and environmental health, desire to project an environmentally friendly image, and willingness to adhere to social norms as a means to contribute to community-based efforts for improving the environment. Swinton, Owens and van Ravenswaay report on corn producers' willingness to pay for reduced risks from herbicide use. Casey and Lynne provide evidence of the impact of social norms on the adoption of water conserving technologies.

New technologies are induced by economic incentives, and research efforts to introduce "green" technologies will intensify with better incentives. The private sector will make an effort to develop technologies that address environmental problems if policy changes either reduce the profitability of existing technologies or if producers are subsidized to adopt new technologies. The importance of incentives in inducing new technology development is further elaborated in Batie and Ervin, Segerson, and in Swinton and Casey. Khanna et al. suggest that private companies may invest in developing new monitoring equipment if there is an increased demand by policymakers or farmers to improve residue monitoring.

The evolving integration of the food supply chain offers new opportunities for private sector involvement in fostering environmental stewardship. Consumer interest in the certification of environmentally sound food and fiber goods production is increasing (van Ravenwaay and Blend). As the supply chain from agricultural input supplier to farmer to processor to wholesaler to retailer becomes more integrated through electronic information flows, it becomes easier to communicate those consumer desires back up the supply chain. Contracts are providing many of the links in the new food and fiber supply chain. With growing consumer demand for environmental attributes, contracts can carry private sector incentives for the adoption of environmental technologies, as Swinton, Chu and Batie illustrate for seed corn production.

Private sector research efforts for the development of environmentally sound technologies will be sub-optimal even if there are significant incentives for the adoption of environmentally friendly technologies. The private sector is interested in environmental technologies that are embodied in new products that can be patented. They may not invest in developing environmental technologies that are not embodied in new products. As a result, there is a significant role for public investment and development activities that will lead to disembodied innovations. Randall provides a philosophical justification for public sector activities. Lovejoy, and Batie and Ervin emphasize the role of public policy when augmenting public sector activities for the production of environmentally sound activities.

Policymakers and the public are interested in simple, transparent policy solutions to complex problems. Browne argues that political feasibility and transparencies are key criteria to assess new environmental regulation policies. Batie and Ervin emphasize the importance of the simplicity of policy designs. These criteria may eliminate some solutions that require a high degree of complexity and sophistication and that are difficult to communicate and understand. Batie and Ervin, and Browne suggest that policy development should not only emphasize economic efficiency, but it should also address issues of public education and the ability to sell solutions to a wide range of individuals with different interests and backgrounds.

Environmental policy objectives should be pursued with a portfolio of incentives that combine penalties, awards, regulations and education. The theory chapters (Batie and Ervin; Segerson; Khanna et al.) emphasize that there are multiple avenues used to obtain policy change that have to take into account efficiency, equity and implementation considerations. Behavior can be modified by incentives, by direct control, or by education. Education and community programs will internalize environmental values and lead to improved social norms that guide environmental behavior. Direct controls (prescribed behavior and technology choices) may be appropriate in situations of much uncertainty and heterogeneity. To be optimal, direct control has to be adjusted across locations and situations. Incentives include subsidies, taxes and the introduction of institutions, such as transferable rights. Ribaudo and Caswell, and Ogg demonstrate how existing federal policies are gradually shifting from direct control toward incentives to adopt more environmentally sensitive practices. Instead of requiring farmers to adopt best management practices, they are providing them with monetary or in-kind support.

Environmental policy intervention may combine measures to reduce pollution generation, exposure and vulnerability to environmental damages. Batie and Ervin, and Khanna et al. suggest that diverse policy tools to control environmental and human health risk from agricultural activities reflect the many dimensions of the health risk generation process. Scientific knowledge and understanding of the processes of human and environmental health risk generation are essential for effective policies. It may be cost effective to combine measures of pollution reduction with policies that protect vulnerable populations (worker safety and clean water policies) through reduced exposures. Swinton, Owens and van Ravenswaay demonstrate the importance of information about human health effects of water quality and the availability of protection technologies for the establishment of effective water quality control policies.

Reliance on public and private willingness to pay for environmental amenities should play a major role in policy design. Three chapters (van Ravenswaay and Blend; Swinton, Chu and Batie; Roka and Main) emphasize the private sector's willingness to pay for environmental amenities as indicators of societal commitment for environmental preservation. Preservation of valuable wildlife habitat may be best achieved by establishing funds to finance acquisition (Roka and Main). Some of these funds could be public but may have mechanisms to accept private contributions, thus revealing willingness to pay for environmental amenities. Policy initiatives should also include institutional support for "green" markets. At the retail level, new grades and standards that support invisible environmental attributes can tap demand that currently may not be manifested (van Ravenswaay and Blend). At the wholesale and processing level, such standards create incentives for processors that enable them to contract with agricultural producers for environmentally assured stewardship (Swinton, Chu and Batie).

Policy design has to take into account the ability to monitor and enforce. The inability to observe pollution patterns may prevent implementation of first-best solutions (such as pollution taxes and trading in pollution permits). Segerson, and Khanna et al. emphasize that policy selection reflects the limitations of the existing information structure and the ability to monitor and enforce. There should be increased emphasis on research and development to improve monitoring capacity. As information gathering capacity and availability improve, so will pollution control policies. With more information, many nonpoint-source problems will become point-source problems. Randall and Lovejoy argue that delegating power to establish and enforce environmental regulations to states and counties will reduce information costs and boost environmental benefits.

Policy design has to take distributional consequences and existing policies into account. Browne emphasizes that environmental policymaking is part of a much greater political process of negotiation between parties. This process involves give and take and reflects the powers of different groups of populations, their commitment and their preferences with regard to environmental issues. Policy impact analysis, therefore, must consider that prospective gainers and losers fully recognize the political implication of the distributional changes. Ribaudo and Caswell, and Ogg suggest that there are times during which environmental objectives cannot be pursued separately but may be incorporated into other agricultural policies. Huffaker and Levin emphasize that any new environmental regulations must fit into the existing legal and regulatory framework. Consequently, it may be more efficient to modify existing policies rather than to overhaul them.

Initiatives to reintegrate livestock and crop production systems should be encouraged. Norris and Thurow, and Carpentier and Bosch demonstrate that animal waste has become a major source of environmental degradation. As U.S. agriculture evolved to separate animal and crop production, livestock farmers struggled to dispose of waste products, and crop farmers spent increased sums of money on the purchase of synthetic chemicals. Little incentive currently exists to introduce animal waste recycling and processing for fertilizer use, which could reduce damages from excessive concentrations of manure nutrients and microbial risks. Norris and Thurow suggest that the environmental impact of industrialization in the livestock sector should be scrutinized in order to develop incentives that encourage a more environmentally friendly, yet modern and efficient, livestock sector.

Environmental policies should be scrutinized to avoid situations in which they are misused as mechanisms to prevent competition. Schmitz et al. (1995) argued that producers with monopoly power could possibly capture regulators through the process of regulation (Stigler, 1971). In this volume, Lovejoy argues that environmental regulation must be used as a vehicle to transfer income to farmers, to limit international trade, and/or to reduce competition in the market. Deepak et al., and Schmitz and Polopolus, also in this volume, provide additional supporting evidence. An important challenge for policy analysis is to develop the capacity to identify and address such situations.

Environmental policy design should recognize that policy failure could endanger attempts to correct market failure. The U.S. Department of Agriculture (Ribaudo and Caswell) and the U.S. Environmental Protection Agency (Ogg) have added increasing flexibility and local involvement to existing environmental policies. Lovejoy, however, argues that existing policies that regulate environmental effects of agriculture are far from optimal. Randall, Batie and Ervin, and Khanna et al. argue that the challenge in the development of environmental policies for agriculture is to find the true balance between excess in free markets and heavy-handedness in government.

The themes discussed above and evoked in the chapters ahead jointly highlight the importance of flexible incentive design to meet the needs of different natural site characteristics, technologies and human preferences. A variety of mechanisms can be utilized to promote adoption of environmental technologies in agriculture—but there is no "one size fits all" approach.

This book is divided into five parts. The Overview describes the current regulatory and institutional framework. It serves as a launching pad for the presentation of policy improvements and of reform in the rest of the book. Three chapters present the conceptual and theoretical foundations for the development of effective flexible environmental policies in agriculture. These are presented in Part II of the book. This foundation is based on the evolving literature in environmental and resource economics. The case studies in Part III make up the majority of the book. They present an array of conceptual approaches with which to analyze the incentives for the adoption of environmental technologies in agriculture. Some of the broad legal, philosophical and political considerations that have to be integrated when enacting policy change are presented in Part IV. Part V builds a bridge from the identification of incentives for the *adoption* of environmental technologies to the creation of incentives for the *innovation* of environmental technologies to serve agriculture of the future. Continued innovation will depend not only on economic incentives, but also on the institutions that affect innovative behavior.

ENDNOTES

1.   All references without dates refer to chapters in this volume.

REFERENCES

Cochrane, W.W. 1993. *The Development of American Agriculture: A Historical Analysis, Second Edition.* Minneapolis, MN: University of Minnesota Press.

Schmitz, A., W.G. Boggess, and K. Tefertiller. 1995. "Regulations: Evidence from the Florida Dairy Industry." *American Journal of Agricultural Economics* 77: 1166–1171.

Stigler, G.T. 1971. "The Theory of Economic Regulation." *Bell Journal of Economics and Management Science* 2: 3–21

2

ENVIRONMENTAL REGULATION IN AGRICULTURE AND THE ADOPTION OF ENVIRONMENTAL TECHNOLOGY

Marc Ribaudo and Margriet F. Caswell
U.S. Department of Agriculture
Economic Research Service
Washington, DC

Agriculture in the United States has been the subject of numerous government incentive programs over the years. These programs have been designed to achieve a wide variety of goals that include supply control, cropland conversion, soil conservation and environmental quality. A variety of incentive mechanisms have been used to achieve these goals. Incentives for protecting and enhancing water quality come from two different programmatic directions. The U.S. Department of Agriculture (USDA) has a long history of promoting natural resource stewardship through various forms of voluntary assistance (carrots), such as education, technical assistance and cost sharing. Water quality programs that have arisen from the Clean Water Act (CWA) and state water quality protection laws can be of the variety (stick), where farmers are forced through command-and-control or other incentive mechanisms to adopt alternative management practices.

The characteristics of nonpoint-source pollution (NSP) and the lack of specific water quality goals have generally led to technology- or behavior-based policies (subsidies and design standards) that are inherently inflexible. Cost-effective control of pollution is best achieved through performance-based instruments that promote flexibility and allow producers to meet goals using their own specialized knowledge.

INTRODUCTION

Agriculture in the United States has been the subject of numerous government incentive programs over the years. These programs have been designed to achieve a wide variety of goals, including supply control, cropland conversion, soil conservation and environmental quality. A variety of incentive mechanisms have been used to achieve these goals. In this chapter, the incentives—used by federal and state policies and programs to influence farmers when adopting management practices that protect water quality—are reviewed.

Incentives for protecting and enhancing water quality come from two different programmatic directions. The U.S. Department of Agriculture (USDA)[1] has a long history of promoting natural resource stewardship through various forms of voluntary assistance (carrots), such as education, technical assistance and cost sharing. Congressional passage of PL 92-500, the Federal Water Pollution Control Act Amendments of 1972, marked the first time that agriculture was noted as a source of pollution. It gave rise to the possibility that farmers might one day have to alter the management of their land in order to address environmental problems. Water quality programs resulting from this legislation could be of the stick variety, where farmers are forced through command-and-control mechanisms to adopt alternative management practices.

Rather than presenting a theoretical treatment of the efficiency of each policy tool, this chapter will describe which incentives are being used and how each has been effective. (Discussions of the efficiency characteristics of various policy instruments can be found in Batie and Ervin, this volume; Segerson, this volume; Khanna et al., this volume). First, the important characteristics of agricultural pollution that influence the performance of policies are reviewed. Second, the USDA's largely voluntary incentive mechanisms are discussed. Third, mechanisms that arise from federal and state water-quality protection laws are enumerated. Finally, we examine possible future policy decisions.

CHARACTERISTICS OF NONPOINT-SOURCE POLLUTION

Nonpoint-source pollution (NSP) has several important characteristics that influence how incentives for controlling pollution may work. NSP loadings depend in part on random variables—such as wind, rainfall and temperature—thus making it a stochastic process. As a result, a particular policy is likely to produce a distribution of water quality outcomes rather than a single outcome (Braden and Segerson, 1993). This, by itself, does not prevent the attainment of *ex ante* efficiency through the use of standard instruments. It does imply, however, that a policy must be designed to consider moments or points of the distribution other than the mean. For example, nearly all soil erosion occurs during the occurrence of extremely heavy rain. Practices that control erosion from average rainfalls but fail with the occurrence of heavy rain will generally not be effective in protecting water resources from sediment inflows.

Characteristics of agricultural NSP vary over geographic area due to the great variety of farming practices, landforms and hydrologic characteristics that are found across even relatively small areas. An effective policy tool should be flexible enough to work in many different circumstances (Braden and Segerson, 1993).

Some important aspects of agricultural NSP are difficult to measure or to observe. The most problematic characteristic, from a policy standpoint, is the inability to observe emissions. NSP infiltrates water systems over a broad front. Changes in ambient water quality can be observed, and aggregate loadings of agricultural chemicals and sediment can be estimated. The sources of these residuals cannot be pinpointed. In addition, monitoring the movement of NSP emissions is often impractical or prohibitively expensive. The inability to observe emissions would not be such an obstacle if there were strong correlations between emissions and some observable aspect of the production process, or between emissions and ambient quality. If such correlations did exist, a policy could then be directed at the production process or at ambient quality. For example, if agricultural chemicals threaten a shallow aquifer that is entirely superimposed with cropland, then a policy could be targeted to regulate chemical use on that cropland. Such correlations, however, are unlikely to occur, and where relationships can be established, they are unlikely to be the same across a range of conditions. Thus, the regulation of NSP involves moral hazard (Malik et al., 1992). Although the regulatory agency can judge the quality of a body of water through biological and chemical measurements, it cannot determine if the observed state of water quality is caused by the failure of nonpoint-source polluters taking appropriate actions or by undesirable states of nature (for example, high levels of rainfall).

Furthermore, production inputs critical for predicting or forming expectations on NSP may also be unobservable or prohibitively expensive to monitor. For example, there is a close correlation between the chemical contamination of groundwater, the soil type and the amount of chemicals applied to that soil. Chemical characteristics and soil type can be observed, but the amount of a chemical reaching an aquifer also depends on the timing and method of application. These activities are generally not observable to a regulating agency without very costly and intrusive monitoring (Segerson, this volume).

INCENTIVES BY THE U.S. DEPARTMENT OF AGRICULTURE

The USDA uses four basic policy instruments, alone or in combination, to achieve conservation and environmental goals. These include: (1) education and technical assistance, (2) subsidies, (3) research programs, and (4) compliance. Each of these policy instruments is designed to encourage a change in the use of certain agricultural production practices. The USDA often focuses this assistance in a particular geographic area through water quality projects.

Education and Technical Assistance

The use of education as a tool for improving environmental quality is based on the assumption that farmers are not fully aware of either the characteristics of alternative production practices that reduce water pollution or of the environmental quality effects of current production practices. In addition, it is assumed that farmers would adopt the alternative practices, for reasons of economics or stewardship, if they had complete information. Common mechanisms for conveying information to farmers include demonstration projects, technical assistance, newsletters, seminars and field days.

The USDA has a long history of using education and technical assistance to promote improved management practices. The results of studies regarding the role education plays when convincing farmers to change management practices have been mixed. This indicates that education assistance is only one of many factors influencing farmers' decisions. Education has been shown to increase the awareness of soil erosion problems on the farm (Gould et al., 1989). A number of studies, however, have indicated that education is a significant factor in the actual adoption of a practice, only when the practice is also profitable (Nowak, 1987; Napier and Sommers, 1994; Camboni and Napier, 1994).

Farmers do respond to education programs when the quality of their own water supply is at stake (Napier and Brown, 1993). The Farmstead Assessment System Program (FARM*A*SYST Program) is an example of a program that teaches farmers how to assess the risks that certain operations around the farmstead can pose to personal health and to the market value of the farm. This program has been effective in getting individuals to take cost-effective actions to remediate and prevent pollution that results from leaking fuel storage tanks, pesticide spills and poor well maintenance. The success of such programs indicates that farmers will take action when their own potential economic or health risks are high (Knox et al., 1995; Anderson et al, 1995).

On the other hand, experience with other education programs indicates that altruism or concern over off-site environmental quality plays only a small role in farmers' decisions to adopt alternative management practices. Part of the problem is that individual farmers are not likely to have a good understanding of the risks they are posing to off-site water users. Producer surveys have consistently shown that farmers have a poor understanding of the relationship between actions at the farm level and local water quality (Lichtenberg and Lessley, 1992; Hoban and Wimberley, 1992; Pease and Bosch, 1994; Nowak et al., 1997). A shortcoming of many education programs is that they concentrate on educating farmers about the merits of new practices, rather than about the impacts of existing agricultural practices on local water quality. Another problem is that agricultural markets are competitive, so it is unlikely that a farmer would voluntarily adopt costly or risky pollution control measures for altruistic reasons alone—even if he were to understand his impact on water quality, especially when the primary beneficiaries are downstream (Bohm and Russell, 1985; Abler and Shortle, 1991; Nowak et al., 1997).

The effectiveness of education as a water quality tool is also conditional on a program manager's ability to identify the most appropriate practices to encourage. The characteristics of NSP make it difficult to identify these practices *a priori*. Program managers must often make an educated guess that the recommended practices, if adopted, would actually produce the desired water quality improvements.

Subsidies

Subsidies that offer farmers an economic incentive to change practices have been widely applied to agricultural NSP. Through a variety of programs, the USDA has provided millions of dollars in cost-share and incentive payments to encourage farmers to adopt specific practices. The assumption underlying the use of incentive payments is that certain pollution reducing practices or technologies will be profitable to the farmer in the long run, but their adoption is impeded by capital requirements and/or transition costs.

The Agriculture Conservation Program (ACP) and the Water Quality Incentives Program (WQIP) were two USDA programs that offered farmers incentive payments to adopt environmental quality enhancing practices. These practices included conservation tillage, contour farming, irrigation water management, nutrient management, integrated pest management, riparian buffer strips and animal waste handling systems. The Environmental Quality Incentives Program (EQIP) replaced the ACP and the WQIP in 1996. Ogg (this volume) discusses this program. The payments offered through these programs, however, differed greatly from the textbook definition of a subsidy. In theory, subsidies that are based on their economic benefits to society should be offered to all who adopt the practice and for as long as the practice is used. In the case of the ACP and WQIP, incentive payments tended to be based on the cost of installing or adopting the practice. Since these subsidy payments were not based on potential economic benefits or on the difference in profits between the current and preferred practices, the adoption of the desired practices was influenced more by economic merits rather than by subsidies.

Evidence suggests that WQIP subsidy rates were inadequate to encourage the long-term adoption of farming practices that were less damaging to water quality. A study by the Sustainable Agriculture Coalition found that WQIP incentive payments were too low in some parts of the country to interest producers to implement management practices identified as necessary for meeting individual project goals (Higgins, 1995). The extra paperwork associated with the program also reduced the incentive to adopt particular practices. Cooper and Keim (1996) found that WQIP incentive payments may have been insufficient for adopting and maintaining practices beyond three years. Adoption rates of 12 percent to 20 percent for the practices of split fertilizer applications, integrated pest management, legume crediting, manure crediting and soil moisture testing could have been achieved for a $0 payment.[2] This suggests that some producers were willing to adopt certain practices without any program incentives because of the profitability of the practice (Cooper and Keim, 1996). At the WQIP offer rate of about $10 per acre, its adoption rate did not exceed 30 percent. Achieving a 50 percent adoption rate for any of the practices would have required a substantial increase in the incentive payments.

Cooper and Keim's results were supported by the findings of a survey conducted in the cornbelt region (Kraft et al., 1996). This survey of farmers' attitudes toward the WQIP found that only 17.5 percent were definitely interested in enrolling. An additional 27.8 percent stated that they might be interested. The average per-acre incentive payment that was requested by those expressing at least some interest in the program was almost $76—much greater than the current maximum level of $25 per acre. Only 18.8 percent of surveyed farmers were willing to accept an incentive payment of $25 per acre or less.

Generally, for a given (limited) financial incentive, practices that enhance net returns will have a higher probability of being adopted than practices that are not profitable or whose benefits accrue largely off-farm. Therefore, traditional cost-share programs can be expected to be less effective in promoting water quality practices designed to provide off-farm environmental benefits. In addition, the success of USDA's subsidy programs to achieve specific water quality goals is more dependent on the program manager's skills in the identification of the correct practices to support than it is on the farmer's business and cultural skills.

The Conservation Reserve Program (CRP) and Wetlands Reserve Program (WRP) have used annual subsidies (rental payments) and other incentives to retire environmentally sensitive cropland from production. Subsidies for land retirement can be viewed as a very blunt policy instrument in which the conservation practice consists of ceasing production. Rental payments have been continuous for the life of the contracts, unlike the short-term incentive payments for conservation practices. Long-term land retirement is an expensive approach, but is more likely to generate significant environmental returns. The CRP has converted a total of 36.4 million acres, about 8 percent of U.S. cropland, to conservation uses since 1985. Net social benefits of the CRP have been estimated to range between $4.2 and $9 billion in present value over the life of the program (Osborn and Konyar, 1990).

Research and Development

A major premise behind investment in agro-environmental research and development is that new and improved farming practices can be developed. They can be developed such that they protect or improve the environment and, at the same time, increase net returns to agriculture. Innovations can be input saving or output enhancing. Either way, a unit of output can be produced with fewer polluting inputs. Innovations from research and development can also reduce pollution flows without affecting input use or production (Parker and Caswell, this volume).

The extent to which research can be an effective strategy for improving environmental quality depends on the scope of existing technological opportunities and the market impacts of new technologies when they are adopted. Without government intervention, input saving or yield increasing innovations will be more attractive to farmers than innovations that merely reduce pollution flows. If widely adopted, however, some new practices may have the perverse result of increasing total input use through an expansion of production at both the intensive and extensive margins (Abler and Shortle, 1995).

A private firm has an incentive to develop profit-enhancing technologies when that company can capture the benefits of these developments. For example, much of the research on precision farming has been done by the private sector. Innovations that only reduce pollution flows may need public research support because the benefits would be widespread and not easily captured by private firms.

Research has had some success in developing complementary technologies that enable farmers to simultaneously achieve improvements in water quality and experience higher net returns. Important examples of such successes have occurred in the areas of integrated pest management, conservation tillage and improved soil nitrogen testing. Unfortunately, federal research funds generally have not been directed to environmentally related topics or to the targeting of innovations that are both profitable and environmentally friendly (OTA, 1995). Since the 1970s, only 10 percent of the work done by federal and state research institutions has been dedicated to natural resource topics, compared to 60 percent for traditional productivity-related topics. Even less research effort has been dedicated to complementary technologies (OTA, 1995).

Compliance

Compliance mechanisms tie the receipt of benefits from unrelated programs to some level of environmental performance. Compliance provisions were enacted in the Food Security Act of 1985 for the purposes of reducing soil erosion, discouraging the drainage of wetlands and discouraging the conversion of fragile grasslands to crop production. Violation of compliance provisions could result in the loss of price supports, commodity loan rates, disaster relief, CRP and Farmers' Home Administration benefits.

Conservation compliance provided a strong incentive to reduce erosion on highly erodible land (HEL). Conservation compliance plans were implemented on over 96 percent of the fields that required such plans (USDA/NRCS, 1996). Compliance resulted in significant reductions in soil erosion. Annual soil losses on HEL cropland have been reduced by nearly 900 million tons. Average soil erosion rates on more than 50 million HEL acres have been reduced to T (the rate at which soil can erode without harming or degrading its long-term productivity). Where conservation plans were fully applied on HEL cropland that was affected by compliance, the average annual soil erosion rate dropped from 16.8 tons per acre to 5.8 tons per acre (USDA/NRCS, 1996).

Evaluations of conservation compliance found significant reductions in soil erosion with minimal or moderate increases in the cost of crop production (Thompson et al., 1989; Dicks, 1986). Regional assessments of conservation compliance varied significantly in costs and benefits. Two of these assessments concluded that conservation compliance was a win-win program that increased farm incomes and reduced soil losses (Osborn and Setia, 1988; Prato and Wu, 1991). Other studies have shown that soil loss reductions were achieved only through decreases in net farm income (Hickman et al., 1989; Nelson and Seitz, 1979; Lee et al., 1991; Richardson et al., 1989; Hoag and Holloway, 1991; Young et al., 1991). The majority of HEL, however, was apparently brought into compliance without a significant economic burden. A national survey of producers who were subject to compliance found that 73 percent expected no decrease in earnings because of compliance (Esseks and Kraft, 1993).

Overall, conservation compliance resulted in a large social dividend, primarily in the form of off site benefits. An evaluation using 1994 data on HEL indicated that the national benefit/cost ratio for compliance was greater than two to one.[3] In other words, the monetary benefits associated with improved water quality, air quality and productivity were twice as great as the program costs to the government and producers (USDA/ERS, 1994). The average annual benefits of conservation compliance were estimated at about $13.80 per acre for water quality alone (USDA/ERS, 1994).

The effectiveness of compliance policies is limited by the extent that farmers are eligible and choose to participate in government programs. For example, USDA commodity programs have not covered vegetables, which traditionally have been chemical intensive crops. The effectiveness of compliance programs also varies with economic conditions. Generally, program participation and benefits decrease when crop prices increase. Ironically, it is precisely during these times that agriculture's pressures on the environment have been the greatest, through changes at the intensive and extensive margins.

The provisions of the 1996 Farm Act offer producers more flexibility in implementing conservation plans. However, producers still risk becoming ineligible for the new production flexibility contract payments, as well as for EQIP and WRP payments, should they fall out of compliance (Osborn, 1996).

U.S. Department of Agriculture Water Quality Projects

There has been an evolution in how the USDA offers assistance for water quality practices. USDA water quality programs and conservation programs were traditionally focused on getting farmers to use good farming practices, and not on the off-site benefits of those practices. In other words, farming practices were recommended on the basis of more efficient input use or of meeting a less clear-cut standard of better stewardship, rather than to ameliorate a particular water quality problem. As a result, USDA policy tools were not targeted to address specific problems.

The USDA began to design focused, watershed scale projects to address specific watershed problems starting with the experimental Model Implementation Program of the 1970s and the Rural Clean Water Program of the 1980s. In the 1990s, under the Water Quality Initiative and WQIP, hundreds of watershed projects were initiated to address water quality problems. These projects offered educational, technical and financial assistance to eligible landowners. An extensive research program was also developed to support these projects. Each project addressed a particular water quality problem, and focused assistance on getting farmers to adopt those practices that were believed to be best suited for addressing that problem.

While these projects succeeded in getting large numbers of landowners to adopt improved management practices, improvements in water quality were documented in only a few areas, such as West Lake in Iowa, Tillamook Bay in Oregon and Snake Creek in Utah (USDA/ERS, 1997). Part of the problem for this lack of documentation was inadequate water quality monitoring at the local level.

FEDERAL WATER QUALITY REGULATIONS

The USDA is not the only government agency whose programs have influenced agricultural production. The selection and use of specific practices and technologies by producers have been and continue to be affected by policies and regulations promulgated by other federal agencies and, increasingly, by state regulations.

Federal water quality legislation began with the 1972 amendments to the Federal Water Pollution and Control Act, now known as the Clean Water Act (CWA). This act initially targeted strong controls, which could be easily identified and monitored for point-source pollution (PSP). A two-tiered regulatory program managed by the Environmental Protection Agency (EPA) has controlled point-source polluters. National Pollutant Discharge Elimination System (NPDES) discharge permits are used as the enforcement tool for this regulation. While this command-and-control approach did improve water quality, the scale of improvements expected by the government and the private sector did not materialize (Adler, 1994).

While the emphasis was on PSP, the CWA also acknowledged NSP. Section 208 of the CWA called for the development and implementation of area wide, water quality management programs to ensure adequate control of all sources of pollutants in any areas in which water quality was impaired. The CWA also directed individual states to develop plans for reducing NSP. This included the development of appropriate land management controls. The 1977 amendments to the CWA further emphasized the role of NSP control in meeting water quality goals.

The Section 208 process was found to have problems (EPA, 1988; Harrington et al., 1985; Cook et al., 1991). A series of hearings held by the House of Representatives' Public Works and Transportation Subcommittee found that technical and financial support for the NSP program was lacking. The subcommittee also discovered that there was a lack of coordination with the PSP program and that data and other information necessary for its implementation were inadequate (Copeland and Zinn, 1986). Consequentially, individual states delayed the development and implementation of Section 208 programs. When these state programs were finally developed, the EPA was not able to readily determine whether they were adequate for achieving the stated water quality goals. The EPA was also not given effective enforcement tools to ensure that states did develop and implement viable management plans (Wicker, 1979).

The Water Quality Act (WQA) of 1987 placed special emphasis on NSP by amending the CWA's Declaration of Goals and Policy (EPA, 1988). Section 319 of the CWA required each state to develop watershed management plans to control and reduce specific types of NSP. Each management plan was required to have a list of best management practices (BMPs) to control NSP, as well as enforcement measures to ensure that these plans were implemented. The WQA also authorized federal loans and granted funds to help individual states develop and implement NSP programs. All states currently have NSP management plans employing a variety of voluntary and regulatory policy tools.

The one sector of agriculture that is regulated directly by the CWA is animal feeding. Under the CWA, animal feeding operations above a certain size (defined as concentrated animal feeding operations, or CAFO) are treated as PSP, and must obtain a discharge permit. They are required to prevent any water runoff from the site except that which could result from 24-hour rainstorm occurrences expected only once every 25 years.

A separate Federal NSP control program that affected agricultural technology choices was implemented for coastal zones. These are land areas that are several counties deep along the Atlantic, Pacific and Gulf coasts. The Coastal Zone Management Act Reauthorization Amendments (CZARA) required specific measures to address agricultural NSP in these zones (EPA, 1993). CZARA required each state, which had an existing coastal zone management program, to submit a plan to the EPA and the National Oceanic and Atmospheric Administration. This plan was to implement NSP management measures that would protect and restore coastal waters. A list of economically achievable measures for controlling agricultural NSP (design standards) was part of each state's management plan. Federal guidelines were broad enough to allow individual states to identify the management measures that were best-suited for their local conditions, thus, the inefficiencies of requiring practices with national standards were avoided. States first tried voluntary incentive mechanisms to promote adoption, but they were able to enforce management measures in the event that voluntary approaches failed. The implementation of these plans is not required until 1999.

The CWA and CZARA only addressed surface water. The Safe Drinking Water Act covered the impacts of agricultural NSP on groundwater quality. The Well Head Protection Program, established in 1986, required individual states to prepare a program for protecting public water wells from contamination from all sources of pollution, including fertilizers and pesticides. Forty-four states prepared such plans, but there were no penalties for individual states that did not prepare a plan.

WATER QUALITY PROGRAMS AT THE STATE LEVEL

Federal water quality laws passed on the responsibility for developing NSP programs to the individual states. States were allowed to use the full range of policy tools—voluntary (education, technical assistance), regulatory (technology and performance standard) and economic incentive (tax, subsidy, trading) instruments—to comply with federal requirements. Previously, individual states developed programs almost exclusively around voluntary approaches that were supported with some cost sharing. In recent years, more states developed programs that contained nonvoluntary elements. The tools that were used and the method by which they were implemented determined the incentives for adopting environmental quality enhancing practices. The following section reviews the five general nonvoluntary approaches to agricultural NSP control currently in use by individual states.

Technology Standards

The most common regulatory mechanisms employed by state water quality programs are technology standards. These generally call for farmers to implement a unique conservation plan that contains recommended BMPs. Individual states can apply this approach either uniformly across the state (nontargeted) or to specific geographic areas within the state (targeted).

Nontargeted technology standards require farmers to adopt conservation plans that incorporate management practices considered representing good stewardship. A few states have developed extensive lists of approved BMPs (for example, Kentucky with a list of 58 practices) while other state lists are currently less specific. All plans must be approved by the state. In the beginning, laws that were directed at crop production generally allowed voluntary adoption but had a regulatory backup. Enforcement was generally through citizen complaint. If a producer had a suitable plan in force, she or she he would not be subject to fines or penalties if someone filed a complaint for damages. Instead, the responsible producer could receive state assistance to alter the plan and to address the specific complaint.

Behavior based standards or regulatory approaches, such as technology standards, are not flexible incentives (Segerson, this volume). By striving for better stewardship instead of refusing to set specific water quality standards or goals, individual states can achieve limited flexibility through administrative means. This usually leads to the acceptance of a wide range of conservation plans that do not greatly constrain farmers.

CAFOs are regarded as PSPs under the CWA and require an NPDES discharge permit to control runoff from the site. A number of states have started using these permits to prevent the over-application of nutrients to cropland by restricting the manner in which they are applied. The permit system is regulatory in nature and is enforced through site inspections as well as through citizen complaints.

A problem with technology standards that rely on citizen complaints for enforcement is that they do not provide adequate incentives for the landowner to implement an efficient amount of pollution control, or for the potential victim to make known the costs of pollution. NSP is characterized by an inability to identify its source and by scattered victims who generally suffer minimal harm. If individual harms are minimal, they may be insufficient to induce citizens to initiate complaints. Even when a complaint is filed, no one can be held accountable for correcting the problem if the source has not been identified.

Another problem is that administrators, who are physically removed from the water quality problem, issue technology standards. The physical and hydrologic linkages between field practices and water quality are difficult to ascertain at any geographic level. Consequently, the practices that are required in statewide conservation plans are based on a best guess of what constitutes good stewardship.

In some states, technology standards have been targeted to specific geographic areas that are defined by a particular water quality problem. In such cases, thorough monitoring of the pollution figures is required. In many cases, laws have been directed at particular problems, such as pesticides in groundwater. In general, producers in such designated areas have to adopt specific BMPs. Enforcement occurs through inspection making the application of targeted technology standards more stringent than that of non-targeted technology standards. By focusing on specific problems in specific areas, better information on what constitutes acceptable management practices can be reasonably provided.

Nebraska, for example, is divided into Natural Resources Districts (NRDs), which are local units of government responsible for the conservation, wise development and proper utilization of natural resources (Bishop, 1994). In 1982, the Nebraska Legislature passed the Groundwater Management and Protection Act that allowed NRDs to establish special control areas to address groundwater quality concerns. In 1986, the Legislature gave NRDs the ability to require farms to use BMPs and to implement education programs to protect water quality. The BMPs for Nebraska were defined as those practices—including irrigation scheduling, proper timing of fertilizer and pesticide application, and other fertilizer and pesticide management programs—that prevent or reduce present and future contamination of groundwater.

The Central Platte NRD used this authority to develop a trigger policy (Segerson, this volume) for addressing a serious and growing problem with nitrate levels in its groundwater. Under Central Platte regulations, areas within the district were divided into three phases based on current groundwater nitrate levels. A Phase I area was defined as having an average groundwater nitrate level between zero and 12.5 parts per million (ppm). Nitrate concentrations in Phase II areas averaged between 12.6 and 20 ppm. Phase III areas had nitrate concentrations averaging 20.1 ppm or greater.

Agricultural practices were restricted according to the level of contamination. In a Phase I area, commercial fertilizer could not be applied on sandy soils until after March 1. Fall and winter applications to sandy soils were prohibited altogether. Phase II regulations included Phase I restrictions, plus the condition that commercial fertilizers could only be applied to heavy soils after November 1 and when an approved nitrification inhibitor was used. In addition, all farm operators using nitrogen fertilizer had to be certified by the state. Irrigation water in Phase II areas had to be tested annually for nitrate concentration. These results, along with the content of fertilizer recommendations and records of nitrate applications and crop yields, had to be filed annually with the NRD. Phase III regulations included Phase II requirements with additional requirements of split fertilizer applications (pre-plant and side-dress) and/or nitrogen inhibitors in the spring. In addition, deep soil analyses were required annually.

An advantage of the application of the Central Platte NRD's approach is that peer pressure could reduce enforcement costs (Randall, this volume). The prospect of having to implement evermore stringent and costly nutrient management practices encouraged producers to monitor and enforce the behavior of each other. It also served to prevent free riding. In this way, producers avoided more costly controls. Groundwater monitoring in the Central Platte NRD has shown a decrease in nitrate levels, indicating that the program is working (Bishop, 1994).

Performance Standards

Technology standards cover the majority of state initiatives. Only Florida is using a performance standard to address an agricultural pollution problem. Emission-based performance standards are not generally suitable for NSP since runoff cannot be easily measured. In Florida, however, the extensive use of drainage structures allows systematic sampling that can identify individual sources of pollution. The Works District Rule is being used in the area south of Lake Okeechobee to reduce the flow of phosphorus into the Everglades (Schmitz et al., 1995). This reduction is being accomplished by placing a maximum allowable phosphorus runoff standard on dairies. The enforcement method is inspection with dairies being allowed to reach the standard in any way.

Performance Taxes

Performance taxes are also being applied to the Everglades in South Florida. The Everglades Forever Act calls for a uniform per-acre tax on all cropland in the Everglades agricultural area. The tax starts at $24.89 per acre per year and increases every four years to a maximum of $35.00 per acre unless farmers exceed an overall 25 percent basinwide phosphorus reduction goal (State of Florida, 1995). The tax creates an incentive for producers to adopt BMPs. It also creates an incentive for producers to apply pressure on recalcitrant neighbors. The number of producers is not so large that free riding should be much of a problem.

This particular tool is flexible in that farmers are not restricted in how they manage their operations to meet the phosphorus reduction goal. The basis upon which the tax is placed, however, is not necessarily consistent with the goal of phosphorus reduction. A more efficient approach may be to directly tax phosphorus loads (Lee and Milon, this volume).

Trading

Trading is a market mechanism for efficiently allocating pollution reductions among different pollution sources with different marginal costs of control. Trading between agricultural PSP and NSP is possible when both sources contribute significant known amounts of the target pollutant in a basin. It is also possible when the costs of reducing loadings from NSP at the margin are less than the costs of reducing PSP. Uncertainty in the performance of agricultural BMPs can be accounted for with a trading ratio, which specifies the units of NSP reduction that can replace a single unit of PSP reduction. By allowing PSP to meet discharge goals by purchasing reductions from NSP, pollution control is achieved at lower cost. PSP/NSP trading requires a large commitment by an individual state in terms of administration costs and basic data acquisition.

The possibility of trading is driven by the Total Maximum Daily Load (TMDL) provisions of the CWA. According to the CWA, if the technology-based, point-source program fails to achieve water quality standards, a second tier of regulations would be implemented. These would be based on the quality of the receiving waters. Federal regulations and EPA guidance for TMDL implementation describe a process whereby regulators establish wasteload allocations (WLAs) for PSP and load allocations (LAs) for NSP and natural sources (Bartfeld, 1993). Together, WLAs and LAs comprise the TMDL, or the maximum discharge of pollutants in the basin, which allows the water quality standard to be met. A necessary component of this process is the identification of all loads and an assessment of the assimilative capacities of the body of water in relation to the water quality standards.

PSP discharge permits are based on the WLAs for the basin. The provisions of the CWA allow regulators to consider the relative costs of control when issuing discharge permits. The law states that if BMPs or other NSP controls make more stringent LAs practicable, then WLAs (PSP controls) can be made less stringent. Thus, the TMDL process provides for NSP control trade-offs (Bartfeld, 1993). The TMDL process, however, does make NSP legally responsible for meeting LAs, just as the NPDES permits do for the WLAs.

North Carolina has adopted a basin-oriented, water quality protection strategy that includes trading. The state is applying the TMDL process to several basins that it has identified as Nutrient Sensitive Waters. One, in particular, is the Tar-Pamlico Basin. Annual reductions in nitrogen discharge allowances have been established for a group of wastewater treatment plants in the basin in order to meet a basin discharge goal. The treatment plants can purchase the right to exceed their discharge allowances at a rate of $56 per kilogram. This payment goes into the Agricultural Cost Share Fund which supports BMPs for farmers in the basin (EPA, 1996). In comparison, the dischargers estimate that the treatment plant upgrades that are needed to achieve the required nitrogen discharge reductions would cost between $250 and $500 per kilogram (EPA, 1996).

This program has been operating since 1992 and has provided incentives for point-source polluters to increase the efficiency of wastewater treatment plants. As a result, these polluters were initially able to meet discharge goals without trades. As the gains from improved operation have been exhausted, however, some trades have occurred (EPA, 1996). A similar system has recently been established for the Neuse River Basin.

FEDERAL VERSUS LOCAL CONTROL

An important issue in developing NSP control strategies is the level of government at which incentives are developed and implemented. Federal water quality laws have passed responsibility for NSP control to the individual states. This has both advantages and disadvantages. A basic principle of the economic theory of federalism is that economic efficiency in the provision of a public good is generally best served by delegating responsibility for the provision of that good to the lowest level of government encompassing most of the associated benefits and costs (Shortle, 1995). The impacts of NSP are usually most pronounced near its point of origin. Contaminated groundwater does not move far from its pollution source. Lakes and small reservoirs are generally affected by local land uses. Likewise, streams and small rivers are impacted by land uses within relatively small watersheds. The impacts of agricultural runoff on water quality are generally most pronounced in small lakes, reservoirs and small rivers (Goolsby and Battaglin, 1993). Also, control of NSP requires regulation of land use that traditionally has been the prerogative of individual states and local governments (Malik et al., 1992).

The characteristics of NSP vary over geographic space due to the great variety of farming practices, landforms, climate and hydrologic characteristics found across relatively small areas of farmland. An efficient, centralized control policy would have to account for many different situations that, in turn, would result in exceedingly high administration costs. The reduction of these costs through national standards could not occur without reduced environmental efficiency. An efficient decentralized policy, however, would not have to account for as much geographic variability.

Decentralized control does not easily address the problem of the interstate transport of pollutants (transboundary issues). While most of the problems from NSP are felt close to the source, some NSP can travel long distances (in major rivers), affecting regional bodies of water such as the Gulf of Mexico or the Chesapeake Bay. The beneficiaries of one state's pollution control policies could therefore reside in other states. There are very few examples in which individual states have come together without Federal prodding to address regional water quality issues. This is despite common goals and the fact that an individual state may not be able to meet water quality goals without better control of interstate pollution. Interstate cooperation would increase the likelihood of a more efficient response to pollution problems.

Turning responsibility for pollution control over to the states can result in quite varied responses to pollution. Individual states react differently to similar pollution problems for a variety of reasons. These differences include socioeconomic characteristics of a state's populous, internal partisanship, organizational capacity and the perceived severity of problems. While individual states may be in a better position than that of the federal government to develop efficient pollution control policies, they do not always have the means or the will. Lester (1994) grouped states by their commitment to environmental protection and their institutional capacity to carry out that commitment. Progressive states are more likely to try innovative policies that go beyond federal mandates and guidelines. Some states have a high commitment to environmental protection but lack the institutional structure or resources to fulfill that commitment. Lester labels them as strugglers. Other states have a strong institutional capacity but a limited commitment to environmental protection (delayers). These states have been slow in implementing federal legislation. The final category contains the states that lack both the will and the means to implement environmental policies (regressive). Lester claims that decentralization of environmental programs would likely be a disaster in the regressive states because those states would be the most likely to provide insufficient protection for their citizens.

CONCLUSION AND FUTURE DIRECTIONS

Agricultural production has produced adverse environmental impacts in some locations. Changes in agricultural practices and technologies may mitigate these impacts and lead to a more sustainable agro-environmental system. The USDA has
traditionally used voluntary policy instruments to encourage the adoption of environmentally preferred technologies. Such voluntary instruments have included
education, financial assistance and technical assistance. The EPA and individual
states have also used regulatory tools (such as technology standards) and economic incentives (such as taxes and trading). Each policy approach has an impact
on agricultural productivity, profitability and environmental quality.

What is evident from the approaches that have been taken is that the characteristics of NSP and the lack of specific water quality goals have led to technology-
or behavior-based policies (subsidies and design standards) that are inherently inflexible. Cost-effective control of pollution is best achieved through performance-
based instruments that promote flexibility and allow producers to meet goals using
their own specialized knowledge (Segerson, this volume). In many cases, programs have been able to achieve some flexibility through administrative means,
such as setting a loosely defined goal of better management that can be achieved
using a wide range of management practices.

If the Clean Water Action Plan (EPA/USDA, 1998) released by the Clinton administration is any indication, there will be more emphasis on achieving specific
water quality goals in the future. This new plan targets NSP for greater control and
calls for establishing specific water quality standards for achieving the goals of the
CWA at the watershed level. It also provides for the use of enforcement mechanisms in ensuring that appropriate management practices will be adopted. This
should result in the U.S. bodies of water becoming more amenable to fishing and
swimming.

Because of their inflexible nature, the technology-based policies that are widely
used today would generally not lead to cost-effective controls for meeting specific
water quality standards (Segerson, this volume; Batie and Ervin; this volume).
Such approaches would put a tremendous burden on regulators to identify appropriate management practices for meeting water quality standards and how they
would be enforced.

A policy framework, such as the one presented by Segerson (this volume), is
one possible approach by which to achieve a more cost-effective control. This approach uses a variety of tools (such as subsidies, education and performance standards) in a trigger policy framework to encourage farmers to meet ambient water
quality standards. Research on the linkages between management practices and
water quality, and the development of water quality models would greatly assist
farmers in linking their actions to water quality.

No general statement can be made about which policy instruments give the
most efficient or cost-effective control. The characteristics of NSP problems vary
tremendously across the country. The choice of policies to control NSP problems
depends on the nature of the environmental quality problems, the information
available to the administering agency on the linkages between farming activities

and environmental quality, farm economics, and societal decisions about who should bear the costs of control. An approach, based on state and locally developed watershed-level control plans, would allow a variety of policy tools to be used and would include both carrots and sticks. Such an approach probably would provide the greatest opportunities for cost-effective control.

ENDNOTES

1. The authors are economists with the Economic Research Service of the U.S. Department of Agriculture (USDA). The views expressed in the chapter do not necessarily reflect those of the USDA.
2. Monetary denominations in this chapter are all in terms of U.S. dollars.
3. Benefit-cost ratios varied widely across regions, due to differences in the social benefits that can be achieved and the costs of implementing a conservation plan (USDA/ERS, 1994).

REFERENCES

Abler, D.G. and J.S. Shortle. 1991. "The Political Economy of Water Quality Protection from Agricultural Chemicals." *Northeastern Journal of Agricultural and Resource Economics* 21(5): 53–60.

_____. 1995. "Technology as an Agricultural Pollution Control Policy." *American Journal of Agricultural Economics* 77(1): 20–32.

Adler, R.W. 1994. "Reauthorizing the Clean Water Act: Looking to Tangible Values." *Water Resources Bulletin* 30(5): 799–807.

Anderson, J.L., F.G. Bergsrud, and T.M. Ahles. 1995. "Evaluation of the Farmstead Assessment System (FARM*A*SYST) in Minnesota," in U.S. Department of Agriculture, *Clean Water, Clean Environment—21st Century, Vol. III: Practices, Systems, & Adoption.* Proceedings of "Clean water, Clean Environment—21st Century: Team Agriculture—Working to Protect Water Resources," Kansas City, MO, (5–8 March).

Bartfeld, E. 1993. "Point-nonpoint Source Trading: Looking Beyond Potential Cost Savings." *Environmental Law* 23(43): 43–106.

Bishop, R. 1994. "A Local Agency's Approach to Solving the Difficult Problem of Nitrate in the Groundwater." *Journal of Soil and Water Conservation* 49(2): 82–84.

Bohm, P. and C. Russell. 1985. "Comparative Analysis of Alternative Policy Instruments," in A.Y. Kneese and J.L. Sweeny, eds., *Handbook of Natural Resource Economics, Vol. 1.* New York, NY: Elsevier Science Publishing Co.

Braden, J.B. and K. Segerson. 1993. "Information Problems in the Design of Nonpoint-source Pollution Policy," in C.S. Russell and J.F. Shogren, eds., *Theory, Modeling, and Experience in the Management of Nonpoint-source Pollution.* Boston, MA: Kluwer Academic Publishers.

Camboni, S.M. and T.L. Napier. 1994. "Socioeconomic and Farm Structure Factors Affecting Frequency of Use of Tillage Systems." Invited paper presented at the Agrarian Prospects III Symposium, Prague, Czech Republic.

Cook, K., A. Hug, W. Hoffman, A. Taddese, M. Hinkle, and C. Williams. 1991. Center for Resource Economics and National Audubon Society. Statement prepared for the Subcommittee on Environmental Protection and presented to the Committee on Environment and Public Works, U.S. Senate, Washington, DC (17 July).

Cooper, J.C. and R.W. Keim. 1996. "Incentive Payments to Encourage Farmer Adoption of Water Quality Protection Practices." *American Journal of Agricultural Economics* 78(1): 54–64.

Copeland, C. and J.A. Zinn. 1986. *Agricultural Nonpoint Pollution Policy: A Federal Perspective.* Study prepared for a colloquium on Agrichemical Management to Protect Water Quality, Washington, DC (21 March).

Dicks, M.R. 1986. "What Will It Cost Farmers to Comply with Conservation Compliance?" *Agricultural Outlook* 124: 27–30.

EPA (U.S. Environmental Protection Agency). 1988. *Nonpoint Sources: Agenda for the Future.* Office of Water , Washington, DC (October).

_____. 1993. *Guidance Specifying Management Measures for Sources of Nonpoint Pollution in Coastal Waters.* EPA 840-B-92-002, Washington, DC (January).

_____. 1996. *Draft Framework for Watershed-based Trading*. EPA 800-R-96-001, Washington, DC (May).

EPA/USDA Plan (U.S. Environmental Protection Agency and U.S. Department of Agriculture Plan). 1998. *Clean Water Action Plan: Restoring and Protecting America's Waters*. Washington, DC (February).

Esseks, J.D. and S.E. Kraft. 1993. *Opinions of Conservation Compliance Held by Producers Subject to It*. Study prepared for the American Farmland Trust, DeKalb, IL.

Florida, State of. 1995. "Everglades Improvement and Management." *Florida Statutes 373.4592*, Tallahassee, Florida.

Goolsby, D.A. and W.A. Battaglin. 1993. "Occurrence, Distribution, and Transport of Agricultural Chemicals in Surface Waters of the Midwestern United States," in D.A. Goolsby, L.L. Boyer, and G.E. Mallard, eds., *Selected Papers on Agricultural Chemicals in Water Resources of the Midcontinental United States*. Open-file Report No. 93-418, U.S. Geological Survey, Reston, VA.

Gould, B.W., W.E. Saupe, and R.M. Klemme. 1989. "Conservation Tillage: The Role of Farm and Operator Characteristics and the Perception of Soil Erosion." *Land Economics* 85(2): 167–182 (May).

Harrington, W., A. Krupnick, and H.M. Peskin. 1985. "Policies for Nonpoint Source Water Pollution Control." *Journal of Soil and Water Conservation* 40(27): 27–32.

Hickman, J.S., C.P. Rowell, and J.R. Williams. 1989. "Net Returns from Conservation Compliance for a Producer and Landlord in Northeastern Kansas." *Journal of Soil and Water Conservation* 44(5): 532–534.

Higgins, E.M. 1995. *The Water Quality Incentive Program: The Unfulfilled Promise*. Research Report, Center for Rural Affairs, Walthill, NE.

Hoag, D.L. and H.A. Holloway. 1991. "Farm Production Decisions Under Cross and Conservation Compliance." *American Journal of Agricultural Economics* 73(1): 184–193.

Hoban, T.J. and R.C. Wimberley. 1992. "Farm Operators: Attitudes About Water Quality and the RCWP," in *Seminar Publication: The National Rural Clean Water Program Symposium*. U.S. Environmental Protection Agency, Report No. 625-R-92-006, EPA, Washington, DC (August).

Knox, D., G. Jackson, and E. Nevers. 1995. "FARM*A*SYST: A Partnership Program to Protect Water Resources," in U.S. Department of Agriculture, *Clean Water, Clean Environment—21st Century, Vol. III: Practices, Systems, & Adoption*. Proceedings of "Clean Water, Clean Environment—21st Century: Team Agriculture—Working to Protect Water Resources," pp. 167–170. Kansas City, MO (5–8 March).

Kraft, S.E., C. Lant, and K. Gillman. 1996. "WQIP: An Assessment of Its Chances for Acceptance by Farmers." *Journal of Soil and Water Conservation* 51(6): 494–498.

Lee, J.G., R.D. Lacewell, and J.W. Richardson. 1991. "Soil Conservation or Commodity Programs: Trade-offs During the Transition to Dryland Crop Production." *Southern Journal of Agricultural Economics* 23(1): 203–211.

Lester. J.P. 1994. "A New Federalism? Environmental Policy in the States," in N.J. Vig and M.E. Kraft eds., *Environmental Policy in the 1990s: Toward a New Agenda*. Washington, DC: C.Q. Press.

Lichtenberg, E. and B.V. Lessley. 1992. "Water Quality, Cost-sharing, and Technical Assistance: Perceptions of Maryland Farmers." *Journal of Soil and Water Conservation* 47(3): 260–263.

Malik, A.S., B.A. Larson, and M. Ribaudo. 1992. *Agricultural Nonpoint Source Pollution and Economic Incentive Policies: Issues in the Reauthorization of the Clean Water Act*. Agricultural Economic Report No. AGES 9229, USDA/ERS, Washington, DC (November).

Napier, T.L. and D.E. Brown. 1993. "Factors Affecting Attitudes Toward Groundwater Pollution among Ohio Farmers." *Journal of Soil and Water Conservation* 48(5): 432–438.

Napier, T.L. and S.M. Camboni. 1993. "Use of Conventional and Conservation Practices among Farmers in the Scioto River Basin of Ohio." *Journal of Soil and Water Conservation* 48(3): 231–237.

Napier, T.L. and D.G. Sommers. 1994. "Correlates of Plant Nutrient Use among Ohio Farmers: Implications for Water Quality Initiatives." *Journal of Rural Studies* 10(2): 159–171.

Nelson, M.C. and W.D. Seitz. 1979. "An Economic Analysis of Soil Erosion Control in a Watershed Representing Corn Belt Conditions." *North Central Journal of Agricultural Economics* 1(2): 173–186.

Nowak, P.J. 1987. "The Adoption of Agricultural Conservation Technologies: Economic and Diffusion Explanations." *Rural Sociology* 52(2): 208–220.

Nowak, P.J., G. O'Keefe, C. Bennett, S. Anderson, and C. Trumbo. 1997. *Communication and Adoption Evaluation of USDA Water Quality Demonstration Projects*. Evaluation Report, University of Wisconsin, Madison, WI in cooperation with U.S. Department of Agriculture, Washington, DC (October).

OTA (Office of Technology Assessment). 1995. *Targeting Environmental Priorities in Agriculture: Reforming Program Strategies*. U.S. Congress, Washington, DC.

Osborn, C. T. 1996. "Conservation and the 1996 Farm Act." *Agricultural Outlook* 235: 22–29.

Osborn, C.T. and K. Konyar. 1990. "A Fresh Look at the CRP." *Agricultural Outlook* 166: 33–37.

Osborn, C.T. and P.P. Setia. 1988. "Estimating the Economic Impacts of Conservation Compliance: A Corn Belt Application." Selected paper presented at the 1988 American Agricultural Economics Association meetings, Knoxville, TN (1 August).

Pease, J. and D. Bosch. 1994. "Relationships Among Farm Operators' Water Quality Opinions, Fertilization Practices, and Cropland Potential to Pollute in Two Regions of Virginia." *Journal of Soil and Water Conservation* 49(5): 477–483.

Prato, T. and S. Wu. 1991. "Erosion, Sediment and Economic Effects of Conservation Compliance in an Agricultural Watershed." *Journal of Soil and Water Conservation* 46(3): 211–214.

Richardson, J.W., D.C. Gerloff, B.L. Harris, and L.L. Dollar. 1989. "Economic Impacts of Conservation Compliance on a Representative Dawson County, Texas Farm." *Journal of Soil and Water Conservation* 44(5): 527–531.

Schmitz, A., W.G. Boggess, and K. Tefertiller. 1995. "Regulations: Evidence from the Florida Dairy Industry." *American Journal of Agricultural Economics 77(5)*: 1166–1171.

Shortle, J. 1995. "Environmental Federalism: The Case of U.S. Agriculture?," in J.R. Braden, H. Folmer, and T. Ulen, eds., *Environmental Policy with Economic and Political Integration: The European Union and the United States*. Cheltenham: Edward Elgar.

Thompson, L.C., J.D. Atwood, S.R. Johnson, and T. Robertson. 1989. "National Implications of Mandatory Conservation Compliance." *Journal of Soil and Water Conservation* 44(5): 517–520.

USDA/ERS (U.S. Department of Agriculture, Economic Research Service). 1994. *Agricultural Resources and Environmental Indicators*. Agricultural Handbook No. 75, Washington, DC (December).

_____. 1997. *Agricultural Resources and Environmental Indicators*. Agricultural Handbook No. 712, Washington, DC (July).

USDA/NRCS (U.S. Department of Agriculture, Natural Resources Conservation Service). 1996. *1994 Final Status Review Results*. Unpublished. Washington, DC.

Wicker, W. 1979. "Enforcement of Section 208 of the Federal Water Pollution Control Act Amendments of 1972 to Control Nonpoint Source Pollution." *Land and Water Law Review* 14(2): 419–446.

3 EVOLUTION OF EPA PROGRAMS AND POLICIES THAT IMPACT AGRICULTURE

A. Clayton W. Ogg
U.S. Environmental Protection Agency
Washington, DC

Environmental policies for agriculture in the United States began with a narrow focus on pesticides. These policies have recently become more flexible and integrated in addressing the serious ecological damages caused by agriculture. Early programs restricting the use of certain widely used pesticides altered the mix of available pesticides and changed the mix of environmental and human health risks from pesticides. More recently, flexible approaches achieve water, wildlife and climate benefits by targeting protection to vital areas and by using the most effective remedies. Under new action plans, the U.S. Environmental Protection Agency (EPA) has addressed the application of agricultural nutrients and developed water quality criteria to serve as goals for reducing nutrient loadings. In implementing the 1996 Farm Bill, the U.S. Department of Agriculture (USDA) has focused incentive payments on the most economically efficient practices, such as nutrient planning and riparian protection. The new emphasis on flexible approaches and effective remedies by both agencies can result in major gains in treating widespread agro-environmental problems.

INTRODUCTION

Agriculture has greatly altered the American landscape. Unfortunately, soil erosion and the extensive use of manufactured nutrients and pesticides by American agriculture have been linked to widespread ecosystem damage and potential health risks to the public (EPA and USDA, 1990; Puckett, 1994). Environmental policies have been designed to address these and other problems associated with agricultural production, that include the following:

- loss of critical ecosystems due to their conversion to crop and pasture uses (Allen, 1994);
- reduction of wildlife populations due to exposure to pesticides and pesticide residues;
- occurrence of pesticide residues in our food and water;
- contamination of our water and air from agricultural nutrients (Mueller et al., 1995; Puckett, 1994; EPA and USDA, 1990; EPA et al., 1998);
- loss of stream flow due to irrigation; and
- sedimentation of our lakes and streams from agricultural erosion and run-off (Ribaudo, 1989).

These six problems and their remedies are interrelated. As an example, the loss of critical ecosystems includes the loss of riparian systems that contribute greatly to several of the other five problems (National Research Council, 1993).

Although the U.S. Environmental Protection Agency's (EPA) early efforts to address these problems were limited to certain pesticide and livestock waste issues, policies evolved over the years to focus on remedies that achieved multifaceted benefits. This evolution has included a shift in EPA policy orientation from a reliance on technology standards to more flexible policy designs. There has also been greater EPA involvement in cooperative activities, particularly with the individual states, the USDA and other federal agencies. The recent Clean Water Action Plan was written by the EPA and by several other federal agencies. It is part of EPA's new environmental policies for U.S. agriculture and demonstrates a high degree of program integration.

Some ecosystems, such as the tall-grass prairies, had the misfortune of occurring on the most productive agricultural land. These systems are virtually gone now except for a few tracts located in parks. Many opportunities still exist, however, to greatly reduce the environmental damage that is associated with agricultural production. The lands most critical for protecting many ecosystems and species of wildlife often constitute only a tiny portion of the cropped landscape. Wetlands are probably the most ecologically valuable lands in the country and have limited value for agricultural production. The above-mentioned riparian lands (critical to the survival of fish, mollusks and amphibians) occupy less than one-half of 1 percent of the cropped acreage in the United States (Lee and Lovejoy, 1994). Such small percentages can offer protection for a significant proportion of the endangered species in the United States. Other widespread problems associated with the application of excess nutrients to cropland (such as water pollution and greenhouse gas emissions) can be reduced through the use of best management practices (Babcock and Blackmer, 1992; Bosch and Napit, 1992; Fleming and Babcock, 1997; Shortle et al., forthcoming; Trachtenberg and Ogg, 1994; EPA and USDA, 1990).

Given the pervasiveness of the six problems mentioned above, the resulting ecological damage and the risks to human health (and opportunities to solve them), one may ask why these problems still persist in the United States. This chapter describes the evolution of the federal policies that have been developed to address these agro-environmental problems. It analyzes the specific problems and opportunities that have been encountered while implementing these policies and evaluates the responses made by involved federal institutions.

EVOLUTION OF AGRO-ENVIRONMENTAL PROGRAMS: 1970–1985

Prior to the 1970s, American agriculture had largely avoided environmental regulation. Regulation of the pesticide industry became a major role of the EPA with the agency's inception in 1970. The main U.S. environmental legislation affecting agriculture up to the present has been the Federal Insecticide, Fungicide and Rodenticide Act (FIFRA) initially legislated in 1947. The FIFRA was amended in 1972, and its amendments provided most of the basis for contemporary pesticide regulation. While taking costs into consideration, this legislation required pesticides to be registered for use only on the condition that they did not pose unreasonable risks to man or to the environment.

Public concerns about the effects of pesticides on non-target species and food safety shaped the 1972 legislation. This legislation has been amended on several occasions since 1972, but the most important policy changes occurred with the 1996 Food Quality Protection Act (FQPA). The FQPA directed the EPA to consider chemically similar families of pesticides and to facilitate the registration of "minor-use" pesticides. The volume of sales of minor-use pesticides was often inadequate to financially justify the significant fixed registration costs.

Since 1972, the number of pesticide cancellations has reflected the many concerns about pesticide risks. The earliest actions focused on pesticides and their effects on non-target species. Bald eagles and other non-target species were being poisoned with various pesticides used in predator and field rodent control. Brown pelican numbers were in rapid decline due to eggshell thinning caused by chlorinated hydrocarbon pesticides.

Cancer concerns became the next focus of the EPA's attention, and several of the most carcinogenic pesticides were removed from use. This included aldrin and chlordane, which were largely removed from agricultural use between 1974 and 1976. As attention shifted to groundwater, ethylene dibromide (EDB), aldicarb and a number of pesticides that tended to leach into groundwater were also removed from use. (EDB cancellations for agricultural use occurred between 1983 and 1985; the use of aldicarb was regulated from 1979 through the present). Food safety concerns that affect agriculture became the next focus of environmental regulations. Several pesticides, including carbon tetrachloride, were the next to face regulatory sanctions due to food safety issues. Another effort, the protection of songbirds, focused on granular forms of certain pesticides.

In dealing with the above risks, the most problematic pesticides faced regulatory sanctions first. Consequently, the most dramatic environmental gains from pesticide regulation were probably realized early in the program's implementation. As regulatory actions have expanded to consider less questionable substances, pest control choices have become more difficult, and the economic value of the remaining pesticides for agriculture has increased. For example, as methyl bromide is phased out to address the ozone layer problem (Deepak et al., this volume) we will likely see the expanded use of a variety of substitute products that may lead to new risk trade-offs. The 1996 FQPA and other recent policy initiatives (discussed later in this chapter) reflect this maturing stage of EPA pesticide regulation.

Another important EPA regulatory program that dates back to the early 1970s is the Clean Water Act (CWA). The CWA launched a massive effort to restore the quality of the nation's waters. This ultimately resulted in an investment of hundreds of billions of dollars by U.S. cities and businesses. Although the CWA did not provide clear direction regarding which opportunities to address in agricultural pollution, it did introduce the National Pollution Discharge Elimination System (NPDS) that regulated large livestock facilities. The NPDES permits issued under the CWA required the utilization of the best available technologies to satisfy the new effluent guidelines. In the case of large livestock operations (more than 1,000 animal units), waste containment facilities capable of handling the runoff from a 24-hour rainstorm expected to occur once every 25 years and any process wastewater were required. The 1,000 animal unit minimum limited this regulation to a very small portion of the total number of livestock operations. Although the implementation of livestock permitting requirements has been delegated to the states in the majority of cases, NPDES did give the EPA the authority to carry out the program should the states fail to do so.

Thus far, the NPDES and other federal regulatory programs have not been used to address the nonpoint-source pollution (NSP) associated with the excessive application of nutrients to cropland (Ribaudo and Caswell, this volume). These problems may not remain within the exclusive realm of incentive-based programs in the future. A new plan, the Clean Water Action Plan: Restoring and Protecting America's Waters, is discussed later. Along with voluntary initiatives, this plan sets goals for an expanded NPDES livestock permitting program. It also calls for regulating the application of nutrients to farmland that contain large livestock operations (EPA et al.).

SAFE DRINKING WATER ACT AMENDMENTS AND CONSERVATION LINKAGES FOR USDA PROGRAMS: 1985–1990

The Safe Drinking Water Act (SDWA) established the current federal/state arrangement whereby individual states were delegated the authority to protect drinking water. Amendments to the SDWA in 1986, however, provided more regulatory authority over agriculture for the EPA. These amendments required that drinking water be monitored to assure that it does not exceed contaminant levels set for 83 chemicals. It also requires that every three years the EPA identify, for a specified period, 25 more contaminants, which are to be monitored in addition to the original 83. It also required the disinfection and filtration of public water supplies, the establishment of a wellhead protection program and other protections that were less relevant to agriculture. These 1986 amendments led to the criticism that the SDWA lacked flexibility. This resulted in major revisions a decade later.

By the mid-1980s, the USDA was drawn into the environmental arena. This initially occurred in response to the plowout of tens of millions of erodible cropland acres during the high crop price years of the 1970s (Ogg, 1992). The EPA was not involved in the development of these soil conservation programs until much later. In retrospect, it evolved that the major soil conservation initiatives of the Food Security Act of 1985 were as important in protecting water quality as they were in conserving cropland productivity (Ribaudo, 1989). In addition, the USDA soil conservation programs that were created in 1985 provided the foundation for some of the nation's major environmental programs that exist today.

Although incentive-based programs had existed for decades in the U.S. Department of Agriculture, its focus and size increased dramatically with the passage of the Food Security Act of 1985. This legislation modified farm policy to address the inconsistencies between natural resource conservation goals and commodity price support programs, which had greatly hampered the farm programs' economic efficiency (Calcicco et al., 1985; Ogg, 1992; USDA/ASCS, 1980). The acreage reduction programs created at that time removed about two-thirds of the nation's highly erodible farmland (in excess of 20 tons per acre per year) out of crop production and enrolled many cropped wetlands (Babcock et al., 1995). In addition, farmers who received commodity program payments had to implement affordable conservation measures (especially reduced tillage) and stop the drainage of wetlands. The 1990 Food, Agriculture, Conservation and Trade Act fine tuned these programs and expanded the focus of the Conservation Reserve Program (CRP) to include water quality concerns.

These highly integrated incentive-based programs represented the first major regulatory efforts to address environmental damage from agricultural activities. The dramatic success of these programs in reducing soil erosion and protecting wetlands was a sharp contrast to the programs of previous decades (Ogg, 1992; USDA/ASCS, 1980). The success of these 1985 and 1990 programs carried forward into the next decade.

FLEXIBLE, DECENTRALIZED PROGRAMS: THE 1990s

Agro-environmental programs administered in conjunction with individual states have become more important during the 1990s. Federal and state agencies have worked together to create flexible solutions to local problems by making use of various incentives as well as by enforcing certain requirements. Four EPA administered regulatory programs now influence farmers' activities that may affect water quality. The specific programs and their legislation are the management measure requirements of the Coastal Zone Act Reauthorization Amendments of 1990 (CZARA); the Total Maximum Daily Load (TMDL) requirements for particular watersheds under the CWA; the changes in source water protection of the SDWA Amendments of 1996; and pesticide registration of the FQPA. During the same period, pesticide regulation became more flexible and more integrated with passage of the FQPA. The potential economic efficiencies sought under these newly constructed EPA programs paralleled the focus on flexibility and economically efficient remedies offered by the new USDA incentive-based programs. These are discussed in the final sections of this chapter.

Coastal Zone Act Reauthorization Amendments

CZARA requires each state to develop plans and enforceable NSP programs for coastal areas or to face phased reductions in their coastal zone grants and the grants under Section 319 of the CWA. The CZARA management measures foster cost-effective practices that address agricultural pollution while assuring that some minimum standard of performance is enforced. The EPA and National Oceanic and Atmospheric Administration (NOAA) identified these management measures, but the CZARA has given each state considerable flexibility to develop alternative measures to suit their specific needs as long as they are equally effective in controlling NSP.

Past interpretations of the CWA limited its application in agriculture primarily to the containment of runoff from confined animal feeding operations. The CZARA provides an important change by addressing the excess application of livestock wastes and other nutrients on cropland in the coastal zone. The management measures identified by the EPA drew on agronomic best management practices developed at state land grant universities. Since individual states are allowed to substitute alternative practices that achieve the same level of environmental improvement, the CZARA approach is extremely flexible. In contrast to the technology standards used in the past, the CZARA management measure approach requires an economically reasonable amount of improvement in solving the problem. An individual state may also identify geographic areas where there are no problems that require treatment. The TMDL program complements CZARA by providing performance standards that assure specified uses of streams.

Although both CZARA and TMDL programs offer great flexibility, neither program necessarily offers incentives that would meet the definition of flexible incentives as discussed by Batie and Ervin in this volume. Individual states may, however, provide those incentives when implementing the respective laws.

The CZARA management measure approach has not been backed by substantially greater federal inducements for states to implement enforceable NSP programs. Broader use of these management measures was part of the Clinton Administration's proposals for amending the CWA. The economic significance of opportunities to improve the efficiency of chemical use were highlighted (EPA, 1994). The more recent action plan (EPA et al.) would increase cost-share funding for EPA's NSP programs and would encourage more use of State Revolving Fund loans for NSP problems.

A major advantage of the CZARA management measures approach has been the opportunity to encourage the widespread use of practices, such as nutrient planning, that are needed to address environmental problems that occur throughout large areas. Nonetheless, the CZARA has not assured that remedies to agricultural water quality problems are adequate. Even though a specified management measure may be very effective throughout large areas, it may not be adequate to address the special needs of some of the most severe problem areas.

Total Maximum Daily Loads

Under the CWA, individual states identify TMDLs from various sources, which are capable of meeting designated uses for streams and other bodies of water. Although the intent is to provide goals for guiding the level of treatment needed for NSP as well as point-source pollution (PSP), few examples of fully implemented TMDL programs are available to date. The EPA recently committed itself to expedite the development of TMDL programs, and new guidelines call on states to implement TMDLs for listed waters using both incentive-based and regulatory programs (EPA et al.).

Greater flexibility in tailoring remedies to specific problems under the TMDL approach offers efficiencies in targeting resources although this flexibility brings with it greater administrative challenges. The EPA is currently attempting to support the efforts of individual states to establish TMDLs by developing nutrient criteria and guidelines and by conducting research to support the identification of problems and goals (EPA et al.).

TMDLs could play a larger role in addressing pollution from agricultural activities in the future. Because both the CZARA and TMDL programs rely on individual states for implementation, the availability of adequate funding is critical to provide the resources and incentives for states to carry out these respective laws. Equally critical is the willingness of federal policymakers to make each state's funding contingent on the development of programs that adequately implement the CZARA and TMDL laws.

The Clinton Administration's Clean Water Action Plan (EPA et al.) calls for a stepped-up implementation of CZARA and TMDL programs, and for making NSP programs enforceable. Increased funding (beyond the $100 million per year currently available to states) will be contingent on the individual states making improvements in their NSP programs. The Action Plan also calls for developing nutrient criteria for bodies of water, and for having the NPDES program address land application of nutrients from large livestock operations. It would also support livestock producers' efforts to manage wastes with increased assistance from the USDA livestock waste management incentive programs (described later in this chapter) as well as through an industry-run fund, the Agricultural Marketing and Promotion Order.

Safe Drinking Water Act Amendments of 1996

The 1996 amendments to the SDWA placed responsibilities on communities for assessing potential sources of drinking water contamination and for seeking remedies to potential problems. Identification of source water areas (that is, the land from which the drinking water originates) that have pollution problems provides another flexible tool for targeting resources to environmental problem areas.

Food Quality Protection Act

In 1996, Congress also amended FIFRA to create flexibility in regulating minor-use pesticides and to expedite the registration of reduced-risk pesticides. The FQPA shifted U.S. pesticide policy toward providing consumers with more information regarding pesticide risks and better protection for the special vulnerabilities of children.

Minor-use pesticides are used on relatively few acres. This has sometimes lead to voluntary cancellation by manufacturers in order to avoid high registration costs. The FQPA set up a program to remedy this situation by having the USDA provide part of the information needed to register these pesticides. The EPA is also charged with providing a separate registration process for pesticides used on minor-use crops.

FQPA also directed the EPA to expedite the registration of reduced-risk pesticides. This provision will reduce the overall risks from pesticides by making safer pesticides available much sooner. To complement these opportunities for the employment of safer pesticide practices, the legislation also called on the USDA to provide information and to assist farmers in the use of integrated pest management (IPM) practices.

Other actions called for by FQPA focused on the provision of pesticide-risk information to consumers so that they could make more informed choices regarding pesticide residues on foods. As consumers consider potential health risks when choosing among products in the marketplace, pesticide-risk information brings economic forces into play to encourage safer pest control practices.

Along with this increased emphasis on flexible incentives for managing pesticides, the FQPA introduced an integrated approach to the regulation of families of pesticides instead of individual pesticides. It focuses on identified human health risks from families of pesticides that share the same biochemical mode of action (for example, organo-phosphate insecticides or triazine herbicides). This approach attempts to address the problem of increased use of substitute pesticides that pose similar risks to ones that have been removed from use. The FQPA focus on families of pesticides, however, raises the regulatory stakes considerably. The loss of a whole family of pesticides would greatly reduce the choice of pest control substances available to producers.

USDA'S NEW INCENTIVE-BASED PROGRAMS

Just as many EPA programs are being modified to focus on economically efficient and flexible remedies, the new USDA programs created by the 1996 Federal Agricultural Improvement and Reform Act (FAIR) offered a parallel focus on economic efficiency and flexibility. The 1996 FAIR Act provided major initiatives for USDA environmental programs and policies to change with respect to addressing environmental problems related to agriculture. The Agricultural CRP became the Environmental Conservation Acreage Reserve Program, and refocused on the CRP to provide long-term protection to environmentally sensitive land. The Environmental Quality Incentives Program (EQIP) consolidated a number of the existing subsidy programs with the goal of maximizing the environmental benefits per-dollar expended (USDA, 1980). In a break with the past, the 1996 statute enhanced local involvement in the implementation of farm programs. Under the EQIP, individual states gained greater autonomy in shaping these programs to suit state and local priorities, so that problem targeting, practice selection and subsidization levels are likely to differ widely among states.

For the first time, the 1996 Farm Bill gave cost-share and incentive programs a directive to efficiently address agro-environmental degradation. The FAIR legislation and the rules developed by U.S. Department of Agriculture (USDA/CCC, 1997a and 1997b) provided much more detailed direction regarding the use of flexible incentives. Each farmer's request to participate in these programs was evaluated based on the payment he or she was willing to accept and the ability of the proposed practice to achieve designated objectives. These new USDA programs are providing an opportunity to apply the theory of flexibility to an actual program. Although current funding for the EQIP program averages only $4 million per state, the Clinton Administration's Action Plan proposes to increase this amount (EPA et al.). The remaining sections of this chapter will describe the 1996 FAIR programs, indicate how they complement other environmental programs for agriculture, and analyze the potential of these new flexible EPA and USDA programs to achieve economic efficiency and environmental gains.

Environmental Quality Incentives Program

EQIP placed new requirements for targeting, planning and cost effectiveness on USDA's cost-sharing and incentive payment activities. Most of the responsibility for implementing these new requirements is held by local and state conservationists who are advised by local work groups and state technical committees. The statute requires that one-half of EQIP's $200 million per-year allotment be applied to livestock-related problems. Although the distinction is not clearly drawn, cost-share payments are provided to encourage the adoption of structural practices (such as livestock waste management facilities) while incentive payments are used to encourage nonstructural practices (such as nutrient plans, IPM, and rotational grazing). Large-scale livestock producers are not allowed to receive cost-share payments for waste management facilities, but they can receive incentive payments for services to assist in nutrient management (USDA/CCC, 1997a).

Several EQIP provisions focus on attaining economic efficiency and on creating program flexibility at the grass roots level, especially with respect to targeting problems and to the selection of appropriate remedies. Most of the funding must be targeted to selected geographic priority areas defined by each state. Some funds are also reserved for statewide concerns that can be addressed using special-emphasis practices (USDA/CCC, 1997a). State and local groups, which include environmental, wildlife and agricultural players, have been set up to develop proposals for priority areas and state wide concerns. The states then rank these proposals according to the importance of the environmental problem addressed and their likelihood of success. For both priority areas and state wide concerns, farmers' bids to receive EQIP assistance are ranked according to the environmental benefits produced for each dollar funded. In a dynamic setting, this requirement should lead to the positive outcome of cost-share investments favoring low marginal-cost activities first (USDA/CCC, 1997a). For the first year, five percent of the EQIP budget was expended on bonus payments to individual states that excelled in the application of EQIP rules.

The State wide Concerns Program offers an opportunity to fill major gaps left by past environmental programs. Under this program, very low marginal-cost activities can receive cost-share assistance over a much larger geographic area than would be included in the state's priority areas. For example, implementing nutrient recommendations costs less than 50 cents per acre and generally can lead to a 15–35 percent reduction in nitrate runoff and leaching (Kuch and Ogg, 1996; Wu et al., 1996). This ability to address the excessive use of nutrients makes EQIP a key player when addressing problems that are basin wide (for example, physteria outbreaks in Eastern estuaries and hypoxia in the Gulf of Mexico) or national in scope (for example, nitrous oxide emissions that may contribute to global warming).

Conservation Reserve Program

Since its creation in 1985, the CRP has focused on protecting highly erodible land from soil erosion. Modifications over the years have allowed the CRP to use priority practices that have utilized natural vegetation to protect riparian areas, grass waterways, contour strips, cropped wetlands and wellhead source water protection areas (USDA/CCC, 1997b). Revised in 1996, the CRP provided a 20 percent bonus to farmers for enrolling riparian areas. This feature, in combination with its continuous enrollment opportunity and vigorous informational campaign, will likely result in substantial enrollments for these ecologically important areas (Lant et al., 1995). Enrollment of two-thirds of the currently cropped riparian areas would exhaust less than 4 percent of the 36 million acre capacity available to the CRP under the FAIR and would provide large benefits in reducing nitrate (Haycock and Pinay, 1993; Jacobs and Gilliam, 1985), herbicide (Rohde et al., 1980; Hall et al., 1993), and sediment pollution (Karr and Schlosser, 1978). One year into the program, Maryland, Tennessee and several other key states have enrolled most of their formerly cropped riparian land.

The CRP also has featured a Conservation Reserve Enhancement Program (CREP) that has captured the interest of many states. CREP has allowed state contributions to the CRP to be part of a program that has targeted specific state and national priorities (USDA et al., 1997). Farmers in Maryland, the first state with an approved CREP, has received a 70 percent bonus for riparian lands, compared to 20 percent offered elsewhere. The state has piggybacked its offers for longer contracts or for permanent easements for riparian lands on the CREP options for farmers.

ECONOMIC EFFICIENCY OF INCENTIVE-BASED PROGRAMS

Realizing major opportunities under the new EPA and USDA programs require economically beneficial remedies for farmers and for the environment. How economically efficient are the new flexible incentive-based programs of the EPA and the USDA from the standpoint of economic theory? It appears that the main economic advantages of their flexibility stem from the diversity of the agricultural sector.

Shortle and Dunn (1986) demonstrated that where there is uncertainty about damage cost functions and asymmetrical production function information, and where farmers have specialized knowledge about their farm operations appropriately specified management practice incentives will generally outperform management practice standards, estimated runoff incentives, or standards. Taken altogether, it seems the ideal policy approach to the amelioration of many agriculturally based environmental problems is a decentralized system of locally specified management practice incentives (Shortle and Dunn, 1986; Zilberman and Marra, 1993). This recent literature supports the highly decentralized approach described above for the implementation of the new EQIP cost-sharing and incentive payments, as well as for the CZARA and TMDL programs.

What guidance is there about how such a system of incentives should operate? The accepted wisdom suggests that aggregate net social benefits (NSB) would be maximized by sorting farmers' bids for funding in descending order by the increment in NSB each bid would generate. Then, beginning at the top, work down the list of funding proposals until one runs out of proposals that generate positive increments in NSB, or until the program's budget is exhausted, whichever comes first. This is the equivalent of maximizing NSB subject to a program budget constraint. Benefit-cost ratios could serve as proxies for contributions to aggregate NSB where all proposals, relative to the allocable budget, are quite small, (Hirshleifer, 1970). All of this conforms to common sense and economic theory, but it is problematic insofar as it is normally difficult to measure the benefits (the environmental damages avoided) that would flow from each proposal. In spite of this difficulty, the EQIP and CRP attempt such a ranking to achieve the highest environmental benefit per dollar (USDA/CCC, 1997a and 1997b).

RESEARCH TO SUPPORT THE NEW PROGRAMS OF THE USDA AND THE EPA

The CRP offers a powerful tool for addressing environmental and ecological problems, but its contracts with farmers last only 10 or 15 years. Future policy decisions regarding what to continue funding could depend heavily on research that track the success of these programs. Research has played a vital role in realizing environmental opportunities and can play an even more vital role in the future by documenting the environmental benefits of CRP, EQIP and the evolving EPA programs (Allen, 1994; Ribaudo, 1989). Research that employs multiple environmental quality and economic indicators (for example, the Center for Agriculture and Rural Development) has much to offer when tracking the success of these programs.

The CRP logically functions as an acreage reserve program when it addresses soil erosion problems. It preserves the productivity of fragile lands when they are not needed so that they can be used when they are needed. Riparian lands and grassed waterways, however, are needed most to reduce pollution when crops are in heavy demand and where cropping is most intensive. Research is needed to identify which practices may prove beneficial so as to merit the consideration of becoming permanent practices.

The opportunities to use EQIP's state wide concern program are well documented in the case of nutrients (Wu et al., 1996). No similar consensus has developed regarding which IPM practices should be supported by incentive payments. Applied economic and interdisciplinary research could support the efforts of state technical committees, extension services, consultants and many others who are likely to play a role in the use of EQIP incentive payments. In particular, research can help to encourage the optimal substitution of technologies, which include chemical recommendations that are provided by consultants. These technological substitutions could provide some of the greatest opportunities toward the advancement of a more sustainable agriculture (Ruttan, 1994).

Research can also support efforts by the EPA and its counterparts in state governments as they provide program managers with more information regarding what loadings of nutrients are considered allowable to support beneficial stream uses (EPA, 1998). This information is particularly important to those who establish priority areas for the newer and more flexible EPA programs and for EQIP. In areas with concentrations of livestock, the costs of addressing phosphorus buildup in the soil can be substantial (Pratt et al., 1997). Lakes differ in their needs, and accurate identification of practical and quantifiable goals for cleaning up lakes can greatly affect treatment costs and methods (Ogg et al., 1983).

CONCLUSION

Early EPA programs regulated the use of pesticides in agriculture, but they provided few incentives or regulations to address other environmental problems related to agriculture. By restricting the use of pesticides that caused the greatest damage to wildlife or the greatest risks to human health, the EPA attempted to minimize those damages. Farmers were forced to rely on alternative pesticides that posed less risk or a different mix of risks. In some cases, these alternative pesticides were more costly and less effective, or their use required more management skill.

As policymakers became more aware of both the environmental problems created by agriculture and the practical opportunities to address those problems, a more flexible set of EPA policies emerged. Pesticide risks may be addressed in the future through greater reliance on incentives that encourage farmers to use IPM practices, and to provide more information and choices for the consumer.

Opportunities to greatly reduce soil erosion, stream sedimentation and further loss of wetlands have been realized in the past decade through the USDA's integration of conservation and income support programs (Ogg, 1992). Other agro-environmental opportunities, which policies are beginning to exploit through flexible incentive-based programs, include the restoration of riparian corridors along streams and the application of nutrients to land at rates more in line with crop needs. These are highlighted in the Clean Water Action Plan, which calls on the EPA to develop nutrient criteria for streams within two years and to include the appropriate application of nutrients to fields as part of its regulation of large livestock facilities. The use of water markets to obtain water from farmers and to restore water to in-stream uses is being pursued by several federal agencies and by state governments (Zilberman, 1997).

Although state EPA programs have established a framework that could provide enforceable mechanisms to support the realization of some of the above opportunities, a lack of funding hinders these efforts. Policymakers continue to place heavy reliance on incentive-based approaches, administered largely by the USDA, to solve many types of problems. Through coordination at the federal, state and local levels, these programs attempt to target important environmental problems and to make efficient use of the scarce resources currently available for incentive payments to farmers.

Applied economic and multidisciplinary research is needed to assure the success of these innovative new programs in meeting water quality, greenhouse gas and wildlife objectives. Researchers must track the multifaceted benefits that emanate from the EQIP and CRP programs in order to ensure that successful programs receive recognition and continued support, and to ensure that remedies are found if some programs are not successful. There is also a need to identify which practices merit cost-share assistance as specified by EQIP, that is, those achieving the highest environmental benefits per dollar. This is particularly important with regard to identifying and ranking the numerous IPM practices, which can vary widely in their benefits and costs (Norton and Mullen, 1994). Finally, close cooperation among agencies is needed to support setting goals for reducing pollutant loadings. This applies to the new USDA programs and other evolving programs. Policymakers need these goals to guide them as they design flexible and efficient programs to deal with potentially costly problems.

REFERENCES

Allen, A.W. 1994. "Wildlife Habitat Enhancement," in Soil and Water Conservation Society, *When Conservation Reserve Program Contracts Expire: The Policy Options,* pp. 18–20. Proceedings of "When Conservation Reserve Program Contracts Expire: The Policy Options." Arlington, VA (10–11 February).

Babcock, B.A. and A.M. Blackmer. 1992. "The Value of Reducing Temporal Input Nonuniformities." *Journal of Agriculture and Resource Economics* 17: 335–347.

Babcock, B.A., P.G. Lakshminarayan, and J.J. Wu. 1995. *Renewing CRP: Results from a Study of Alternative Targeting Criteria.* Iowa State University, Center for Agriculture and Rural Development, Briefing Paper No. 95-BP-6, Ames, IA.

Bosch, D.J. and K.B. Napit. 1992. "Economics of Transporting Poultry Litter to Achieve More Effective Use as Fertilizer." *Journal of Soil and Water Conservation* 47: 342–346.

CARD (Center for Agriculture and Rural Development). 1996. *RAPS 1996: Agricultural and Environmental Outlook.* Iowa State University, Center for Agriculture and Rural Development, Ames, IA.

Colacicco, D., A. Barbarika, Jr., and L. Langner. 1985. "Conservation Benefits of the USDA's 1983 Payment-in-kind and Acreage Reduction Programs." Unpublished Mimeo, Washington, DC.

EPA (U.S. Environmental Protection Agency). 1994. "President Clinton's Clean Water Initiative: Analysis of Benefits and Costs." Office of Water, EPA 800-R-94-002, Washington, DC.

EPA et al. (U.S Environmental Protection Agency, U.S. Department of Agriculture, U.S. Department of Commerce, U.S. Department of Defense, U.S. Department of Energy, U.S. Department of Interior, U.S. Department of Justice, U.S. Department of Transportation and the Tennessee Valley Authority. 1998. *Clean Water Action Plan: Restoring and Protecting America's Waters.* (14 February).

EPA/USDA (U.S. Environmental Protection Agency and U.S. Department of Agriculture). 1990. "National Water Quality Inventory, Report to Congress." Washington, DC.

Fleming, R.A. and B.A. Babcock. 1997. "Resource or Waste? The Economics of Swine Manure Storage and Management." Center for Agriculture and Rural Development, Working Paper No. 97-WP 178, Iowa State University, Ames, IA.

Hall, J.K., N.L. Hartwig, and L.D. Hoffman. 1983. "Application Mode and Alternate Cropping Effects on Atrazine Losses From a Hillside." *Journal of Environmental Quality* 12: 336–340.

Haycock, N.E. and G. Pinay. 1993. "Groundwater Nitrate Dynamics in Grass and Poplar Vegetated Riparian Buffer Strips During the Winter." *Journal of Environmental Quality* 22: 273–278.

Hirshleifer, J. 1970. *Investment, Interest, and Capital.* Englewood Cliffs, NJ: Prentice-Hall.

Jacobs, T.C. and J.W. Gilliam. 1985. "Riparian Losses of Nitrate from Agricultural Drainage Waters." *Journal of Environmental Quality* 14: 472–478.

Karr, J.R. and I.J. Schlosser. 1978. "Water Resources and the Land Water Interface." *Science* 21: 229–224.

Kuch, P.J. and C.W. Ogg. 1996. "The 1995 Farm Bill and Natural Resource Conservation: Major New Opportunities." *American Journal of Agricultural Economics 78*: 1207–1214.

Lant, C.L., S.E. Kraft, and K.R. Gilliam. 1995. "Enrollment of Filter Strips and Recharge Areas in the CRP and USDA Easement Programs." *Journal of Soil and Water Conservation* 50: 193–200.

Lee, J.G. and S.B. Lovejoy. 1994. "Land Area Available and Characteristics for Riparian Buffers to Control Agricultural Nonpoint Source Pollution: A GIS Approach." Unpublished Mimeo, Purdue University, West Lafayette, IN.

Mueller, D.K., P.A. Hamilton, D.R. Helsel, K.J. Hitt, and G.C. Ruddy. 1995. "Nutrients in Groundwater of the United States—An Analysis of Data Through 1992." U.S. Geological Survey, Water Resources Investigations, Report No. 95-4031, Denver, CO.

National Research Council, National Academy of Science. 1993. *Soil and Water Quality, An Agenda for Agriculture.* Washington, DC: National Academy Press.

Norton, G.W. and J. Mullen. 1994. "Economic Evaluation of Integrated Pest Management Programs: A Literature Review." Virginia Cooperative Extension Publication No. 448–120, Virginia Polytechnic Institute and State University, Blacksburg, VA (March).

Ogg, C.W. 1992. "Addressing Environmental Needs in Farm Programs." *Agricultural History* 66: 273–278.

Ogg, C.W., J.D. Johnson, and K.C. Clayton. 1982. "A Policy Option for Targeting Soil Conservation Expenditures." *Journal of Soil and Water Conservation* 37: 68–72.

Ogg, C.W., H.B. Pionke, and R. Heimlich. 1983. "A Linear Programming Economic Analysis of Lake Quality Improvement Using Phosphorus Buffer Curves." *Water Resources Research* 19: 21–31.

Pratt, S., R. Jones, and C.A. Jones. 1997. Livestock and the Environment: Expanding the Focus. Texas Institute for Applied Environmental Research, TIAER PR 96-03, Tarleton State University, Stevenville, TX (June).

Puckett, L.J. 1994. *Nonpoint and Point Sources of Nitrogen in Major Watersheds of the United States.* U.S. Geological Survey based on Water Resources Investigations Report No. 94-4001, Department of Interior, Washington, DC.

Ribaudo, M.O. 1989. Water Quality Benefits from the Conservation Reserve Program. U.S. Department of Agriculture, Report No. AER-606 USDA/ERS, Washington, DC.

Rohde, W.A., L.E. Asmussen, W.W. Hauser, R.D. Wauchope, and H.D. Allison. 1980. "Trifluralin Movement in Run Off from a Small Agricultural Watershed." *Journal of Environmental Quality* 9: 37–42.

Ruttan, V.W. 1994. "Constraints on the Design of Sustainable Systems of Agricultural Production." *Ecological Economics* 10: 209–219.

Shortle, J.S., W.N. Musser, W.C. Huang, B. Roach, K. Kreahling, D. Beegle, and R.M. Fox. Forthcoming. "Economic and Environmental Potential of the Pre-sidedressing Soil Nitrate Test." *Review of Agricultural Economics* 17: 25–35.

Shortle, J.S. and J.W. Dunn. 1986. "Agricultural Pollution Control Strategies." *American Journal of Agricultural Economics* 68 : 668–677

Trachtenberg, E. and C. Ogg. 1994. "Potential for Reducing Nitrogen Pollution through Improved Agronomic Practices." *Water Resource Bulletin* 30: 1109–1118.

U.S. Congress. 1972. *Federal Water Pollution Control Act Amendments.* 92nd Congress, PL 92-500, Washington, DC (October).

_____. 1978. *Federal Insecticide, Fungicide, and Rodenticide Act Amendments.* 95th Congress, PL 95-396, Washington, DC (30 September).

_____. 1985. *Food Security Act of 1985.* 99th Congress, PL 99-198, Washington, DC (23 December).

_____. 1996. *Safe Drinking Water Act.* 99th Congress. PL 104-182, Washington, DC (December).

_____. 1996. *Federal Agriculture Improvement and Reform Act of 1996.* 104th Congress, PL 104-127, Washington, DC (4 April).

_____. 1996. *Food Quality Protection Act.* 104th Congress, PL 104-170, Washington, DC (3 August).

USDA/ASCS (U.S. Department of Agriculture, Agricultural Stabilization and Conservation Service). 1980. *National Summary Evaluation of the Agricultural Conservation Program: Phase I.* Washington DC.

USDA/CCC (U.S. Department of Agriculture, Commodity Credit corporation). 1997a. *EQIP, Final Rule*.7 CFR Part 1466, RIN 0578-AA19, Washington, DC.

_____. 1997b. *EQIP, Final Rule.* 7 CFR part 1410, RIN 0560-0125, Washington, DC.

USDA et al (U.S. Department of Agriculture, Commodity Credit Corporation and the State of Maryland). 1997. *Memorandum of Agreement Between the U.S. Department of Agriculture, the Commodity Credit Corporation, and the State of Maryland.* Signed at Spaniard's Point Farm, Queen Anne's County, MD (20 October).

USDA/SCS (U.S. Department of Agriculture, Soil Conservation Service). Various Issues. National Resource Inventory, Washington, DC.

Wu, J.J., P.G. Lakshminararyan, and B.A. Babcock. 1996. "Impacts of Agricultural Practices and Policies on Potential Nitrate Water Pollution in the Midwest and Northern Plains of the United States." Iowa State University, Center for Agricultural and Rural Development, Working Paper No. 96-WP 148, Ames, IA

Zilberman, D. 1977. "Water Marketing in California and the West." Paper presented at the seminar on Economic Policies and Water Use in Colombia, Bogota, Colombia (25 August).

Zilberman, D. and M. Marra. 1993. "Agricultural Externalities," in G.A. Carlson, D. Zilberman and J.A. Miranowaski eds., *Agricultural and Environmental Resource Economics,* pp. 221–267. New York, NY: The Oxford University Press.

4 ENVIRONMENTAL CONSERVATION STRATEGIES: WHAT WORKS AND WHAT MIGHT WORK BETTER

Stephen B. Lovejoy
Purdue University, West Lafayette, IN

While environmental conservation as a component of agricultural policy dates back to the 1930s, it remained a neglected stepchild until the more recent Farm Bills of 1985, 1990 and 1996. The last three farm bills have spawned massive programs to induce, cajole and force farmers and ranchers to adopt and implement better conservation technologies. These recent strategies have included regulation and targeting in addition to, or combined with, the old standard of buying the cooperation of farmers and ranchers. While some of these programs have been very popular, the actual conservation impacts have often been less spectacular than anticipated. Many of the programs have not been implemented in a way that maximizes society's environmental objectives. Research, which suggests why these programs were effective or ineffective at altering the behavior of farmers and ranchers, will be reviewed in this chapter. The programs' impact as conservation tools will be explored. In addition, the potential impacts of the new Environmental Quality Incentives Program (EQIP) are examined, and some suggestions for alternative conservation strategies that may be more effective and efficient are delineated.

INTRODUCTION

Many in agriculture have long advocated conservation practices and have suggested that farmers and ranchers should adopt production technologies and systems that improve environmental quality and conserve resources. Traditionally, the general public has been more concerned about the price of food and the overall production of food and fiber. Recent events suggest that the general public now want a great deal more from the agricultural sector than just a cheap and plentiful supply of food. The public also wants clean water, clean air, sufficient wildlife, good habitat, scenic landscapes and recreational opportunities. It now expects farmers and ranchers to provide these amenities and, in many cases, is willing to help by providing financial and technical assistance.

The general public is neither homogenous nor consistent in its preferences and demands. Some environmental laws have been passed while others have failed. Some individuals suggest they want ecolabeling but are unwilling to pay for it. Some see a wetland as a valuable ecosystem and still others see it as a mosquito habitat. With these multiple criteria and changing preferences, policymakers and program implementers face a considerable challenge in the application of the appropriate weights to these various amenities.

With these multiple desires as a base, recent policies attempted to induce farmers and ranchers to adopt management systems that better satisfy the public's demands. Some programs, for example the Conservation Reserve Program (CRP) and the Wetlands Reserve Program (WRP), used financial incentives, such as cost sharing. Other approaches involved education, technical assistance, peer pressure and community recognition (such as being named Conservation Farmer of the Year). Still other programs involved mandatory compliance. In general, these programs were popular with the agricultural community. They funneled billions of dollars into the hands of farmers and ranchers, and participation was largely voluntary. The few regulatory-type programs that were implemented were minimally enforced, which led to very few sanctions.

Although soil conservation as a component of agricultural policy dates back to the 1930s, it was not seriously addressed until the Farm Bills of 1985, 1990 and 1996. These three farm bills have spawned massive programs to induce, cajole and force farmers and ranchers to adopt and implement better conservation practices. Recent strategies have included regulation—compliance, wetland conservation and the Environmental Conservation Acreage Reserve Program (ECARP)—and targeting (the CRP and WRP) in addition to, or in combination with, the old standard of buying the cooperation of farmers and ranchers. While some of these programs have been very popular, the actual conservation impacts have often been less spectacular than anticipated. Many of the programs have *not* been implemented in a way that maximizes society's objectives of achieving a less environmentally degrading agricultural sector (Ribaudo and Caswell, this volume; Ogg, this volume).

Presumably, these new and revised programs are based upon U.S. Department of Agriculture (USDA) experience with previous water quality projects and their research related to what works and what does not work. Effective programs are difficult to design because of complex spatial, temporal and political parameters. The experience and research of the USDA in implementing and evaluating similar programs should lead it to a design process similar to that used by a private company when that company introduces a new product or service. If the past is indicative of the future, there are reasons to doubt the value of these programs. An examination of some of these projects and the research evaluating their results follow.

PAST AND CURRENT PROGRAMS: A REVIEW

Rural Clean Water Program

The Rural Clean Water Program (RCWP), initiated in 1980, established 21 projects selected on the basis of their diversity of agricultural nonpoint-source pollution (NSP). Agriculture was shown to be the major source of pollution in 18 of these projects. The program was a demonstration of the effects on water quality from implementing agricultural best management practices (BMPs). Each project was required to submit plans that included project goals, methods to be used to obtain these goals and a program that would monitor water quality to determine the success of the program. Ten years were allowed for implementing the projects, and five years were allocated to obtain signed contracts. If the circumstances dictated, this time frame was modified. Two examples of such circumstances were the early achievement of the project's goals or the termination of the project because of low levels of water quality impairment. The types of water uses most commonly impaired were recreational, drinking and commercial fishing.

Piper et al. (1989) analyzed the successes, or lack thereof, in each project. Although less than one-third of the projects resulted in improvements in water quality, the researchers concluded that the program should be integrated into a larger and more comprehensive plan. These authors also suggested that, since the program had been demonstrative in nature, it should be used as the foundation of future programs to reduce and eliminate NSP. The authors for future projects suggested the following important considerations:

- Economic benefits must be gained through the implementation of the program. These benefits may be gained by any number of parties, such as recreational users, commercial fishermen or those who experience increased land values because of an increase in the aesthetic quality of the area.
- BMPs must be adopted if water quality is to improve. This leads to the question of enforcement.
- Benefits must be greater than both the costs of engineering and running a program. Benefits are defined to include social benefits that are difficult to measure quantitatively.

These criteria sidestep a primary issue: The benefits from such NSP programs may not accrue to those who bear the costs of the programs (Lovejoy and Hyde, 1997). As Piper et al. (1989) concluded, the program should be implemented in conjunction with a larger regional program. It should also be expanded to include neighboring bodies of water (streams, estuaries, reservoirs). These requirements would lead to a wider-based more focused program than the RCWP. The extent to which such an expanded program would lead to actual water quality improvement remains an empirical question.

Water Quality Initiative

The 1989 Water Quality Initiative (WQI) was the USDA's response to public concern for the nation's surface water and groundwater quality. It was a coordinated effort to protect the nation's water from contamination by agricultural chemicals (Sutton et al., 1994). The goal of the program was to give agricultural workers the knowledge, technology and funding to address the environmental concerns related to agriculture.

Under the WQI, several projects were implemented to help reduce agricultural NSP. Two types of these projects, Hydrologic Unit Areas and Water Quality Demonstration Projects, are discussed below. One of the main features of these programs was the cooperative nature of their administration. Three separate agencies of the USDA were involved along with local authorities and farmers participating in the projects.

Hydrologic Unit Areas

Under this program, the USDA provided farmers with financial and technical assistance to help them meet water quality goals that were set forth by their respective states. Initially, 37 different areas were selected. The selection process was based on three considerations: (1) the significance of agriculture in the pollution process; (2) the amount of agricultural pollution that came from a group of designated pollutants (for example, pesticides, nutrients or animal wastes); and (3) the extent to which the area conformed with other water quality programs. The WQI included a plan to initiate 275 such projects within five years, but this goal was never achieved.

Water Quality Demonstration Projects

Twenty-four projects were selected under this program. The goal for each project was to show the extent to which selected practices could be effective in reducing agricultural NSP. The projects were chosen for their diversity in agricultural, soil and geological conditions. Two other goals of the projects were to show farmers cost-effective methods for reducing NSP and to accelerate the adoption of technology that had been developed but not yet widely implemented. The USDA provided leadership and funding for the farmers who participated in these projects.

Results and Recommendations

There is little substantial information on the success of the WQI program since it was begun only five years ago. It has been reported, however, that the 16 original demonstration projects have reduced nitrogen and phosphorus usage by 6.7 million pounds and 4 million pounds, respectively. An interim evaluation study (Sutton et al., 1994) concluded that:

Except for three or four projects, it will be difficult to link practice installation to measured improvements in water quality . . . The reasons for this are inadequate monitoring networks and a low emphasis on tracking chemical use and land management, along with weather conditions and the fact that projects last only five years. (p. 15)

The evaluation led to many recommendations. The most important of these recommendations was that a project should begin with clear objectives and unbiased methods of measuring pre- and post-project performance levels for each objective. For the tens of millions of dollars expended, however, this recommendation seems inadequate.

Programs in the Chesapeake Bay

There have been several programs that fight pollution in the Chesapeake Bay ranging from individual state programs to regional compacts among states. In 1983, the District of Columbia and the states of Maryland, Pennsylvania and Virginia signed the first Chesapeake Bay Agreement that allowed environmental research of the Bay. Because it was considered to be a national treasure, Chesapeake Bay was targeted under interstate regulation. This regulation came in response to the Environmental Protection Agency's (EPA) work in determining the causes of degradation in the Bay. The general environmental movement of the 1970s was the impetus for this research on the Bay. The most noteworthy pollutants from NSP were nitrogen and phosphorus. Sediment from harbors in Maryland and Virginia also was targeted by the agreement.

A second Chesapeake Bay Agreement was signed in 1987. It established 29 distinct goals to be achieved. The most relevant of these was the creation of a plan to reduce the controllable levels of nitrogen and phosphorus by 40 percent by 2000 from the 1985 base. Schuyler (1993) prepared a report on the key elements of this program along with an evaluation of its success thus far.

This second Chesapeake Bay Agreement established several control measures to reduce pollutants. One of these measures was a reduction of pollution caused by animal manure from feedlots, animal production facilities, land applications and other sources. Manure became a key target for efforts in the Chesapeake Bay since it contained both phosphorus and nitrogen. The EPA appropriated $7 million annually to this program. This appropriation was matched, dollar-for-dollar, by participating state funds. These states also allocated monies from other sources to control NSP loadings. Thus far, this program has achieved only limited success. Only 12 percent to 14 percent of the targeted lands have been treated with BMPs and only 10 percent to 12 percent of animal wastes have been controlled. The results of a watershed model have suggested that only 12 percent of nitrogen and 8 percent of phosphorus have been controlled.

As with other programs, the lack of enforcement measures may have contributed to the lack of success. Schuyler (1993) stated: ". . . if voluntary cooperation does not increase any time in the next few years, the programs will have no option but to go regulatory." (p. 222)

Summary

A common theme runs throughout this examination of some of the previous environmental programs. They have been, at best, only somewhat successful in dealing with NSP because of the incongruence of costs and benefits. As stated earlier in this chapter, those who gain benefits may be different from those who pay the costs. The farmer who bears the costs of changing inputs or management practices may gain little from the changes that these practices have upon the environment. Instead, the beneficiaries of these changes are the commercial fishermen, swimmers, boaters and others who do not use the water because of NSP.

In the future, all programs should take into account the incentives factored in by the relevant decision-makers as well as the question of who benefits and who pays. These programs may employ economic incentives, such as linking available funds to a program's success, or they may require civil or criminal actions (Braden and Lovejoy, 1989). Either way, the future of the nation's waters is dependent on policies that are successful in reducing the pollutant loadings. Future programs must focus more on the results and less on the process. The goal is to protect and enhance the quality of the nation's waters so that everyone can continue to enjoy the benefits of clean water.

NEW PROGRAMS: NEW CRITERIA AND PAST KNOWLEDGE?

The late 1990s was an era of renewed discussions about solutions to the conservation problem. Programs that are voluntary, targeted and locally based have been developed. While some of the acronyms are different, some of the program incentives differ little from previous programs. Can we expect these new programs to perform any better than those of the past?

In the late 1980s, several studies suggested that the implementation of environmental regulatory programs for agriculture was less than optimal. The reasons often cited for program failures include a lack of enforcement, hastily developed farm plans, perverse incentives and lack of clear objectives.

The National Resources Conservation Service (NRCS) has always maintained that it should provide technical assistance to landowners and operators and should not just be an enforcement agency. More recent statements from the agency suggest that the two characteristics essential for a good agricultural conservation policy are the targeting of programs and voluntary programs.

Targeting Programs

Lovejoy and Lee (1995) illustrated that the CRP was not an optimal program in terms of per-ton costs of erosion saved or from obtaining the maximum water quality benefits per-dollar spent. While few expect government programs to always be optimal, one could expect most program goals to be achieved. This has not occurred. One reason the CRP failed to meet its goals was that the perform-ance criteria of these goals were not well specified and that there was no system for prioritizing these criteria.

An analysis of CRP targeting options indicated that selecting 7.2 million acres could result in costs ranging from $255 million to $334 million per year, whereas water quality benefits could range from $211 million to $572 million, depending on the acres selected (Lovejoy and Lee, 1995). This emphasis was consistent with efforts in Congress during the spring of 1997 to shift the emphasis of the CRP program from soil erosion in the Plains region to water quality in the Midwest, Southeast and Mississippi Delta regions. The re-enrollment of CRP acres, how-ever, seems to be concentrated more in the mountain states and in the southern plains states. These areas will not maximize the CRP program benefits based on any environmental or conservation criteria.

Another targeting concept is to identify cropland acres that abut streams so that they can be planted to some type of permanent vegetation. Lee and Lovejoy (1995) indicated that only slightly more than 2 million acres were both cropland and were situated within 100 feet of a stream. The benefits of retiring these ripar-ian acres might be much more substantial than all other CRP benefits even though these areas represented only 5 percent of the acres currently enrolled in the CRP. More than 1 million (50 percent) of these CRP-enrolled riparian acres are located in just 12 states. These 12 states are located in the Upper Great Plains, the Lake States or the Corn Belt.

The present riparian buffer strip program (which provides a bonus rental pay-ment) may be a positive step, but it still does not address the issue of targeting those acres that could yield the greatest environmental benefit. Even with a bonus plan for riparian acres, geographic targeting is still necessary. While these results are well known, recent negotiations between Congress and the USDA indicate that the concept of targeting solely for environmental protection and/or improvement is still not fully accepted.

Voluntary Programs

While voluntary programs are politically attractive, evidence from the RCWP, the WQIP and other programs is far from conclusive about the actual environmental impacts of such programs. The USDA counters that the programs clearly illustrate landowners' and operators' adoption of conservation practices and these results are the criteria by which the programs should be judged. The USDA argues that the adoption and use of conservation practices is sufficient since this will certainly lead to environmental improvements, even though the USDA's own research questions that causal link. The question of whether the programs will actually lead to the increased adoption of conservation practices has been raised recently.

Napier and Johnson (forthcoming) examined a locally based program that was designed to make farmers aware of pollution problems, their role as agricultural producers and the practices that they could utilize to farm in a more environmen-tally sound manner. However, neither awareness gained from the local project nor familiarity with federal conservation programs had any impact on conservation behaviors. The researchers found no statistically significant difference in behavior between those more or those less familiar with the pollution problem. This outcome belied the millions of dollars spent in the watershed to educate and promote conservation behavior. This finding was consistent with the USDA evaluation of the WQI Demonstration Projects (Nowak et al., 1997) that also found no association between exposure to the project and attitudes toward water quality problems.

Another recent study evaluated the impacts of increased financial and technical assistance upon conservation behavior in side-by-side (with and without) watersheds (Napier and Johnson, 1998). The authors found that the millions spent on the experimental watershed only served to stimulate the same behaviors found in the control watershed. They said:

> The major policy implication of the conclusions drawn from these study findings is that future conservation programs should be evaluated in the context of changes in conservation behaviors among target populations rather than the number of activities enacted, local projects sponsored, or positive attitudes among target populations . . . Proponents of the Darby Creek program will be hard pressed to prove they significantly effected long-term conservation behaviors even though they expended large amounts of public and private resources to accomplish that end. (p. 83)

Again, this is consistent with the findings of Nowak et al. (1997) that concluded that the WQI Demonstration Projects did not significantly influence the adoption of BMPs.

Design Principles Need Change

While recent projects may sound improved, the design principles have not changed appreciably over the past several decades. The rhetoric may suggest targeting, but the reality does not always match. While new voluntary programs are being developed, they are often modeled after old voluntary programs that did not perform well. While there is talk about local control and community-based projects, it is still the federal agencies that normally decide on project criteria, the projects to be funded and the money to be spent. If we truly want an agricultural sector that supplies us with all the goods and services we desire, it is time to try some new ideas.

IDEAS TO TRY

We could attempt to form a National Trust made up of relevant land-use and environmental groups. This concept was highlighted in Lovejoy (1996) as the Agriculture Environmental Enhancement Trust. This Trust would be funded with the dollars from the CRP and WRP. It would have a Board of Directors with 50 percent of its members from land-user interests and 50 percent from environmental organizations. Local and state groups would propose land-retirement projects to the Trust. Expenditures by the Trust would have to be approved by a two-thirds majority vote by the Board of Directors, thereby eliminating some of the more parochial projects. The Board of Directors would have to decide which proposals are most important and which would do the best job of protecting the environment at the lowest cost. This would force explicit choices to be made and priorities to be set and would lead to greater effectiveness of our tax dollars.

Although we have the Natural Resources Foundation, as specified in the 1996 Farm Bill, it is a far cry from the type of institution proposed (Lovejoy, 1996). The Agriculture Environmental Enhancement Trust proposed here would be structured so there would be implicit goals and objective setting, as well as true compromises between land users and the environmental community. This structure would establish some quasi-market-type functions that would lead to better outcomes and greater environmental quality.

Another idea would be to allow the government to sell land-use rights that it had previously appropriated (Lovejoy, 1991). The federal government could auction holdings, such as drainage rights to wetland acres, which would promote a market for wetlands that could respond to the changing desires by the public. These public desires would include those for wetland amenities, food production and housing. This would ensure that consumers and producers have a method by which to determine how much of the various commodities they want, based on those commodities that they most value.

The above proposition implies that the federal government would auction drainage rights it appropriated under Section 404 of the Clean Water Act (CWA) and the Swampbuster program. In such an auction, local or national environmental groups could purchase drainage rights to environmentally valuable wetland areas. They would have incentives to raise money for this purpose. Those same groups, however, may not be so eager to purchase drainage rights to small pockets of wetlands in the middle of a cornfield because of the meager environmental benefits.

Commercial land developers might purchase some of these drainage rights by outbidding environmental groups, but that would constitute very few acres. The dollars generated by these developer purchases could be utilized to fund projects to restore wetlands in other environmentally sensitive areas. For those acres that are neither environmentally valuable nor attractive to developers, the present fee-simple landowners (for example, farmers) could bid a nominal amount ($1 per acre) and purchase those drainage rights.

This type of structure would force us to make choices concerning the value, including its environmental benefits, of a piece of property for various uses. In addition, it would initiate the establishment of a market for wetlands that would adjust to the changing preferences of the public and to the changing knowledge concerning the benefits derived from wetlands. Another benefit suggested by Lovejoy (1991) is "that it will allow members of society, through an observable market, to place values on wetland functions, and avoid having public bureaucrats or university scientists do so." (p. 419)

Although we have not been successful in the past, we could try to do more targeting based upon environmental criteria (Lovejoy and Lee, 1995). We could establish the primary goals of CRP to be environmental, especially off-site, where the focus would be (for example, water quality and wildlife habitat). While some moves have been made in that direction, targeting remains a debated issue, particularly the political dimensions of this form of redistribution.

In addition, conservation projects could be based on more long-term planning and could include concepts such as permanent easements. Temporary programs like the CRP often cost nearly as much as the outright purchase of the land, but they only produce the desired land-use changes for a limited period. For most environmental amenities, the assumption that such goods and services would always be desired might be warranted. While the 1996 Farm Bill limits easements to a period of 30 years, we should be moving in the other direction. If environmental quality is our goal, there is no evidence that a series of temporary programs would maximize anything except federal largess.

It is possible to focus upon more recently introduced community-based approaches to environmental protection. Changes in institutional roles and responsibilities, however, would be needed if community-based environmental protection (CBEP) were to achieve its potential. Federal agencies would need to play new roles such as leader, convenor, mediator and educator (Randall, this volume).

For CBEP to work, collaborative partnerships are needed among federal, state and local agencies, in addition to private parties and communities. While this may sound like the EQIP, the USDA does not seem to be changing its roles substantially. After all, who is still deciding (1) which projects are acceptable, (2) which projects will be funded, and (3) what protection is most important?

In related actions, the EPA is suggesting greater involvement by state and local governments and local groups in environmental protection. While this might be reminiscent of the concept of Environmental Federalism, in reality it is quite different. Also, the EPA is threatening certain states by suggesting that they are not prepared for the job of protecting the environment. Inside Washington, DC, apparently only the EPA is capable of saving us from ourselves.

IMPLICATIONS FOR FUTURE PROGRAMS

The examination of present initiatives illustrates their shortcomings and suggests improvements to consider for future programs. The following are a few suggested areas to consider when formulating policy and programs for attenuating the impacts of agricultural production upon environmental resources:

- We should continue to develop new technologies to meet society's ever-changing environmental needs and wants. The question of whether this means precision farming, new crops or production systems that have not yet been conceived is beyond the scope of this chapter (Khanna et al., this volume).
- We need to investigate the reason(s) some farmers are not using crop residue management systems or conservation tillage. What are the constraints to the adoption of these environmentally sound practices? Lack of management skills? There are many solutions if this is the problem. Is it the unique combination of soils, climate and crops? If this is so, technologies

must be developed to help produce the environmental quality and crops that we all want. Is it tradition? If that is the constraint, it will be solved as those farmers and ranchers retire, or as they find that they cannot compete with other producers who have higher net returns.

- In this era of tight budgets and demands for accountability, we must better target our conservation dollars. The focus should be on creating the greatest environmental benefit for each dollar expended. Environmental funding should not be focused on political pork, bureaucratic turf or tradition in the form of *laissez-faire* attitudes.
- Greater innovation and creativity is needed in developing programs to deal with these problems, which include routine evaluations to determine what works and what does not work. It seems incongruent that we have decades of experience with some of these programs and know so little about how to design successful programs for the future.
- Perception of the problem needs to be altered. Most of our emphasis seems to lie on the manner in which we can reduce degradation of the environment and protect farmers from unduly constricting rules and regulations. Instead we should be asking: "How can we assist farmers in providing the environmental enhancements desired by the public?"

CONCLUSION

Farmers and ranchers in this country have shown that they respond well to clear signals about what is important (for example, increasing production of commodities). Unfortunately, we have not sent the same unambiguous signals about the provision of environmental amenities. Restructuring our programs and policies to utilize more market and quasi-market signals would encourage farmers and ranchers to supply all the goods and services demanded by consumers.

REFERENCES

Braden, J. and S.B. Lovejoy, eds. 1989. *Agriculture and Water Quality: International Perspectives.* Boulder, CO: Lynne Rienner Publishers.

Lee, J.G. and S.B. Lovejoy. 1995. "Riparian Buffers: A 2,000,000 Acre Environmental Enhancement." Department of Agricultural Economics, Staff Paper No. 95-8, Purdue University, West Lafayette, IN (March)

Lovejoy, S.B. 1991. "Wetlands: Sell to the Highest Bidder?" *Journal of Soil and Water Conservation* 46(6): 418–419.

_____. 1996. "Environmental Enhancement in Agriculture: The Case for a National Trust." *Journal of Soil and Water Conservation* 51(3): 202–203.

Lovejoy, S.B. and J. Hyde. 1997. "Nonpoint-source Pollution Defies U.S. Water Policy." *Forum for Applied Research and Public Policy* 12(4): 98–101.

Lovejoy, S.B. and J.G. Lee. 1995. *CRP in 1995: Fewer Acres, More Targeting?* Department of Agriculture Economics, Staff Paper No. 95-8, Purdue University, West Lafayette, IN (April).

Napier, T.L. and E.C. Johnson. Forthcoming. "Awareness of Operation Future Among Landowner Operators in the Darby Creek Watershed of Ohio." *Journal of Soil and Water Conservation.*

_____. 1998. "Impacts of Voluntary Conservation Initiation in the Darb Creek Watershed of Ohio." *Journal of Soil and Water Conservation* 53(1): 78–84.

Nowak, P., G. O'Keefe, C. Bennett, S. Anderson, and C. Trumbo. 1997. "Communication and Adoption Evaluation of USDA Water Quality Demonstration Projects." USDA Agricultural Research Service, Water Quality Information Center, National Agricultural Library, Washington, DC.

Piper, S., R. Magleby, and C. Young. 1989. *Economic Benefit Considerations in Selecting Water Quality Projects: Insights from the RCWP*. Economic Research Service, Staff Paper No. AGES 89-18, USDA/ERS, Washington, DC.

Schuyler, L. 1993. "Nonpoint Source Programs and Progress in the Chesapeake Bay." *Agriculture Ecosystems and the Environment* 46: 217–222.

Sutton, J., D. Meals, and R. Griggs. 1994. *Interim Evaluation of USDA Water Quality Projects*. National Water Quality Enhancement Program Notes, No. 64, North Carolina State University Extension Service, Raleigh, NC.

5 FLEXIBLE INCENTIVES FOR ENVIRONMENTAL MANAGEMENT IN AGRICULTURE: A TYPOLOGY

Sandra S. Batie and David E. Ervin
Michigan State University, East Lansing MI and the Henry A. Wallace Institute for Alternative Agriculture, Greenbelt, MD

Flexible incentives are incentives that do not dictate how environmental objectives are to be achieved, and they are important tools in managing agro-environmental problems. They are, however, a means to an end and not and end in themselves. Successful implementation of these incentives depends on clear, enforceable performance standards. Furthermore, the best flexible incentive approach appears to be one that involves a combination of instruments that fit local, social, economic and environmental conditions. It is important to recognize, as well, that the flexible incentive approach can impose substantial transaction costs and can require a high level of both producer and agency human management skills. Thus the policy challenge is to find effective ways to lower the costs of using flexible incentives and of expanding the management capacity of those farmers and ranchers who can deliver high environmental values over the long run.

INTRODUCTION

In response to increasing political desires for smaller, less intrusive and less expensive government, policymakers are searching for new approaches to environmental management. Despite the broad recognition that federal environmental legislation and regulation has resulted in significantly cleaner air and water during the past three decades, many complain that these rules and regulations are too complex, frequently contradictory, involve too many federal agencies, and can be both expensive and inflexible. There are also accusations that much environmental legislation embodies a command-and-control philosophy that requires certain technologies and thwarts the use of less expensive and more innovative methods of achieving environmental goals.

The interest in more flexible approaches for environmental management does not equate with a retreat from environmental quality goals. It is clear that the public does not want to roll back the improvements in environmental quality that has been achieved during the past two decades. Indeed, the results of public opinion polls indicate that a majority prefers existing or higher standards for drinking water quality, wetlands conservation and endangered species protection (USDA/NRCS, 1995).

To date, agriculture has been exempt from many of the land, air and water regulations that affect other industries. Because of public concerns with agro-environmental problems, there is growing interest in the less traditional, flexible incentive approaches when dealing with these problems by both government and private sectors. The seemingly simple concept of designing and implementing incentives to achieve agro-environmental goals in a flexible and cost-effective manner is, in reality, quite difficult. Not only does the concept involve choosing from a typology of flexible incentives, but it must also answer the following questions: Why are flexible incentives desirable? What are the goals of the flexible incentives? If flexible incentives are to be used for agro-environmental problems, who should have flexibility? How can flexible incentives be assessed?

These questions present four themes that highlight the complexities of flexible incentives:

- A flexible incentive is a *means* to an *end* and not an end in itself. Its successful implementation depends upon clear performance objectives.
- A flexible incentive panacea does not exist; the best flexible incentive approach probably would involve a combination of instruments that fit local, social, economic and environmental conditions.
- A flexible incentive approach imposes substantial transaction costs. Institutional reform and innovation are necessary to lower these costs and to spread the use of management systems.
- A high level of both producer and agency human management skills are required to implement a flexible system in an effective and low-cost manner.

TYPOLOGY OF FLEXIBLE INCENTIVES

Flexible incentives refer to environmental management tools that specify objectives but usually do not dictate how the environmental objective is to be achieved. They give agricultural producers discretion over what technologies they would use in order to reach a performance standard, such as ambient water quality in a watershed. An example of a performance standard that specifies a desired outcome is a 10 milligram per liter agricultural nitrogen runoff. Performance standards leave a producer flexible to find his own least-cost pollution prevention or amenity creating strategy. Thus, a performance standard specifies what needs to be accomplished, but it does not specify a certain technology as the method by which to reach that standard.[1] Performance standards may come from local, state or national government programs, voluntary industry agreements, international trade pacts, common law precedents and other sources.

Ideally, with a flexible incentive typology, the agency responsible for securing the objectives has the discretion to target certain areas that will generate the greatest long-run improvement for a given social expenditure. For example, a performance standard might be administered on a watershed basis as opposed to an individual farm basis, such that each producer within the watershed would not be required to attain the same level of compliance. By exempting certain operators, it may be more likely that a least-cost combination of pollution reduction across all producers could be obtained.

Well-designed flexible incentives imply the provision of comprehensive and consistent incentives such that environmental improvements in one area are not offset by degradation in other ecosystem resources. They also provide the impetus for the adoption of technologies, so that producers can ultimately meet environmental objectives that apply to the industry or subsector.[2] These incentives may stem from a variety of sources that include the desire for profit maximization, stature, improved public relations, or from perceived liability of existing or foreseeable government policies, rules and regulations.

Flexible incentives pertain to instruments for pollution control, pollution prevention and the supply of positive environmental services, such as landscape amenities and wildlife habitat. These instruments include economic incentive schemes, such as charges for effluent discharges or ambient conditions above a threshold minimum; subsidies, such as grants, tax allowances or deposit refund systems; strategies that link government payments to environmental compliance; and market creation, such as those associated with emissions trading or ecolabeling. These instruments for pollution control also include moral suasion; peer pressure; education and technical assistance; green certificate awards; and regulations that impose performance standards but permit unrestricted technology choice and the trading of pollution rights. Space limitations preclude a discussion of all possible flexible incentives for agro-environmental management. Major conceptual approaches with potential flexibility, however, are described below and are summarized in table 5.1.

Charges

There are three types of charges—effluent, ambient and input. Effluent charges are fees levied on the producers who are responsible for discharging pollutants into the air, water, soil or for the generation of noise (Hanley et al., 1997). Ambient charges are penalties placed on producers who are responsible for causing the concentration of pollutants in the medium of interest (such as water) to exceed specified standards. Input charges are taxes placed on production inputs, such as those that exist on fertilizers and pesticides in California, Iowa and Michigan. By increasing the cost of agricultural inputs, one possible outcome of input charges is the substitution of safer methods of production for the more harmful products. The success of such charges depends upon the size of the tax and the strength of the

TABLE 5.1 Typology of Flexible Incentives

Conceptual Approach	Potential Flexibility

Charges:
  Effluent
  Ambient
  Input
	Charges levied on pollutants into air, water or soil, on the generation of noise or for exceeding standards of pollutant concentration, or on production inputs.

Subsidies	Financial assistance given to promote pollution prevention and/or pollution control.
Education and Technical Assistance	Assistance and/or education on pollution problems and solutions.
Compliance Rewards	Environmental performance requirements provided as a condition for continued eligibility for other government program participation.
Deposit Refunds	Incentives to recycle, reclaim or properly dispose of potential pollutants.
Marketable Permits	Provision of tradable permits for predetermined levels of pollution.
Ecolabeling	Market labels asserting environment protecting production processes or products.
Performance Bonds	Posting of a financial bond that is forfeited with unacceptable pollution behavior.

Other:
  Contracts
  Assigned Liability
	Other mechanisms to promote pollution prevention and/or pollution control.

Source: Authors' compilation.

linkage between the input and the desired environmental condition (Khanna et al., this volume). These charges create economic incentives for producers to consider, at least partially, when they encounter the environmental effects that remain outside the normal market channels. They are due to the non-rivalry or nonexclusive characteristics of flexible incentives (Randall, this volume). With the imposition of such charges, the property rights to the environmental resource shift in favor of those who cause the degradation and, in this way, producers assume part of the cost of environmental improvement (Vatn and Bromley, 1997). This input charge approach has been discussed in textbooks and articles but has rarely been used in agriculture at a level that actually alters polluter behavior.

Subsidies

Subsidies are financial incentives for producers to create positive environmental services, such as prevention, control or remediation of pollution. Subsidies may take the form of grants or cost sharing, low interest loans and tax allowances. The basic difference between subsidies and input charges depends on who has the property rights with respect to agro-environmental pollutants and services (Bromley, 1996). For example, does the off-farm public have the right to a high quality environment such that polluters must have penalties levied on them by officials? Conversely, do producers have the right to pollute and be compensated through subsidies when they are required to reduce pollution? If they are large enough and are not directed toward ambient performance conditions, subsidies can cause the industry to expand and consequently can exacerbate pollution problems. Input charges are more likely to have the opposite effect.

An example of a flexible subsidy involves cost sharing for pollution prevention. This example assumes that specific prevention technologies are not mandated as conditions of cost sharing to the exclusion of others. Ideal cost-sharing subsidies are based on their economic benefits to society, and are offered as long as the practices are used and paid to all that adopt the practices (Ribaudo and Caswell, this volume). These conditions are not usually met with most agro-environmental cost-sharing subsidies.

Rental payments, which temporarily or permanently retire land from production, are the most common forms of environmental subsidies in agriculture. Land retirement schemes have been the publicly preferred approach for natural resource conservation, pollution control and positive environmental services in U.S. agriculture. Indeed, of the nearly $3 billion in annual federal spending allocated to conservation and environmental programs in agriculture, about 70 percent has been apportioned for land rental expense. The Conservation Reserve Program (CRP) is the most utilized type of subsidy program, followed by the Wetland Reserve Program (WRP). Retirement contracts preclude almost all commercial uses of the land and, therefore, provide little flexibility to producers when meeting the environmental objective, except that participation in the program is voluntary. In this way, retired producers do not satisfy the definition of flexible incentives given above. Yet, the approach may be flexible from the point of view of the management agency when assembling a portfolio of strategies.

Cost sharing (for pollution prevention) and control practices (on cropland and livestock operations under the Environmental Quality Incentives Program (EQIP) of the 1996 Federal Agricultural Improvement and Reform (FAIR) Act) are other types of subsidy programs (Ribaudo and Caswell, this volume). Other subsidy incentives, gleaned from the FAIR provisions, are available to assist states when purchasing farmland development rights for the protection of landscape amenities.

Tax allowances, another form of subsidy, include income tax, sales tax and property tax reductions in exchange for producers who choose land-use practices that improve environmental quality. Income tax provisions may provide tax reductions for the adoption of conservation practices, but they probably do not provide the main motivation for such decisions. There are examples of property tax reductions designed to encourage environmental protection. One such example is found in Pepin County, Wisconsin, where per-acre property tax credits are linked to the adoption of an approved conservation plan.

Education and Technical Assistance

In instances in which farmers or ranchers inadvertently create environmental risks, education and demonstration can provide incentives for the adoption of environmental technologies. This adoption can be accomplished by increasing the awareness of such problems, the knowledge of technologies, the awareness of technical and financial assistance and the awareness of market opportunities that are created by the demands for environmental quality. Since the 1930s, the provisions of education and technical assistance (ETA) have been at the core of federal conservation programs. This is a voluntary approach that is a direct result of the Depression era philosophy of helping financially needy farmers reduce soil erosion. The underlying rationale is that, once producers are aware of environmental problems within their operations, they will seek to reduce these problems. Cost sharing would also be made available to these producers, encouraging them to adopt the approved practices.

In one sense, the ETA approach epitomizes the notion of flexibility. Producers can voluntarily receive environmental advice and counseling from informed officials on concerns and their possible solutions. Evidence indicates that ETA alone exerts insignificant effects while cost sharing has been influential in the adoption of the conservation practices (Lovejoy, this volume). Cost sharing usually is tied to a restricted set of practices that is approved for use within the state or local area. New guidelines under EQIP promise to increase the flexibility of ETA and of cost sharing.

Another ETA example is the recently announced U.S. Department of Agriculture (USDA) program, which has a goal of using integrated pest management (IPM) practices on 75 percent of the cropland by the year 2000. This program is designed to reduce the deleterious impacts of pesticides on humans and on the environment. The USDA budget for this program contains a request for about $10 million in new resources that could create new ETA efforts. The IPM program could be successful in reducing the use and harmful effects of pesticides if its practices were profitable. Studies have shown, however, that education alone is not effective in promoting the adoption of practices that are unprofitable (Camboni and Napier, 1994) unless the farmer's immediate environment or health is at risk (Napier and Brown, 1993).

The key role of profitability is not surprising to economists, but the elements that need to be considered when assessing profitability must be explored more fully. Of the various production and environmental practices observed in use among farms and ranches, profitability is not a uniform determination. In the past, for example, many farmers immediately adopted the reduced tillage technologies, but the extent of use has now leveled off at below 50 percent of cropland. Profitability depends upon a variety of factors that include existing capital stock and technologies, management skill (human capital), natural resource conditions, relevant input and output prices, and rules and regulations.

Compliance Rewards

Compliance rewards rely on the existence of other government subsidies in order to achieve environmental objectives. They establish environmental performance standards as conditions for the continued eligibility of producers for government programs and benefits, such as commodity program payments. Compliance requirements are not considered true regulations if they are attached to a program that is voluntarily implemented by producers. In concept, compliance schemes could be flexible incentives, if the producer were to have wide latitude in selecting the practices that would meet the standard. If the compliance strategy required the use of a restricted set of technologies that may or may not fit the farm, then it would not be a flexible approach. In practice, the conservation and wetlands compliance provisions of the 1985 Food Security Act (FSA) began with little flexibility. With few exceptions, all operators of highly erodible lands and owners of unfarmed wetlands had to comply. Over time, more flexibility for producers was incorporated by varying the standards according to the natural resource situation and by permitting a wider range of practices. FAIR continued the trend of incorporating more flexibility (Ribaudo and Caswell, this volume). Weaknesses of the compliance strategies, as practiced, are that the foregone benefits that stem from noncompliance do not necessarily correspond with the incentives necessary to meet the standards. The incentives vanish when the program payments cease.

Deposit Refunds

Deposit refund systems provide incentives to recycle or properly dispose of potential pollutants. Deposit refunds are most applicable in agriculture for the proper disposal of pesticide containers. Purchasers of pesticides pay an additional fee that is refunded when the empty container is returned to a designated disposal or recycling site. Several states have deposit refund incentives.

Marketable Pollution Permits

Marketable pollution permits are based on a predetermined level of effluent emissions or ambient concentrations that are acceptable within a watershed or airshed. Permits allocate this acceptable amount among the producers in the watershed or airshed who are allowed to trade them for money with others. Because the permits are scarce, they have value. Permit trading allows producers to develop least-cost pollution abatement strategies. Experiments in marketable trading permits, particularly point-source pollution (PSP) and nonpoint-source pollution (NSP) trades, are ongoing in several parts of the nation—although actual trades have been limited. An example of a marketable pollution permit program that has not been successful is one established in 1989 in the Tar-Pamlico River Basin in North Carolina. No trades have yet occurred for this basin due to a variety of reasons that include performance standards, which can be met more easily with new technology than with trades, imprecise property right expectations and bureaucratic obstacles. The primary advantage of marketable permits is that the system delegates the responsibility for calculating the costs and benefits of pollution control to the firms that hold or desire to obtain those permits. This responsibility, however, often entails large transaction costs for information discovery that hinders trading (Stavins, 1995).

One successful U.S. tradable permit program, which has unquestionably reduced the costs of compliance with the Clean Air Act, is the program for controlling aerial sulfur emissions. Almost all of the trading, however, has been between large polluters. Furthermore, the deregulation of the railway rates and the subsequent drop in the delivered price of low-sulfur coal has made it difficult to ascertain how much of the improvement in air quality was directly related to the trading program (Tietenberg, 1990).

Ecolabeling

Market creation can include the ecolabeling of products grown or raised by production processes with certain environmental benefits (van Ravenswaay and Blend, this volume). These labels inform consumers that the product contains some level of environmental performance. Such labeling schemes possess flexibility because they use decentralized market systems to convey information. Consumers are free to choose and reward the type and level of environmental performance that they wish, and the farmer or firm is free to meet the demand in the most efficient way. To be effective, labeling schemes may require public action in order to define the content of the label and to ensure its validity. These are services that deliver non-rival and nonexclusive benefits.

More than 20 countries in addition to the European Union have adopted ecolabeling programs. The oldest program is Germany's Blue Angel Seal that was established in 1988 and is now applied to over 3,500 products. There are still many difficulties associated with ecolabeling. It is frequently hard to discern exactly what is being promised. The potential for consumer deception is high. Products may be more environmentally accountable in one use or location than in another.

Companies may face liability if a product fails to live up to its expectations. Eco-labeling may have mixed results for businesses, consumers and the environment. Still, there is evidence that, for some companies, ecolabeling provides both market advantage and improved environmental outcomes. As an example, Gerber's Baby Food Products is a private company that strives to have zero incidence of pesticide residue. Perhaps more consumers will purchase greater quantities of these products than they will of those products without ecolabeling.

There is also a private industry code, referred to as ISO 14000, that has been established. It is still under development at the industry-driven International Organization for Standardization centered in Geneva, Switzerland. The objective of ISO 14000 is to provide a common approach to the ecolabeling of products and environmental management. The code addresses a company's entire range of activities that include product design, planning, training and operations. Although compliance with ISO 14000 is currently voluntary, there may be strong market forces that ultimately make it profitable to comply.

Performance Bonds

One mechanism for assigning liability to a potential environmental polluter is a performance bond. This mechanism operates such that an agricultural producer or processor must post a bond that would be forfeited if his or her production practices were judged to have caused pollution above acceptable levels. Ideally, the amount of the bond would be based on the potential environmental impact and would be negotiated between the polluter and regulator. Such instruments have been used by certain states as conditions to the importation of nonindigenous species that have a potential for commercial and environmental damage (OTA, 1995a). Performance bonds are also occasionally used in the transport of pesticides or manure on public highways.

Other Flexible Incentives

Contracts between private parties, such as processors and growers, may also influence the adoption of environmentally conserving technologies. Thus, a flexible incentive might be one that alters contractual obligations so as to provide incentives or to remove barriers to the adoption of conserving technologies (Chu et al., 1996). Cases exist in which the contractor desires improved environmental quality (for example, for public relations or for consumer demand reasons) and uses the contract as a means to achieve these results.

Incentives could also be created through legislation or common-law doctrine that would assign liability for certain environmental outcomes or acts. Examples exist in which regulation, the threat of future regulation, or the assignment of liability has provided the incentive for businesses to have self-regulation. Often these self-regulating standards exceed existing government-imposed standards and frequently do so at lower costs (Batie, 1997). Liability may also come from the threat of private lawsuits.

Interpreting the Typology

There are many flexible incentive options. An instrument that is a panacea has not yet emerged from federal, state or local experiences because incentive instruments are means to an end and are not ends in themselves. Therefore, these instruments must fit the particular environmental problem and must include its associated socioeconomic and political conditions. Although we have discussed the types of instruments individually, multiple approaches can be used for any problem in any area. Mandatory regulatory penalties for environmental performance below a minimum standard can be combined with voluntary subsidy approaches to encourage performance above the minimum standard.

WHY ARE FLEXIBLE INCENTIVES DESIRABLE?

In concept, it is desirable that flexible incentives accommodate a diverse mix of natural resource conditions, heterogeneous production systems and socioeconomic factors that vary over space and time. In other words, spatially and dynamically flexible incentives are necessary to solve the standard economics cost effectiveness problem under uncertainty.

There is good physical and biological evidence and emerging environmental economic theory to support the need for such dexterity in the treatment of the highly variable environmental management situations typical to agriculture (Barbash and Resek, 1996; Antle and Just, 1992; Mueller et al., 1992; Opaluch and Segerson, 1991). The joint distribution of natural resource and production parameters over the natural landscape creates a wide diversity of conditions that are not amenable to uniform strategies (Antle et al., 1996). For example, Cooper and Kiem (1996) estimated that variable cost-share rates are necessary to reach selected water pollution goals at least cost. Moreover, there is clear evidence of the integrated relationships between soil, water and air quality as affected by agricultural production systems (NRC, 1993). For example, improvement in the levels of soil organic matter can serve to trap the runoff and leaching of fertilizers and pesticides from farmland.

Dynamic influences are as important as spatial influences but are not adequately designed into agro-environmental management programs. Key factors shifting the temporal demand and supply of environmental quality are usually not incorporated into programs. These factors include infrequent, extreme episodic events; learning due to monitoring information or regulation research and development innovation; and rising environmental prices from income growth (Ervin and Schmitz, 1996).

One way costs are reduced is through innovations that are induced by the response to an environmental regulation. Polluters can redesign the polluting system (for example, the farm) so as to achieve improvements in both profits (through reduction in costs or enhancement of revenues) and environmental performance. Thus, flexible incentives should be thought of in a dynamic context in which the search for solutions leads to asset replacement and technological development; this, in turn, lowers compliance costs and leads to the development of new products and processes. Lowered compliance costs are often referred to as innovation offsets.

The evidence to support or refute the notion of innovation offsets is incomplete (Palmer et al., 1995; Porter and van de Linde, 1995). More study is necessary to understand the conditions that may prompt or deter such innovations—such as the role of incomplete information markets and other missing incentives. Nonetheless, the maintained hypothesis is that flexible dynamic incentives offer the best chance of increasing long-run profits. The technological innovation and redesigned systems that can result from whole-farm planning, which integrates production and natural resource processes with marketing opportunities, is one example (Chambers and Eisgruber, 1998; Jones, 1998).

It is important to note, however, that flexible or other forms of economic incentive mechanisms cannot be shown superior *a priori*, either on a first or second best basis, to some other regulatory path (Russell and Powell, 1996). Superiority is often presumed, but that presumption usually rests on an incomplete accounting of costs and an underappreciation of the complexity of institutions that are necessary to administer incentive-based systems. Economists articulate the virtues of flexible environmental management approaches that allow producers with specialized knowledge of their production systems to reallocate inputs and outputs in lowest cost fashion. Effluent charges have been a favorite pedagogical device. In reality, the accrual of savings from the implementation of these analytically elegant systems requires different charges for each spatially distinct environmental market. Furthermore, adjustments in charges over time must be made for shifts in pollution, supply, environmental quality, demand and inflation.

When public transaction costs for information gathering and enforcement are combined with private compliance costs, it is not obvious which policies are the most cost effective (Carpentier, 1996). Indeed, the nontrivial administrative costs of implementing such flexible systems may explain, in part, why there have been so few cases of effluent charges in agriculture. Furthermore, some recent experiments to implement pollution-trading regimes have not performed as expected because of high transaction costs (Stavins, 1995). NSP and PSP trading strategies are frequently handicapped by both the lack of clearly defined limits on the rights to pollute by nonpoint polluters and the lack of ambient performance standards.

While flexible incentives have many desirable attributes, their uses in agriculture have been rare. (The exception to this generalization has been the use of subsidies.) Even in the case of industrial pollution, the use of flexible incentives has been limited. Also, when these incentives have been used, they frequently have not changed pollution behavior. For example, effluent charge rates and product charge rates are frequently set too low to provide true incentives. While there has been an increasing role for economic incentive instruments, much of the motivation has occurred because of the revenue raising properties associated with some of them (Pearce and Turner, 1990). It is a fair statement to say that much of what we know about flexible incentives is obtained from theoretical discussions and analysis—there are few actual experiences to examine.

APPLICABILITY TO AGRO-ENVIRONMENTAL PROBLEMS

Achieving agro-environmental goals with flexible public policies requires clear, specific and measurable objectives. They also require information on sources of pollutants, the movement of pollutants through time and space, and the impacts of pollutants on various environmental attributes of concern. Unfortunately, none of these requirements are well met in agriculture.

Measurable Objectives

Ambient environmental objectives have not yet been applied to most agro-environmental problems, and their absence has profound implications for the design of flexible incentives. Without clear, specific and measurable objectives, incentive-based environmental programs will flounder and fail (Davies and Marousek, 1996). Such a prognosis would not be so stark if technology design standards or land retirement strategies could be shown to consistently and reliably correlate with desired environmental outcomes. Unfortunately, while the use of some design standards (for example, buffer filter strips) score better than others, the use of many design technologies and untargeted land retirements do not necessarily correlate with the achievement of a given environmental goal (NRC, 1993).

Despite 60 years of U.S. federal conservation and environmental programs for farmers and ranchers, few performance objectives have been set for the industry. This outcome can be traced to the dominant use of voluntary payment approaches (subsidies for land retirement and best management practices) that began during the Great Depression. The environmental objectives of the vast majority of federal programs that affect the industry are generally couched in terms of the use (or non-use, in the case of toxic pesticides) of particular management technologies. Alternatively, other federal efforts have established codes of good practice for farmers who receive public subsidies, such as compliance strategies. For example, the CRP was established with an objective of retiring 40 million acres of cropland in order to control supply and government expenditures, and to meet certain environmental goals. This enrollment goal, which depended upon the specific lands that were enrolled in the program, did not automatically translate into specific environmental improvements. In the early stages, the environmental performance of the CRP was criticized because some enrolled lands had little conservation value.

While the USDA estimates that the CRP has achieved significant environmental benefits (USDA/ERS, 1994), the estimates generally rely on simulated effects rather than on actual data. Few studies have measured actual environmental responses on the ground. Because no ambient performance goals were set and no pre-CRP baseline conditions were measured, most measures of improved environmental conditions have to be considered unverified estimates, at best, and anecdotal, at worst. Improvements in wildlife numbers and diversity are exceptions because they have received considerable monitoring (OTA, 1995b).

The FSA also established soil erosion and wetlands compliance provisions that required participating farmers to meet conservation standards if they desired to remain eligible for commodity program payments. The objectives of the compliance provisions were to achieve these conservation standards on all highly erodible cropland and wetlands (about 150 million acres) on participating farms. In effect, these compliance provisions only established codes of good practice for farmers. Like the CRP, specific ambient environmental objectives were not set.

The goals and objectives that guided the CRP and compliance standards were mostly renewed in the FAIR Act. Congress reauthorized the existence of the CRP, but this time it had the dominant goal of maximizing environmental benefits per-dollar spent. Supply control purposes were removed—which was consistent with other commodity, market deregulation measures. (Recently announced enrollments suggest that the environmental cost effectiveness of the CRP will improve.) A more inclusive measure of environmental benefits was used. Revised enrollment procedures dropped the average rental rate by about 20 percent. The compliance provisions also continued. Their scope, however, was restricted, and the removal of some program payments, such as crop insurance, weakened their leverage. If agricultural program payments were to be phased out as planned in the FAIR Act, then the incentives to maintain erosion and wetlands conservation standards would disappear entirely by the year 2002.

It appears that traditional federal programs have incorporated limited flexibility and few explicit objectives. Federal funding now predominantly supports land retirement that is driven by enrollment goals. It has little apparent flexibility for producers, but it has discretion for agencies. The traditional pattern appears to be shifting with the FAIR initiatives and with several experiments that are underway in the United States. EQIP may be the prototype for future federal agro-environmental programs. It has focused, in part, on an emerging environmental priority for agriculture—the management of livestock wastes. This program will employ new procedures that purportedly give producers more flexibility in designing waste control and disposal systems. The prototype EQIP program does not have explicit environmental objectives. Rather, the emphasis appears to be on the process. Thus, the program operation could easily slip into the technology design tradition.

Information on Source, Movement and Impacts of Pollutants

While there are federal agro-environmental polices that have measurable attributes; they tend to be focused on practices and land coverage rather than on ambient quality conditions. This tendency is due to the diffuse NPS nature of many agro-environmental problems and from political precedent. Setting an objective of meeting specific water quality standards, such as nitrate or pesticide concentrations in water, presumes that the technology exists to trace those contaminants to their source. This requirement, to relate source to impact, is complicated by diverse farm businesses, topography, practices and stochastic weather events. Currently, a sufficient level of scientific sophistication is not available to deal with most agro-environmental problems on an ambient quality, performance standard basis. While simulation models can substitute for empirical studies in many situations, they are also hampered by the lack of validated data. Even when the technology does exist, the achievement of specific quality objectives through voluntary payment approaches still requires adequate funding. That presumption is less likely in a tight fiscal climate. Thus, the land coverage objectives may be seen as better policy objectives for most agro-environmental problems.

Performance Standards in Practice

An exception to the generalization that performance standards are missing in agro-environmental federal legislation is the Clean Water Act (CWA) of 1972. This Act required Confined Animal Feeding Operations (CAFOs) with more than 1,000 animal units to have discharge permits (Norris and Thurow, this volume). CAFOs are, in effect, treated as PSPs. They are easier to identify and therefore easier to relate to ambient impacts. The stipulated performance standard, however, is no discharge even after a rainfall occurrence. This standard exceeds that which is applied to municipal waste. The CWA discharge permit requires that a CAFO be able to store an amount of rainfall that would be equivalent to a 24-hour rainstorm that is expected to occur once every 25 years. Thus, to satisfy the compliance requirements, this generally would mean the construction of a lagoon. This strict no-discharge standard limits the choice of treatment technologies (many of which produce some discharge) and also effectively eliminates CAFOs from participating in pollution-trading markets. Thus, the CWA's use of a performance standard is not a flexible incentive. This inflexibility may explain why CAFO standards have not been implemented in a uniform fashion across the United States.

A second example of implemented federal agro-environmental performance standards is embodied in the Coastal Zone Management Act Reauthorization Amendments (CZARA). These Amendments require that each coastal state submit approved coastal zone plans, which identify design standards that are best suited to solve NSP problems. Individual states can use voluntary incentive mechanisms, but they are required to impose mandatory measures if these fail to achieve the appropriate levels of protection. Coastal waters include the oceans, the Great Lakes and the watersheds that drain into them. Thus, the potential impact of CZARA is large. The implementation of CZARA provisions for agriculture is still in its early stages. After the failure of earlier efforts that imposed strong technology design standards, most states currently appear to be adopting slower processes that favor voluntary approaches.

Some states are experimenting with direct, ambient environmental objectives for agriculture that will encourage flexible responses. For example, Oregon has established total maximum daily loads (TMDLs) for nitrates and phosphorus in rivers and streams that fail to meet standards for certain designated uses. Once the TMDLs are set, the State Department of Agriculture works in concert with local political agencies to reduce water pollution, first using voluntary payment measures (Wolf, 1996). In the Oregon experiment, all farmers within targeted watersheds were responsible for developing their own sets of management practices. These practices were then evaluated by local governing agencies for consistency with TMDL goals. Because of the difficulty in linking farm and ranch practices to ambient conditions, local agencies were using landscape performance standards, such as minimum residue on tilled acres and no tailwater irrigation discharges into streams. If producers within the watershed failed to meet these standards, the state agency would then intervene and impose civil fines to secure compliance.

Oregon authorities view performance standards, such as TMDLs and landscape conditions, to be more flexible and efficient than narrowly defined land management practices. The principal advantages are that the imposition of the standards (along with the threat of penalty) sets in motion public and private research and development processes that can lower the costs of meeting the target (that is, innovation offsets) in addition to creating value for waste reduction processes and products.

Nebraska also uses performance standards for groundwater in the Central Platte Region that is also backed by agricultural practice restrictions. Nebraska's performance standard program would be flexible only if producers were farming over aquifers that have not yet exceeded the threshold of groundwater nitrate levels. Where aquifer nitrate levels exceed specific thresholds, farmers are restricted to certain agricultural practices.

Florida has established performance standards for phosphorus runoff from dairies that flow into Lake Okeechobee. These dairies are free to meet the phosphorus standard using any method they desire. If they fail to maintain compliance with the governing water quality programs, however, state action could follow.

Political sentiment for using more direct control measures to achieve agro-environmental goals appears to be spreading as agricultural production concentrates in larger operations. Problems with the adequate enforcement of CAFO permits also influence public sentiment for more protection. A review of state water quality programs identified 23 states that could place constraints on agricultural activities through penalty mechanisms (Ribaudo, 1997). Most of these programs were focused on particular pollutants or water resources, which suggest that performance objectives were likely to be included in these state programs. Whether these standards will allow for flexibility in response, however, remains to be seen.

The Challenge

The challenge of using flexible incentives to solve agro-environmental problems is multifaceted. Effective use of incentives requires clear, measurable and enforceable goals as well as information on the linkage between the environmental impacts and the location and characteristics of the farm source. In addition, incentives should be designed to encompass land changes when they are required to obtain the desired outcomes. Clearly, there are both research agenda and policy implications found in the paucity of performance goals; missing knowledge that links source with impact; a lack of knowledge about adoption behavior; and a lack of research and development processes to stimulate low-cost or profitable technological innovations.

There is considerable scientific and anecdotal evidence showing that changes in land use can reduce runoff and leaching rates from individual farms. Such reductions have considerable promise for improving the environmental performance of individual farms (NRC, 1993), but they do not necessarily translate into overall improvements in environmental quality. This is due to factors, such as topography, storm occurences and pollutant levels. Flexible incentives can be used to encourage reductions in runoff and leaching and to better link these reductions with environmental outcomes. To the extent that agro-environmental programs accomplish these outcomes, the more they will fall into the true definition of cost-effective flexible incentives.

WHO SHOULD HAVE FLEXIBILITY?

Effective public policy consists of intentions, rules and enforcement (Bromley, 1996). Intentions in agro-environmental policy are conveyed by objectives such as performance standards. Even if performance standards are set, there remains the task of structuring and enforcing the rules. Appropriate and effective flexible incentives get the rules right in order to attain the environmental objectives. Getting the rules right means that the various actors involved—such as agencies, farmers and ranchers, processors and contractors—are faced with rules and a property right structure that clearly assigns responsibilities for outcomes. Translating the conceptual notions of flexibility into actual effective programs will be difficult because of the need for institutional change. A first step in anticipating those hurdles is to review the perspectives of the major actors in agro-environmental management.

Consider the following agro-environmental problem: an excess of phosphorus in a watershed. Assume that the environmental objective is to limit the TMDL of phosphorus into the water such that the ambient concentration does not exceed 10 milligrams per liter during a given time period. Examples of potential measures include:

- Imposing regulatory penalties on observable effluent concentrations above permissible levels and providing subsidy payments for achieving observable effluent concentrations below these levels.
- Imposing taxes on phosphate fertilizer that are paid by input suppliers or producers.
- Private liability assignment for effluent concentrations above the threshold value.
- Technical assistance and cost sharing provided by agencies and universities to facilitate the adoption of desirable management practices when effluents are unobservable.

What measure or mix of incentive instruments would work best? The answer depends upon the criteria that we use to define what is best. We assume the primary criterion is minimizing the long-run expected social cost of achieving the objective, which includes both private and public expenses.[3] The search for the best incentives to minimize long-run costs, however, requires an examination of the kinds of flexibility needed to achieve dynamic cost minimization.

Pragmatically, three different sets of agents require flexibility. They are (1) governmental agencies that target and tailor management approaches to priority areas that reflect the integrated environmental systems of multiple media—air, water and land; (2) farmers and ranchers who design and implement the lowest cost technologies for their operations that link production and environmental management; and (3) processors and contractors who meet the environmental objectives that markets or government programs may require. To solve the dynamic decision problem, the incentives for these agents must also permit adaptation to external shocks over time.

Government Agencies

Any shift to more flexible approaches will encounter bureaucratic inertia that is motivated, in part, by the political cost of redistributing program benefits and by agency attempts to retain control over program resources. Traditionally, programs have been designed to broadly spread benefits so that they satisfy diverse political interests and generate continuing support for reauthorization and funding (Browne, this volume). Despite these forces, there is a discernible trend toward programs that target problems and grant more latitude in the selection of management measures.

There are five areas in which greater flexibility for government agencies would benefit agro-environmental management:

- Perhaps the most obvious need for flexibility by agencies is the freedom to select geographical targets or other measures (for example, filter strips) that merit special program efforts. The enrollment of CRP lands under the FAIR dictum, "to maximize environmental benefits per dollar spent," illustrates this aspect. Enrollment will be guided by an environmental benefits index (EBI), which accounts for different types and degrees of environmental problems in comparison to the requested rental rate. Although some environmental dimensions are still absent from the formula, the EBI is a significant change from past procedures. Original enrollment procedures called for the selection of lands from a pool of eligible cropland that had only to satisfy some minimal criteria, such as degree of erodibility. There is a chance that enrollments could revert to previous patterns because the EBI and rental rate processes are not fully transparent. If such a reversion could be avoided, however, simulated estimates indicate that a targeted land retirement program would have net benefits (Ribaudo et al., 1994).
- Agencies also require the discretion to design incentives that will assure adequate participation and the most cost-effective protection over time. An example of restricted flexibility is the WRP that will enroll lands in a fashion that maximizes net benefits. This program mandates a fixed proportion of contracts of particular length rather than it allows agency officials to offer landowners an unrestricted range of options. Another example about to gain public attention is the execution of contracts under the EQIP program. It is unclear whether the responsible agency will be able to pursue contracts that assure the maximum economic life of pollution control measures. In an ideal world, the agency should have the ability to im-

pose incentives (or penalties) and to change objectives (or quality targets) with the rising environmental demand or with declining compliance cost (supply). Furthermore, there is an economic rationale for providing subsidies to those farmers who are expected to remain in business over the life of the pollution prevention investment.

- A flexibility dimension for agencies is the discretion to match instruments to the type and level of problem. For example, does the agency have sufficient latitude and funding to select the set of lowest cost measures for controlling nitrate pollution into groundwater? Does it have to employ land retirement when the technical assistance for soil nutrient testing would suffice at less cost? The flexibility to coordinate agency efforts is important as well because some 40 federal programs for agro-environmental management currently exist and there is little coordination in their execution. Different departments, and even different agencies within a department, may be unaware of related and potentially useful efforts.

- Agencies need the discretion to foster the adoption and development of management practices and systems that fit specific farm and ranch operations and whole natural resource systems, such as watersheds. This need is perhaps the central perception of needed flexibility for agro-environmental management. A key to successful approaches may be the ability of the agency to foster collaborative, public/private partnerships because sufficient public resources will likely not be available to address each problem in each location.

- The agency ideally should be empowered to stimulate public and private research and development that would deliver lower cost or more efficacious measures than research that is currently available. This desirable flexibility feature is virtually absent from the present set of agro-environmental management programs.

Farmers and Ranchers

What kinds of flexibility might be most helpful to producers when meeting environmental objectives? Two general types should be highlighted:

- The successful application of environmental management technologies that keep costs low or even improve profits appears to hinge heavily on operator management (Ervin and Graffy, 1996; OTA, 1995a). For example, those operators, who took early advantage of reduced tillage technologies that simultaneously lowered production costs and reduced erosion and water runoff, had more formal education (Rahm and Huffman, 1984). Farmers and ranchers, therefore, require the opportunities and the resources needed to invest in specific human capital to most efficiently manage wastes through redesigned production systems or innovative marketing strategies. Public education and technical assistance programs are vehicles that provide these opportunities but have traditionally been oriented to practice-by-practice approaches. Emerging trends suggest that private firms are increasingly providing production and environmental management services that may augment human capital or substitute for producer management and that may address whole-farm planning.

- To take full advantage of enhanced management skills, operators require the flexibility to organize the resources and outputs of the whole farm or ranch as an integrated production and marketing system. Given their intimate knowledge of system interrelationships, this flexibility maximizes the opportunity that farms will not only meet profit objectives but that they will meet environmental requirements as well. Whole farm or ranch plans are not new. Early farm management education efforts stressed the development of plans driven by operator objectives that encompassed all farm production resources. The extension of those planning concepts to account for on-farm natural resources and off-farm environmental impacts is a new concept. Federal and state environmental management agencies are experimenting with the use of this new, broader approach to whole farm plans. It is too early, however, to assess the potential success of these efforts.

Processors and Contractors

The growth of contracted agricultural production in the United States is well documented (Drabenstott, 1994). This growth is most evident in poultry, hogs, vegetables, fruits and some grains. These contracts usually specify product quality aspects, production practices, delivery dates and quantities as well as prices (Chu et al., 1996). Contracts could be used to provide incentives for improved environmental performance. In many cases, however, the use of contracts in this manner requires substantial changes in contract provisions. Because many contracts reward producers for high yields, producers (or their consultants) are loath to experiment with new practices or systems. If such experimentation were to result in below average yields, even for only one year, the contract might not be renewed. Thus, while contracts could allow flexible approaches to achieve environmental outcomes, many actually would have implicit disincentives for such flexibility. Few contracts specify enforceable, environmental performance standards (Chu et al., 1996). The future will probably see more contract specified performance standards for producers, particularly with respect to food safety.

HOW CAN FLEXIBLE INCENTIVES BE ASSESSED?

The success of flexible incentives can be assessed with five criteria: environmental effectiveness, economic efficiency, administrative efficiency and practicability, political feasibility, and equity.

Generally, flexible incentives need to be designed in order to get the rules right so that they encourage choices that result in the desired outcome. There is substantial anecdotal comment from the industry that the specification of clear, simple performance standards that reduce uncertainty may be of prime concern. Furthermore, progress toward the outcome needs to be monitored, and information must be shared with those whose actions influence the outcome. Because NSP is affected by stochastic climatic events, any given policy will result in a distribution of possible outcomes rather than a single outcome (Braden and Segerson, 1993). Consequently, the design and evaluation of such policies are particularly difficult. This difficulty is compounded by the problem of tracing pollution back to its source or correlating pollution with one aspect of the production process (Ribaudo and Caswell, this volume).

Achievement of desired goals is only one criterion. A major reason for favoring flexible incentives is that, at least conceptually, environmental goals can be obtained at minimum cost. (Strictly, this criterion is a subset of economic efficiency, that is, cost effectiveness.) Conceptual gains obtained from increased flexibility can easily flounder on the rocks of transaction costs, particularly those costs that are associated with monitoring and enforcement. Different flexible incentives will have different transaction costs and require different levels of administrative capacity than others. For example, ambient charges are more complicated than many educational projects, and marketable permit trading requires the development of a new institution. Taxing products usually does not require a new institution.

There are also many administrative criteria by which to judge flexible incentives. These include information requirements, management costs, the propensity to preclude opposition and compatibility with existing institutions (such as the existing agency rules or trade agreements). Other criteria can be added to this list, such as the ease with which existing rules can be modified, as new information becomes available.

Finally, some flexible incentives are more politically acceptable than are others (Browne, this volume). Political acceptability may be dependent on perceived changes in property rights, the incidence and magnitude of any costs, the legitimacy of the enforcing authority or on the acceptability of the environmental objective. Some flexible incentives will be seen as more equitable and fair than others and, therefore, will have greater political acceptability.

CONCLUSION

The following four conclusions emerge from this investigation of flexible incentives.

First, flexible incentive instruments are the means or tools that are used to achieve environmental objectives in agriculture—they are not ends in themselves. This conclusion is obvious but deserves emphasis as a starting point. Without clear targets to guide the application of flexible incentive schemes, the search for efficient instruments may bear little fruit. Despite the increasing efforts to specify clear environmental objectives for agriculture during the past decade; there are very few examples of flexible incentive programs that have clear measurable performance objectives.

Agriculture, as the nation's largest single land user, impacts the health of ecological systems in a myriad of ways (NRC, 1993; OTA, 1995b). Because environmental pressure can shift geographically and among resources, meaningful progress will be difficult until the specified objectives capture the totality of those impacts. The science of defining the full spectrum of impacts has grown considerably over the last decade (NRC, 1993; OTA, 1995a; USDA/ERS, 1994), but it is still less than adequate to finely tune flexible incentives. Nevertheless, enough solid evidence has accumulated to improve the definition of performance objectives and thereby to help in the development of flexible incentives. Yet agro-environmental policies have ventured only slightly away from the largely untargeted policies that were born in the Depression. The absence of clear performance standards hampers the evolution to flexible incentive approaches. Indeed, until there are more and better defined performance standards, coupled with consistent and effective enforcement, scientific research will probably not be well focused on the source movement impact knowledge gaps.

Second, our review of possible flexible incentive instruments shows a wide range of options, with no obvious panacea. Economic incentives (charges or subsidies), regulation through clear performance standards, tax provisions, compliance schemes and marketing (ecolabeling) strategies can all qualify. It is highly likely that some combination of instruments, rather than any single approach, will be necessary to achieve most agro-environmental objectives in a flexible, low-cost fashion. The best set of instruments will depend upon the particular environmental problem, the different types of operators that must be induced to change behavior (for example, good, indifferent and bad actors), input and output markets, and the operative political condition.

Third, the nature of agro-environmental problems, as revealed by emerging science, paints a picture of complex site-specific natural resource relationships with local production systems. Thus, even with clear performance objectives, there is no one-size-fits-all prescription for the cost-effective design of flexible incentives to address these problems. There appears to be considerable potential in designing incentives that encourage site-specific approaches that match operator abilities and goals with environmental conditions. Rapidly increasing numbers of watershed-based projects around the country imply that the feasibility of such incentive systems is growing.

The complexity of the production environment system that appears to necessitate such decentralized, flexible approaches also implies high transaction costs for agencies and producers. Therein lies a key trade-off. The implementation of flexible incentive programs for site-specific management, which lowers producer costs and improves environmental performance, will require substantial administration, information and other transaction costs. Economists have been notably poor at including such costs in environmental management system design (Russell and Powell, 1996; Khanna et al., this volume). Analysis that includes an explicit accounting of transaction costs will help illuminate the trade-offs for existing institutions and will help short-run policy decision making. The problem should be cast in a dynamic learning context.

The long-run challenge is to explore institutional changes that can reduce these transaction costs. Such changes can be discovered and adopted through experimentation, adaptation and innovation in all levels of government and the private sector. The path to flexible approaches would be more clear if there were reforms that would help institutions: (1) overcome the inertia associated with existing agency traditions and regulations; (2) design new contractual arrangements that would encourage and enable experimentation at the farm level; and (3) garner new knowledge that would lower the transaction costs of implementing and enforcing flexible incentive approaches.

Finally, the crucial role of human capital (management) when designing and implementing flexible incentive schemes cannot be over emphasized. Quite simply, improved management can be expected to lower transaction and compliance costs and to increase the probability of achieving and sustaining performance standards. This proposition applies to agency personnel as well as to farmers and ranchers. Tackling the seemingly intractable NSP problems with whole-farm system approaches will require agency staff to design programs under considerable scientific uncertainty. These programs need to send clear signals, yet they must allow flexibility for producers. Such an imposing task necessitates quality input from all disciplines—physical, biological and social—in addition to effective multidisciplinary collaboration. Producers have an equal, if not greater, challenge. Stories of innovative management that simultaneously lower compliance costs and improve economic performance through innovation offsets are increasingly common. Those innovators, however, probably represent only a very small proportion of producers. The policy challenge is to find effective and low-cost ways of expanding that management capacity in farmers and ranchers who can deliver high environmental values over the long term.

ENDNOTES

1. This definition differs from that suggested by Segerson, (this volume). Her article explores these differences.
2. A natural question is whether the recent reform of agricultural programs to decouple payments from particular crops and yield levels provides flexible incentives via market prices. The evidence on this question is clear. Commodity program reform will remove some distortions to plant certain crops but does not correct the root causes of environmental problems—missing markets. Simulations of agricultural program reform generally show that the shift to market prices will cause both reductions and increases in pollution loadings, with a modest net environmental improvement likely (Ervin et al., 1991; Miranowski et al., 1991; Hrubovcak et al., 1990). Actual experience from New Zealand supports the simulated effects.
3. Some sense of the surety of the outcome (extreme risk aversion) can be added if that is relevant to the problem, such as controlling highly toxic compounds. Conditions on the incidence of impacts can be included; however, we will not analyze the distributional effects here because they frequently depend on specific provisions of policy measures. We will not delve into political ramifications; instead, we will leave that discussion to Browne and to others.

REFERENCES

Antle, J., C. Crissman, J. Wagenet, and J. Huston. 1996. " Empirical Foundations for Environment Trade Linkages: Implications of an Andean Case Study," in M. Bredahl, N. Ballenger, J. Dunmore, and T. Roe, eds., *Agriculture, Trade and the Environment: Discovering and Measuring the Critical Linkages.* Boulder, CO: Westview Press.

Antle, J. and R. Just. 1992. "Conceptual and Empirical Foundations for Agricultural Environmental Policy Analysis." *Journal of Environmental Quality* 21: 307–316.

Barbash, J. and E. Resek. 1998. *Pesticides in Groundwater: Distributions, Trends and Governing Factors.* Ann Arbor, MI: Ann Arbor Press.

Batie, S. 1997. "Environmental Issues, Policy and the Food Industry," in L.T. Wallace and W.R. Schroder, eds., *Perspectives on Food Industry Government Linkages*, pp. 235–236. Boston, MA: Kluwer Academic Publishers.

Braden, J.B. and K. Segerson. 1993. "Information Problems in the Design of Nonpoint-source Pollution Policy," in C.S. Russell, and J.F. Shogren, eds., *Theory, Modeling and Experience in the Management of Nonpoint-source Pollution*, pp. 1–36. Boston, MA: Kluwer Academic Publishers.

Bromley, D. 1996. "The Environmental Implications of Agriculture." Paper presented to the Organization of Economic Cooperative Development conference, *The Environmental Benefits of Agriculture*. Helsinki, Finland (September).

Camboni S.M. and T.L. Napier. 1994. "Socioeconomic and Farm Structure Factors Affecting Frequency of Use of Tillage Systems." Invited paper presented at the *Agrarian Prospects III* symposium, Prague, Czech Republic (September).

Carpentier, C. 1996. "Value of Information for Targeting Agro-pollution Control: A Case Study of the Lower Susquehanna Watershed." Unpublished Ph.D. dissertation. Department of Agricultural and Applied Economics, Virginia Polytechnic and State University, Blacksburg, VA (May).

Chambers, K.S. and L. Eisgruber. 1998. "Green Marketing as Green and Competitive," in S.S. Batie, D.E. Ervin and M. Schulz, eds., *Business-led Initiatives in Environmental Management: The Next Generation*, p. 25–36. Department of Agricultural Economics, Michigan State University, East Lansing, MI.

Chu, M., S. Swinton, S. Batie, and C. Dobbins. 1996. *Agricultural Production Contracts to Reduce Nitrate Leaching: A Whole Farm Analysis*. Department of Agricultural Economics, Michigan State University, Staff Paper No. 96-78. East Lansing, MI.

Cooper, J. and R. Kiem. 1996. "Incentive Payments to Encourage Farmer Adoption of Water Quality Incentive Practices." *American Journal of Agricultural Economics* 78: 54–64.

Davies, T. and J. Mazurek. 1996. *Industry Incentives for Environmental Improvement: Evaluation of U.S. Federal Initiatives*. Global Environmental Management Initiative, Washington, DC (September).

Drabenstott, M. 1994. "Industrialization: Steady Current or Tidal Wave?" *Choices* 9: 4–7.

Ervin, D., K. Algozin, M. Carey, O. Doering, S. Frerichs, R. Heimlich, J. Hrubovchak, K. Konyar, I. McCormick, T. Osborn, M. Ribaudo, and R. Shoemaker. 1991. *Conservation and Environmental Issues in Agriculture*. Agricultural Economic Service, Staff Report No. 9134, USDA/ERS, Washington, DC.

Ervin, D. and E. Graffy. 1996. "Leaner Environmental Policies for Agriculture." *Choices* Fourth Quarter: 27–33.

Ervin, D. and A. Schmitz. 1996. "A New Generation of Environmental Programs for Agriculture?" *American Journal of Agricultural Economics* 78(5): 1198–1206.

Ervin, D. and J. Tobey. 1990. "European Agricultural and Environmental Policies: Sorting Through the Incentives." *Is Environmental Quality Good for Business? Problems and Prospects in the Agricultural, Energy and Chemical Industries*. Paper presented at the American Enterprise Institute, Washington, DC (June 11–12).

Hanley, N., J.F. Shogren, and B. White. 1997. *Environmental Economics in Theory and Practice*. New York, NY: Oxford University Press.

Hrubovcak, J., M. Le Blanc and J. Miranowski. 1990. "Limitations in Evaluating Environmental and Agricultural Policy Coordination Benefits." *American Economic Review* 80(2): 50–72.

Jones, M. 1998. "Whole Farming Planning as Industrial Ecology," in S.S. Batie, D.E. Ervin and M. Schulz, eds., *Business-led Initiatives in Environmental Management: The Next Generation*, p. 157–164. Department of Agricultural Economics, Michigan State University, East Lansing, MI.

Lovejoy, S. 1997. "Conservation Strategies: What Works and What Might Work Better." Paper presented at *Flexible Incentives for the Adoption of Environmental Technologies in Agriculture* conference, Gainesville, FL (8–10 June).

Miranowski, J., J. Hrubovcak, and J. Sutton. 1991. "The Effects of Commodity Programs on Resource Use," in R. Just and N. Bockstael, eds., *Commodity and Resource Policies in Agricultural Systems*. New York, NY: Springer-Verlag.

Mueller, D.K., P.A. Hamilton, D.R. Helsel, K.J. Hitt, and B.C.Ruddy. 1995. *Nutrients in Ground Water and Surface Water of the United States: An Analysis of Data through 1992*. U. S. Geological Survey, Water Resources Investigations, Report No. 95-4031, U.S. Government Printing Office, Washington, DC.

Napier T.L. and D.E. Brown. 1993. "Factors Affecting Attitudes toward Groundwater Pollution Among Ohio Farmers." *Journal of Soil and Water Conservation* 48(5): 432–438.

Norton, G.W. and J. Mullen. 1994. *Economic Evaluation of Integrated Pest Management Programs: A Literature Review*. Virginia Cooperative Extension, Publication No. 448–120, Blacksburg, VA.

NRC (National Research Council, Board on Agriculture). 1993. *Soil and Water Quality: An Agenda for Agriculture*. Washington, DC: National Academy Press.

Opaluch, J. and K. Segerson. 1991. "Aggregate Analysis of Site-specific Pollution Problems: The Case of Groundwater Contamination from Agricultural Pesticides." *Northeast Journal of Agricultural Resource Economics* 20(4): 83–97.

OTA (Office of Technology Assessment). 1995a. *Agriculture, Trade, and the Environment: Achieving Complementary Policies*. OTA-ENV No. 617, U.S. Government Printing Office, Washington, DC (May).

_____. 1995b. *Targeting Environmental Priorities in Agriculture: Reforming Program Strategies*. Congress of the United States, OTA-ENV No. 640, U.S. Government Printing Office Washington, DC (September).

Palmer, K., W.E. Oates, and P.R. Portney. 1995. "Tightening Environmental Standards: The Benefit-cost of the No-cost Paradigm?" *Journal of Economic Perspectives* 9(4): 98–118.

Pearce, D.W. and R.K. Turner. 1990. *Economics of Natural Resources and the Environment*. Hemel Hempstead, U.K., Harvester Whetsheaf.

Personal Communication. 1996. M. Wolf, Water Quality Program Director, Oregon Department of Agriculture Salem, OR (July).

Porter, M.E. and C. van der Linde. 1995. "Towards a New Conception of the Environment Competitiveness Relationship." *Journal of Economic Perspectives* 9(4): 122–138.

Rahm, M.R. and W.E. Huffman. 1984. "The Adoption of Reduced Tillage: The Role of Human Capital and Other Variables." *American Journal of Agricultural Economics* 66(11): 405–412.

Randall, A. 1983. "The Problem of Market Failure." *Natural Resources Journal* 23(1): 131–148.

Ribaudo, M. 1997. "Managing Agricultural Nonpoint-source Pollution: Are States Doing the Job?" Draft paper for the Resource Policy Consortium Meeting, Washington, DC (June).

Ribaudo, M. and M. Caswell. 1997. "U.S. Environmental Regulation in Agriculture and Adoption of Environmental Technology." Draft paper prepared for the USDA/ERS, Washington, DC.

Ribaudo, M., C.T. Osborn, and K. Konyar. 1994. "Land Retirement as a Tool for Reducing Agricultural Nonpoint Source Pollution." *Land Economics* 70(5): 77–87.

Russell, C. and P. Powell. 1996. " Choosing Environmental Policy Tools," in M. Livingston, ed. *Environmental Policy for Economies in Transition: Lessons Learned and Future Directions*. Resource Policy Consortium Proceedings, Greeley, CO.

Segerson, K. 1988. "Uncertainty and Incentives for Nonpoint Pollution Control." *Journal of Environmental Economics and Management* 15: 87–94.

Stavins, R. 1995. "Transaction Costs and Tradable Permits." *Journal of Environmental Economics and Management* 29: 133–148.

Tutenberg, T.H. 1990. "Economic Instruments for Environmental Regulation." *Oxford Review of Economic Policy* 6(1): 17–34.

USDA/ERS (U.S. Department of Agriculture, Economic Research Service). 1994. "Agricultural Resources and Environmental Indicators," *Agricultural Handbook 705*. U.S. Government Printing Office, Washington, DC (December).

_____. 1995. *Rural Conditions and Trends* 6: 6–10.

USDA/NRCS (U.S. Department of Agriculture, Natural Resources Conservation Service.) 1995. "National Survey of Attitudes Towards Agricultural Resource Conservation." Unpublished Gallup Report (February).

Vatn, A. and D. Bromley. 1997. "Externalities—A Market Model Failure." *Environmental and Resource Economics* 9: 135–151.

6 FLEXIBLE INCENTIVES: A UNIFYING FRAMEWORK FOR POLICY ANALYSIS

Kathleen Segerson
University of Connecticut, Storrs, CT

The difficulties inherent in controlling nonpoint-source pollution (NSP) have led to an interest in the use of flexible incentive mechanisms as a means of reducing agricultural pollution. A framework for analyzing alternative flexible incentive mechanisms is presented in this chapter. The framework includes sole reliance on cost-sharing incentives and mandatory instruments (such as taxes and regulations) as special cases. It also shows how alternative mechanisms, discussed in other chapters in this volume, fit into an overall model of farmer incentives. It demonstrates, however, the potential gains from creating incentives to reduce NSP by using both subsidy and mandatory instruments as complements in a policy package. A brief discussion of recent examples of policies that are based on this approach is presented.

INTRODUCTION

Given the relative success of efforts that have been designed to reduce water pollution from concentrated or point sources of pollution (PSP)—such as factories and sewage treatment plants—attention has turned to the control of more diffuse or nonpoint sources of water pollution—such as agricultural runoff and leaching. Agriculture is thought to be a major contributor to water pollution in many areas, and in some areas, significant improvements in water quality can be achieved only with reductions in pollution that stem from agricultural sources (EPA, 1992 and 1995).

Control of agricultural pollution is complicated by a number of factors that have not been important impediments to the control of PSP. First, agriculture has historically enjoyed a favored-industry status relative to other industries, such as manufacturing. Concerns about the perceived decline in the agricultural sector, in general, and in the number of family farms, in particular, have led to policies that are designed to boost the agricultural sector by increasing agricultural income. Many of these policies, in fact, exacerbate the environmental impacts of agricultural production (Just and Bockstael, 1991). Despite the recognition that agriculture is a major polluter in some areas, there has been little political will to impose policies that would entail net costs for farmers. Unlike manufacturing, where mandatory pollution controls have been used extensively, the historical focus in agriculture has been on education and cost-sharing programs.[1] The one exception is the regulation of large-scale livestock operations, which are regulated as PSP under the Clean Water Act (Norris and Thurow, this volume).

Second, the diffuse nature of agricultural runoff and leaching makes it difficult to observe or monitor emissions and to control those emissions directly. Instead, policies must focus on indirect approaches to emission control through the control of other variables that are somehow (although generally imperfectly) related to emissions. In other words, policies must be based on variables that serve as proxies (generally, imperfect proxies) for emissions.

Third, because there is considerable variability in the physical characteristics of farms (for example, topography, proximity to bodies of water and the leaching characteristics of the soil), generally, the relationship between the proxy variable and emissions will be different for each farm. Thus, the effectiveness of controlling the proxy will vary across farms.

These difficulties, inherent in controlling nonpoint-source pollution (NSP), have led to an interest in the use of flexible incentive mechanisms as a means of reducing agricultural pollution. This chapter presents a framework for analyzing flexible incentives. We begin by examining what is meant by the term flexible incentives. In the following section, we briefly review some of the theoretical literature on designing economic incentive mechanisms to control NSP, given its unique characteristics (in particular, the lack of observability of emissions and heterogeneity across firms). The conclusion of this review is that the theoretically efficient instruments that have been discussed in previous literature could be difficult and costly to implement in practice. As a result, there is a need for alternative policy approaches that recognize the unique characteristics of NSP but involve lower information and/or transaction costs. After reviewing the literature, we present a framework for assessing different types of incentive policies, how they might be related to each other, and how they might be used as substitutes or complements in the design of an overall policy package. The framework includes sole reliance on cost-sharing incentives and on mandatory instruments (for example, taxes or regulations) as special cases. It also demonstrates, however, the potential gains from creating incentives to reduce NSP by using both subsidy (*carrot*) and mandatory (*stick*) instruments as complements in a policy package.

For ease of exposition, the discussion throughout this chapter is focused on incentives to reduce agricultural sources of water pollution. However, the framework presented is sufficiently general to apply in other contexts in which the goal is to meet an environmental quality standard. With slight modifications, for example, the framework could be applied to incentives for private landowners to maintain wildlife habitats for the protection of endangered species (Roka and Main, this volume). It could also be applied to incentives for manufacturers/private landowners to reduce emissions of air pollutants (Segerson and Miceli, 1998; Deepak et al., this volume; Huffaker and Levin, this volume).

THE MEANING OF FLEXIBLE INCENTIVE

As can be seen from the various chapters in this volume, the term flexible incentive means different things to different people. One possible interpretation is based on the distinction between price and quantity instruments that has been used in the literature on PSP.[2] Price-based instruments set a target for behavior or environmental quality and then use economic or financial incentives (such as taxes or subsidies) to induce firms to meet the target. For this reason, price instruments are often termed economic incentive instruments. In contrast, quantity-based instruments set requirements for behavior or environmental quality (such as the use of certain pollution abatement equipment or limitations on allowable emissions) and then impose penalties for failure to comply with those requirements. Because the firm is required (rather than induced) to meet the standards for behavior or environmental quality, quantity-based instruments have not been viewed as incentive-based policies.

Under the above definition, any kind of subsidy falls under the incentives umbrella, which includes subsidies that entail very little flexibility when meeting pollution reduction targets. For example, a subsidy to induce firms to adopt a particular pollution abatement technology would be considered an incentive, even though use of that particular equipment or practice may not be the least-cost means of meeting the pollution control target (Norris and Thurow, this volume). Such policies provide incentives for adoption but do not provide flexibility in terms of how pollution reduction will be achieved. Thus, not all incentives are flexible.

Similarly, not all policies that grant flexibility necessarily involve economic incentives. Batie and Ervin (this volume) employ a definition of flexible incentives that emphasizes flexibility over incentives. According to their definition, flexible incentives are environmental management tools (for example, economic instruments) that specify objectives but allow choices as to response. Because of the need for measurability, these objectives are usually defined in terms of emissions or ambient concentrations of pollutants, rather than in terms of human exposure or susceptibility. Under this definition, mandatory performance standards would qualify as flexible incentives since they specify an end but allow flexibility regarding the means by which that end is met (Carpentier and Bosch, this volume; Lee and Milon, this volume; Casey and Lynne, this volume). This is true despite the fact that mandatory performance standards do not involve direct economic incentives (for example, taxes or subsidies) to induce firms to meet the standard.[3] Instead, firms are required to meet the standard. Tradable emission permits would also qualify as instruments that specify ends but not means (Deepak et al., this volume; Randall, this volume), yet both of these instruments would have been considered quantity-based instruments under the above definition. Baumol and Oates (1988) noted that, in the context of multiple firms, the price versus quantity literature implicitly assumes that the quantity instruments incorporate tradable permits. These permits ensure that, under quantity instruments, the aggregate emissions will be allocated efficiently across firms.

Regardless of the definition used, the objectives of flexible incentives are to establish performance standards and then to impose policy instruments (for example, taxes, subsidies, or penalties for noncompliance) that ensure those standards will be met. Because the ultimate goal is to achieve some level of environmental performance at the lowest possible cost, the standards that are set must be performance-based rather than behavior-based. Behavior-based standards for using certain types of pollution control equipment or certain production processes or inputs will not generally lead to efficient outcomes because the overall pollution reduction goal will not be met at least cost.

In designing a policy based on flexible incentives, policymakers must decide whether the performance standards will be mandatory or target standards. The difference between the two standards is dependent upon where the burden or responsibility for meeting the standard lies. Under mandatory standards, the burden is placed on the firm to ensure that the standard is met. If the standard is not met, the firm will face penalties for noncompliance. In contrast, under target standards, the burden is on the regulator to come up with ways to induce firms to meet the standard. If the standard is not met, no legal action can be taken against the firm. It is then incumbent on the regulator to redesign the incentive policies if the standard is to be met. Historically, the U.S. Department of Agriculture (USDA) has relied on target standards to meet land retirement and water quality goals. For example, both the Conservation Reserve Program (CRP) and its successor, the Environmental Quality Initiative Program (EQIP), are based on this approach (Ogg, this volume; Lovejoy, this volume). If the programs fail to meet their goals, the burden of responsibility will be placed on the policymakers rather than on the firms.

The above discussion highlights the fact that environmental quality ultimately depends on the decisions of two different parties—regulators/policymakers and firms. Batie and Ervin (this volume) note the need for granting flexibilities in the decision making of both parties. Again, the required flexibility is dependent upon whether the standards are mandatory or are simply targets. Hence, it is necessary to know where the burden for meeting the standards belongs. With mandatory standards, flexibility for firm-level decision making is crucial since these decisions will determine how the standard is to be met. In contrast, under target standards, flexibility for regulators is necessary to ensure efficient policy design. For example, increased flexibility in the allocation of EQIP funds under the 1996 Federal Agricultural Improvement and Reform (FAIR) Act will lead to increased efficiency (Ogg, this volume; Norris and Thurow, this volume; Lovejoy, this volume).

ECONOMIC INCENTIVES FOR NONPOINT POLLUTION CONTROL

The literature on regulating NSP has recognized that traditional economic incentive instruments such as Pigouvian emissions taxes, which are often advocated by economists in the context of PSP, cannot be used to induce efficient pollution abatement for NSP (Braden and Segerson, 1993; Tomasi et al., 1994). This situation stems from the fact that individual emissions for NSP are not observable. They cannot, in general, be inferred from observations on ambient water quality because of the influences of stochastic variables, such as weather. When multiple firms contribute to pollution within a given watershed, the lack of observability of emissions implies that it is not possible to identify the source of ambient pollutants. In addition, heterogeneity across sites, in terms of both pollution and production characteristics, implies that least-cost pollution control strategies will be site-specific—that is, they will vary across firms.

The inability to use standard emissions-based taxes has led economists to suggest alternative economic incentive policies for controlling NSP (Griffin and Bromley, 1982; Shortle and Dunn, 1986; Segerson, 1990a; Tomasi et al., 1994). One such alternative is the use of taxes on inputs, such as pesticides or fertilizers. Taxes on pesticide and fertilizer inputs have been used to control NSP in Europe (Kumm, 1990; Dubgaard, 1990), although the main goal in such programs has been to raise revenue to finance other pollution control expenditures rather than to alter farmer behavior. Generally, input taxes do not ensure that water quality standards are met at least cost because they do not account for heterogeneity across firms or for differences in the way that inputs are used (for example, the timing or method of application). In addition, because of low elasticities, high tax levels are generally necessary to induce the required farmer response (Dubgaard, 1990).

The disadvantages of using input taxes have led some economists to suggest other incentive instruments that provide farmers with more flexibility when choosing the means of meeting the specified water quality standard. For example, Segerson (1988) suggested the use of a tax on ambient water quality (ambient tax) under which farmers would pay a tax if ambient water quality were to fall below a given standard and would receive a subsidy if the standard were exceeded.[4] This tax can be designed to induce farmers to choose cost minimizing abatement strategies for meeting water quality standards.

Hansen (1998) has shown that a tax based on damages, rather than on ambient water pollution concentrations, could also induce cost minimizing abatement choices. Hansen's tax mechanism is essentially equivalent to a strict liability rule in which farmers pay for the damages that result from water pollutants. The information requirements for setting the tax rate are lower for this mechanism than they are for the ambient tax. The determination of the total that is due, however, still requires the evaluation of damages. In addition, Hansen (1998) notes that both ambient and damage-based taxes are susceptible to coalition formation and proposes a modification that reduces the likelihood that coalitions will be formed.

Xepapadeas (1991) proposed a system of random fines, under which all firms are rewarded for meeting a water quality goal. If the goal were not met, then one firm (chosen at random) would be fined. This mechanism would only induce efficient abatement decisions if firms were sufficiently risk-averse (Herriges et al., 1994). Drawing from the literature on work incentives, Govindasamy et al. (1994) suggested the use of rank order tournaments for NSP. Under this middle ground policy, some firms are penalized if the water quality target is not met, but the targeted firms are chosen based on some observable measure (input use or abatement effort) rather than randomly chosen. The authors demonstrated that, under certain conditions, a nonpoint tournament would replicate the efficiency properties of a Pigouvian tax, although they noted a number of conditions under which the tournament would fail. Finally, Shortle and Abler (1994) suggested the use of firm-specific input taxes designed to elicit both truthful revelation of private information about firm type and efficient pollution abatement decisions.[5]

While policies such as taxes on ambient water quality provide farmers with flexible incentives to meet water quality targets, they have been criticized for a number of reasons. A major criticism relates to the amount of information required to implement these policies (Cabe and Herriges, 1992; Batie and Ervin, this volume). For example, first-best outcomes under an ambient tax require that the regulator know all of the pollution and production characteristics of the firm so that the level of the tax can be set correctly. When site characteristics vary considerably across firms, a greater information burden is imposed on the regulator. In addition, when watersheds include multiple polluters, the general form of the tax requires that each farmer know the pollution and production characteristics of all of the other farms in the watershed. Without this information, the desirable properties of the incentive mechanism may not hold. In fact, Hansen's (1998) damage-based tax and the nonpoint tournaments proposed by Govindasamy et al. (1994) are both responses to the high information requirement of an ambient tax. Shortle and Abler (1994) also noted the information intensity of their proposed tax mechanism and suggested that this limits its practicality.

Other criticisms of these proposed mechanisms exist as well. For example, in the context of multiple polluters in which marginal damages are nonlinear, ambient taxes have been criticized for being distortionary, that is, varying across firms. Similarly, random fines have been viewed as politically infeasible (Govindasamy et al., 1994; Herriges et al., 1994).

The administrative and information requirements of these first-best incentive instruments have led some authors to advocate the use of second-best policies, such as input taxes, despite their limitations (Shortle and Abler, 1994; Helfand and House, 1995; Wu et al., 1995; Wu and Babcock, 1996b). This response is based on the premise that the imposition of some mandatory instrument is necessary in order to induce NSP control. In the following section, an alternative approach is suggested, under which a mandatory approach is part of an overall policy package.

A FRAMEWORK FOR ANALYZING FLEXIBLE INCENTIVES

The previous sections of this chapter suggest that the discussion of flexible incentives for the control of agricultural pollution has incorporated a number of different ideas. These ideas include the use of subsidies that encourage the adoption of certain pollution reduction activities (for example, land retirement or the adoption of more environmentally friendly production techniques). They also include the choice between first-best mandatory instruments (for example, ambient water quality taxes or mandatory performance standards) and second-best alternatives (for example, input taxes). In this section, a simple model is presented that demonstrates how these different pieces of the discussion can be put into a unifying framework. This framework highlights the fact that the different pieces represent alternative policy approaches that can be used in isolation or in combination in an overall policy package. In addition, it demonstrates the possible gains and resulting implications for policy from taking a broader perspective rather than focusing on individual policy instruments.

The development of this framework is motivated by the consideration of a particular type of policy package that might be used to control agricultural sources of water pollution.[6] As noted by Batie and Ervin (this volume) and others, while flexible mechanisms (such as ambient taxes or legal liability) can induce cost minimizing pollution abatement decisions in principle, such policies would be difficult to implement in practice. Ideally, one would want to induce theoretically optimal outcomes without having to implement one of these mandatory instruments. One possible way of doing this is through the use of a trigger policy. Under such a policy, performance standards would be set. These would represent target, rather than mandatory, standards in the sense that a failure to meet such standards would not be punishable by legal sanction or fines. Technically, compliance with these standards would be voluntary, but failure to meet them could trigger the imposition of mandatory instruments (for example, taxes, regulations and legal liability) that are designed to ensure that the standard is met. This would require that the standards be quantifiable and measurable (Batie and Ervin, this volume) so that failure to meet the standard could be easily detectable.

If the threat of mandatory instruments were explicit, farmers would know the consequences of not taking the abatement actions necessary to meet the standard voluntarily. As Goodin (1986) noted, such a program is not truly voluntary in that the firm is essentially choosing the lesser of two evils. Nonetheless, the explicit threat shifts the burden of meeting the standard onto the farmers. The threat of costly mandatory instruments would provide an inducement to meet standards voluntarily. In addition, the cost of meeting the standard voluntarily could be partially offset by subsidy payments designed to increase the likelihood that voluntary compliance would be the least-cost strategy for the farmer. With this last additional feature, the policy package would include both *carrot* and *stick* approaches to pollution control.

Similar policy packages have been analyzed in the context of PSP. For example, Segerson and Miceli (1998) examined a policy context in which the regulator and the firm negotiate a voluntary agreement in the presence of a threat to impose mandatory regulation should a voluntary agreement not be successful. They asked how the agreed upon level of pollution abatement was likely to compare to the level that would have been imposed if the agreement had failed. Stranlund (1995) compared the use of mandatory and voluntary programs to meet a specified environmental target but viewed the two as alternatives rather than as complements (that is, no threat of mandatory regulation was imposed under the voluntary approach). Wu and Babcock (Forthcoming) extended Stranlund's model to the case of NSP, again viewing the two approaches as alternatives. Segerson (1998) extended the model developed in Segerson and Miceli (1998) to the NSP context, explicitly considering a policy package that combined the *carrot* and *stick* approaches. This extension provides the basis for the framework presented here.

The choice and possible payoffs to the farmer when he or she is faced with the policy package described above are depicted schematically in figure 6.1. The regulator takes the initial move in the process. He sets a performance standard for the firm, such as a target level of ambient water quality. Because of potential measurement difficulties associated with exposure-based performance standards, we assume that the standard is based on ambient concentrations of pollutants.[7] The choice of the target standards could be based on a balance of the benefits and costs of water quality improvements or on some other criterion—such as ensuring that the body of water is conducive to supporting some desirable activity or species.

Since the main focus of this chapter is how to induce farmers to reduce pollution, we do not explicitly consider the regulator's choice with regard to the level of the standard. In a more general analysis (Segerson, 1998), the decisions and payoffs of the regulator would be considered as well. Ogg (this volume) emphasizes the need for Congress to provide correct incentives for individual states so that they can efficiently implement programs such as EQIP. To consider regulatory incentives, the farmer's choice (figure 6.1) would have to be embedded into the regulator's choice of alternative means of implementation.

Given a target, the policy goal is to induce farmers to undertake abatement activities that ensure the target is met. In practice, since water quality is dependent on both the abatement activities of the farmer and other stochastic variables (such as weather), the farmer might not necessarily be able, or willing, to ensure that the standard is met at every point in time. When water quality varies randomly, the target could be defined in terms of the average water quality level over some period of time or in terms of meeting the standard a certain percentage of the time.

Once the standard is set, the farmer must decide whether or not to undertake the abatement activities that are necessary to meet the standard. If the farmer were to choose to meet the standard voluntarily, the net cost to the farmer would be $C - S - B$, in which C is the cost of meeting the standard, S is the amount of the subsidy (if any) received by the farmer to offset some of those costs, and B is any benefit that the farmer receives from meeting the standard voluntarily.

The value of C would depend on the farmer's choice of technology or practices to reduce water pollution. Such pollution reduction practices could include: changes in crop mix; use of filter or buffer strips; adoption of reduced tillage; changes in the timing and method of fertilizer and pesticide applications; substitution toward more environmentally friendly pesticides; adoption of integrated pest management; improvements in manure storage facilities; implementation of precision farming; investment in more efficient irrigation technologies; and land retirements.

Recent advances in environmentally friendly farming technologies, such as precision farming, could presumably lead to reductions in C. Hence, a focus on the promotion of these technologies as a flexible solution to NSP (for example,

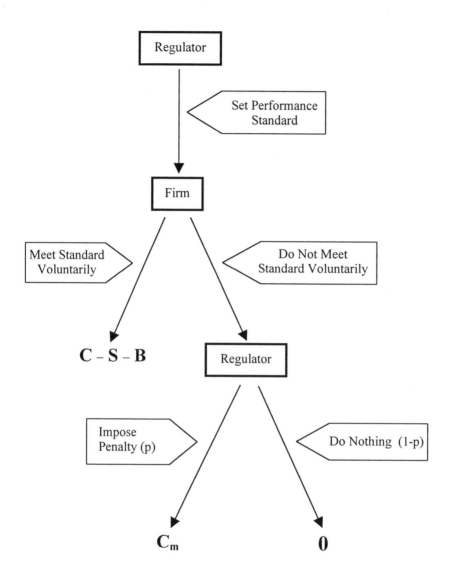

FIGURE 6.1 The Choice and Possible Payoffs to the Farmer when Faced with a Policy Package

Khanna et al., this volume; Parker and Caswell, this volume) would correspond to a focus on reducing C. In the absence of any subsidy, farmers who choose to meet the standard voluntarily would have an incentive to choose the production techniques that insure that the standard is met at the lowest possible cost, since any cost savings would directly benefit them. Also, in choosing the least-cost strategy, a farmer could respond to his own particular situation—that is, to his own pollution and production characteristics—and would choose his strategy accordingly.

The amount of the subsidy, S, is determined by government policy. The regulatory agency or the legislature can decide what, if any, subsidy it can offer to farmers to help defray the costs of meeting the standard voluntarily. Researchers who focus on the use of subsidies as a means of achieving water quality goals (Ogg, this volume) are concerned primarily with the appropriate design of S. As noted above, tying the subsidy to the adoption of specific pollution abatement practices is not likely to be efficient since it would discourage both innovation and incentives for the adoption of least-cost practices (Norris and Thurow, this volume). Instead, the subsidy should be provided for whatever practices the farmers choose as a means of meeting the standard. For example, a cost-sharing subsidy under which the government pays a fixed percentage of the total abatement costs (regardless of the specific activity that led to those costs) would not distort the farmer's pollution abatement decisions. He or she would still have an incentive to choose the abatement strategy that is least-cost, given the specific farm characteristics. In addition, the required cost share can be less than 100 percent, given the threat that failure to meet the standard would trigger mandatory instruments. In other words, it is possible that the farmer would choose to meet the standard voluntarily even if the amount of the subsidy were less than the abatement costs incurred (that is, even if $S < C$), given the possible imposition of mandatory instruments.

In some cases, a farmer could receive some direct benefits, B, from meeting the standard voluntarily. For example, the farmer might realize some public relations or reputation benefits (either with customers or within the community) as well as personal benefits (based on moral or ethical premises) from voluntarily practicing environmental stewardship (Swinton et al., this volume; Casey and Lynne, this volume). Similarly, adoption of certain pollution abatement practices could allow the farmer to use ecolabeling (van Ravenswaay and Blend, this volume) in the marketing of its products. To the extent that the demand for the product is sensitive to the way in which it is produced, a voluntary improvement in environmental quality could translate into an increased demand for the farmer's product. This could, in turn, increase profitability. Such a benefit could offset some (or possibly all) of the costs associated with undertaking the pollution abatement activities. These marketing opportunities could be promoted as a flexible means of reducing agricultural sources of water pollution (van Ravenswaay and Blend, this volume). Of course, if the benefit from increased demand were greater than the cost of the abatement activities, the farmer should have an incentive to invest in pollution abatement without the need for any government inducement. There is certainly anecdotal evidence that some farmers have done this. They have used their environmental practices as successful marketing tools. In such cases, any subsidy for the adoption of these practices would generate a windfall for the farmers.

Finally, the farmer could realize on-farm benefits from pollution control, which would also serve to offset part of the associated costs. Examples of these benefits include (1) production-oriented gains—such as reduced soil loss and increased efficacy of input use (for example, nutrients and pesticides) from reduced losses through runoff (Parker and Caswell, this volume) or innovation offsets—and (2) environmental gains—such as improvements in drinking water for the farm family or reduced farmer exposure to agricultural chemicals. Empirical studies suggest that farmers value improved environmental quality and reduced health risks and would be willing to pay for these improvements (Swinton et al., this volume).

If the farmer were to choose to meet the standard voluntarily, then the policy objective would have been met, and there would be no need for further action by policymakers. If the farmer were to choose not to meet the standard voluntarily, then he would face the threat of mandatory policies. Note, however, that the threatened mandatory policy instrument must be credible—it must be a threat that the legislature or regulator would be willing to impose. Policies that involve excessive costs would not be viewed by farmers as credible and would not induce farmers to seek voluntary compliance.

When the farmer is deciding whether to meet the standard voluntarily, there could be some uncertainty about whether the mandatory instruments would actually be imposed if the voluntary approach fails. In figure 6.1, we capture this uncertainty with the parameter p, which represents the probability that mandatory instruments would be imposed if the standard were not met voluntarily. There are alternative interpretations of p. For example, imposition of a mandatory policy could require action by the legislature and, given the nature of the political process and the numerous factors that affect environmental policy decisions (Browne, this volume), it is uncertain whether the legislature would have the political will necessary to carry out the threat. In this case, p represents the probability that the legislature would have the political will to impose mandatory instruments or that imposition would be a legislative priority that would actually be acted upon. Alternatively, if the regulator had the authority to impose these policies, then p could be chosen. In particular, the regulator could choose p to be equal to one, implying that the threat would be certain. Setting p equal to one could also represent a case where the legislature establishes a voluntary program that would stipulate the imposition of mandatory instruments if the voluntary standards were not met. Finally, p = 1 could also represent a situation in which mandatory instruments were actually imposed, but an exception would be granted for farmers who choose to meet the standard voluntarily. This is similar to the approach taken under Project XL for PSP (Davies and Mazurek, 1996).

If the threat were not actually imposed, then the farmer would incur no cost. Alternatively, if the mandatory instruments were imposed, then the farmer would incur the associated cost, C_m. The magnitude of C_m would depend on the specific policy that would be imposed. As noted previously, there are a number of different mandatory instruments that could be imposed on farmers. These include input taxes; ambient water quality taxes; direct regulation of farming activities; legal

liability for environmental damages and contamination; and loss of other government benefits. As noted above, the choice among alternative (mandatory) policy
instruments has been the focus of the economic literature on incentive policies,
and it continues to be an important component of the debate over the use of flexible incentives (Deepak et al., this volume; Carpentier and Bosch, this volume).

These policy instruments differ in terms of the total costs, C_m, that they impose
on farmers. The differences in total costs stem both from the different incentives
that they create to choose least-cost abatement strategies and from differences in
direct tax payments. As noted above, input taxes would not necessarily induce
farmers to meet the standard at least cost, but in theory, ambient taxes would.
Thus, the input tax that is necessary to ensure that the standard is met would impose a larger abatement cost on farmers than would the necessary ambient tax. In
addition, input taxes would involve net tax payments to the government. In equilibrium, an ambient tax that is imposed only if the ambient water quality were
below the standard would not result in net tax payments. Similarly, under direct
government regulation, farmers would incur the costs of compliance with the
regulation but would not face any tax liability.

The more efficient the instrument, the closer its compliance cost will be to the
minimum. As noted above, first-best or cost minimizing instruments can often
involve high transaction costs (Batie and Ervin, this volume; Carpentier and
Bosch, this volume). When these transaction costs are included, the total cost under the mandatory approach will likely exceed the cost of meeting the standard
voluntarily, even when efficient regulatory instruments are used. Thus, in all
cases, the cost of the mandatory instrument must be at least as great as the cost
incurred under the voluntary approach; that is, $C_m \geq C$, given that C is the minimum cost of meeting the standard. In many cases, we expect it to be greater than
the minimum cost, that is, $C_m > C$.

The fact that we do not include B in the farmer's payoff under the mandatory
policy implies an assumption that possible benefits from pollution abatement stem
from voluntary abatement but not from the response to mandatory instruments.
This is likely to be true for public relations benefits. On-farm benefits from improved drinking water or reduced chemical exposure, however, would be realized
regardless of whether the abatement were voluntary or not. If these benefits were
substantial, they would be included in the payoff as well. Distinguishing between
the two types of benefits and incorporating the latter type into the payoff under
mandatory instruments, however, would neither change the basic structure of the
problem nor change the qualitative results. Similarly, S is not included in the
farmer's payoff since it is implicitly assumed that the mandatory instrument will
not include the payment of a subsidy. If the mandatory policy were to include subsidy payments, then C_m would be interpreted as the farmer's cost, net of any subsidy payment received.

Given the payoffs in figure 6.1, the farmers would be induced to meet the standard voluntarily if and only if the expected cost of meeting it were less than (or
equal to) the expected costs of not meeting it, that is, if and only if

(1) $C - S - B \leq pC_m.$

Clearly, given p, the regulator should always be able to induce participation in the voluntary program by setting the subsidy, S, sufficiently high. In particular, the farmer should choose to participate whenever S is set such that

(2) $\quad S \geq C - pC_m - B$.

As noted above, if $p > 0$, then even with $B = 0$, it would not be necessary for the regulator to set S at a level that would cover the entire cost of abatement for the farmers. In particular, equation (2) implies that the minimum S necessary to induce participation is less than C if $p > 0$ or $B > 0$. In addition, the larger the p (that is, the stronger the threat of mandatory controls), the smaller is the required minimum subsidy.

Note that if $p > 0$ and C_m were sufficiently large relative to C, which might be true if the mandatory policy were to involve large tax payments, then farmers could choose to meet the standard voluntarily even without a subsidy, that is, even with $S = 0$. In other words, even in the absence of a subsidy, the farmer might prefer to incur the (certain) costs of meeting the standard voluntarily rather than to take the chance of incurring even higher costs were the mandatory instrument imposed. Thus, even though the policy package could be designed so that the mandatory instruments would never be imposed in equilibrium, the choice of which mandatory instrument would be imposed were the threat carried out could affect the incentives for farmers to participate voluntarily in meeting the standard.

The choice of mandatory instruments could also affect the information that the regulator must have to set a subsidy high enough to induce participation. For example, Segerson (1998) demonstrated that, if the threatened mandatory instruments were to yield cost minimizing abatement decisions and no net tax payments (so that $C_m = C$), the minimum subsidy rate that the regulator must offer to induce participation would be independent of farm type. Thus, the regulator does not need specific information about farm characteristics to induce cost minimizing abatement. If the threatened mandatory instruments were to impose costs that were higher than those costs that would have been incurred with voluntary participation, then the minimum subsidy rate would vary with farm type.

Finally, note that figure 6.1 includes as special cases the sole reliance on both voluntary programs and on mandatory instruments. In particular, if $p = 0$, then the framework would depict the use of a voluntary approach without any threat of imposition of mandatory controls. In this case, farmers would choose to meet the standard voluntarily, that is, they would choose to participate in the voluntary programs, if and only if $S > C - B$. Thus, the government must offer cost sharing sufficient to cover the total net costs that would be incurred by the farmer. This provides an explanation for the large subsidy rates that are embodied in the CRP and EQIP (Ogg, this volume). Alternatively, if the regulator does not offer the farmer the opportunity to meet the standard voluntarily, then the model would depict a case of sole reliance on mandatory measures.

The framework, depicted in figure 6.1, is based on a single firm making production decisions that affect a given body of water. As noted above, in many contexts many firms would affect the ambient water quality of a given body of water.

In that context, the policy problem would be complicated by the fact that the regulator would not be able to associate specific pollutants with specific sources or farms. In particular, if an ambient water quality standard were not met, the regulator could not easily identify the farms that had undertaken suitable pollution abatement and the ones that had not. The above approach, however, could still be used. In this case, imposition of the mandatory instruments would take the form of a group penalty that would be imposed on all of the farmers for failure of the group, as a whole, to meet the standard (Randall, this volume). Segerson (1998) illustrated that, if the threatened mandatory controls were to induce cost minimizing abatement without net tax payments, then the regulator could set a subsidy rate that would induce the farmers in the watershed to collectively make pollution abatement decisions to ensure that the standard would be met voluntarily. This outcome is most likely to occur in small watersheds where farmers can communicate with their neighbors about their pollution abatement plans. Thus, even in the context of multiple farmers who contribute to a common watershed, it is possible to design a voluntary program that would induce cost minimizing abatement and avoid free riding. Group penalties of this type could also create incentives for within group monitoring.

CONCLUSION

Much discussion about flexible incentives for the reduction of agricultural pollution has focused on individual pieces of the policy puzzle (such as the design of subsidies to induce certain abatement activities), the development of technologies (such as precision farming, which allows heterogeneous farmers to achieve water quality goals at lower costs) or the choice between first- and second-best mandatory instruments. However, there is some evidence that policymakers are beginning to take the broader perspective, embodied in the framework outlined above, and are designing policies that involve the simultaneous use of *carrot* (voluntary) and *stick* (mandatory) approaches.

In the context of PSP, several studies have noted the increasing use of the combined approach. For example, a 1996 survey by the Commission of the European Communities (CEC, 1996) reports evidence of the increasing use of voluntary agreements within its member states. In many cases, background threats of the imposition of legislation or liability have provided the impetus for participation in the voluntary programs. Similarly, in the United States, the U.S. Environmental Protection Agency (EPA) and other regulatory agencies have undertaken several initiatives that are based on the increased reliance on voluntary agreements with the threat or certainty of possibly harsher measures if the agreements are not successful. Examples include Project XL, the Common Sense Initiative and the 33/50 Program. These programs have been the subjects of a number of recent studies. For example, Davies and Mazurek (1996) provided a description and evaluation of these programs and other programs based on industry incentives. Arora and Cason (1995) presented an empirical study of the factors that affect participation in the voluntary 33/50 Program. Segerson and Miceli (1998) developed a conceptual framework that was utilized to analyze the environmental quality implications of voluntary agreements that are used to control PSP.

States have begun to experiment with similar policies in the context of NSP (Ribaudo and Caswell, this volume; Batie and Ervin, this volume; Lee and Milon, this volume). Examples include (1) the Coastal Zone Management Reauthorization Amendments, which impose mandatory measures for those firms that fail to meet target levels of protection; (2) the Everglades Forever Act, under which a tax on cropland will automatically increase unless phosphorous reduction goals are met basin wide; (3) Lake Okeechobee, Florida, where failure by local dairies to maintain compliance with water quality programs (which are designed to meet phosphorous standards) precipitates state action; (4) the State of Oregon, where ambient water quality standards and civil fines have been established to ensure compliance if producers do not participate in plans to achieve those standards; and (5) the Oregon Salmon Restoration Program, under which certain salmon will be listed as endangered species unless voluntary measures are successful in restoring habitats and populations. Such policies recognize the advantage of holding farmers responsible for reductions in agricultural pollution. At the same time they offer farmers the opportunity to meet water quality goals voluntarily (with the possibility of cost sharing to reduce the financial burden of meeting those goals). These policies ultimately provide both flexibility and incentives that encourage farmers to adopt more environmentally friendly production practices. It remains to be seen whether, in practice, these policies are capable of ensuring compliance with environmental objectives as well.

ENDNOTES

1. These policies generally involve the use of subsidies that induce farmers to participate in voluntary programs designed to meet those targets. There has, however, been some regulation of agricultural activity. A notable example is the regulation of pesticide use. For a general description of U.S. Department of Agriculture (USDA) policies, see Ogg (this volume), Reichelderfer (1990) and Ribaudo (1998). For other discussions of the use of subsidies to reduce agricultural nonpoint-source pollution, see Norton et al. (1994), Lohr and Park (1995), Bosch et al. (1995), Wu and Babcock (1995), Hardie and Parks (1996), Wiebe et al. (1996), Wu and Babcock (1996a), Babcock et al. (1996), and Cooper and Keim (1996). Regulatory approaches to controlling agricultural pollution are discussed in Anderson et al. (1990) and Dubgaard (1990).
2. Weitzman's (1974) seminal paper analyzes price versus quantity instruments in the presence of uncertainty. A similar distinction exists in the literature on legal liability. For example, Cooter (1984) notes that strict liability is a price-based liability rule while a negligence standard corresponds to a quantity-based instrument.
3. The firm does face an economic incentive to meet the standard in the least-cost way since any cost savings translates into increased profits for the firm. In addition, one can view the imposition of penalties for failure to meet the standard as an economic incentive instrument.
4. Ambient taxes can be used in the context of surface water pollution. The corresponding policy instrument for groundwater contamination is legal liability for damages (Segerson, 1990b and 1995).
5. This discussion of alternative incentive mechanisms is not meant to be exhaustive. Rather, it is intended to illustrate the types of policies that have been proposed. Other mechanisms that we have not discussed include those described in Xepapadeas (1992, 1995), Laffont (1994) and Smith and Tomasi (1995).
6. The framework is equally applicable to either surface water pollution or groundwater contamination, although the specifics of policy design could vary in the two contexts. In addition, as noted above, it can be applied to other environmental contexts, such as air pollution and land use.

7.   Alternatively, the regulator could set a goal for a variable that is correlated with (and, hence, serves as a proxy for) emissions, such as industry size. This is the approach embodied in, for example, the Dairy Buyout Program.

REFERENCES

Anderson, G.D., A.E. de Bossu, and P.J. Kuch. 1990. "Control of Agricultural Pollution by Regulation," in John B. Braden and Stephen B. Lovejoy, eds., *Agriculture and Water Quality: International Perspectives.* Boulder, CO: Lynne Rienner Publishers.

Arora, S. and T.N. Cason. 1995. "An Experiment in Voluntary Environmental Regulation: Participation in EPA's 33/50 Program." *Journal of Environmental Economics and Management* 28(3): 271–286.

Babcock, B., P.G. Lakshminarayan, J. Wu, and D. Zilberman. 1996. "Public Fund for Environmental Amenities." *American Journal of Agricultural Economics* 78 (4): 961–971.

Baumol, W.J. and W.E. Oates. 1988. *The Theory of Environmental Policy.* Cambridge, MA: Cambridge University Press.

Bosch, D., Z. Cook, and K. Fuglie. 1995. "Voluntary Versus Mandatory Agricultural Policies to Protect Water Quality: Adoption of Nitrogen Testing in Nebraska." *Review of Agricultural Economics* 17(1): 13–24.

Braden, J.B. and K. Segerson. 1993. "Information Problems in the Design of Nonpoint-source Pollution," in Clifford Russell and Jason Shogren, eds., *Theory, Modeling, and Experience in the Management of Nonpoint-source Pollution*, pp. 1–35. Boston, MA: Kluwer Academic Publishers.

Cabe, R. and J.A. Herriges. 1992. "The Regulation of Nonpoint-source Pollution Under Imperfect and Asymmetric Information." *Journal of Environmental Economics and Management* 22(2): 134–146.

CEC (Commission of the European Communities). 1996. *On Environmental Agreements.* Communication from the Commission to the Council and the European Parliament, Brussels, Belgium.

Cooper, J.C. and R.W. Keim. 1996. "Incentive Payments to Encourage Farmer Adoption of Water Quality Protection Practices." *American Journal of Agricultural Economics* 78 (1): 54–64.

Cooter, R.D. 1984. "Prices and Sanctions." *Columbia Law Review* 84: 1523–1560.

Davies, T. and J. Mazurek. 1996. "Industry Incentives for Environmental Improvement: Evaluation of U.S. Federal Initiatives." *Global Environmental Management Initiative.* Washington, DC: Resources for the Future.

Dubgaard, A. 1990. "Programs to Abate Nitrate and Pesticide Pollution in Danish Agriculture," in J.B. Braden and S.B. Lovejoy, eds., *Water Quality and Agriculture: An International Perspective on Policies.* Boulder, CO: Lynne Rienner Publishers.

EPA (U.S. Environmental Protection Agency). 1992. *Managing Nonpoint-source Pollution.* Office of Water (WH-553), U.S. Environmental Protection Agency, EPA-506/9-90, Washington, DC.

_____. 1995. *National Water Quality Inventory: 1994 Report to Congress.* U.S. Environmental Protection Agency, Report No. 841-R-95-005, Washington, DC.

Goodin, R.E. 1986. "The Principle of Voluntary Agreement." *Public Administration* 64: 435–444.

Govindasamy, R., J.A. Herriges, and J.F. Shogren. 1994. "Nonpoint Tournaments," in C. Dosi and T. Tomasi, eds., *Nonpoint Source Pollution Regulation: Issues and Analysis.* Boston, MA: Kluwer Academic Publishers.

Griffin, R. and D.W. Bromley. 1982. "Agricultural Run Off as a Nonpoint Externality." *American Journal of Agricultural Economics* 64(3): 547–552.

Hansen, L.G. 1998. "A Damage Based Tax Mechanism for Regulation of Nonpoint Emissions." *Environmental and Resource Economics* 12(1): 99–112.

Hardie, I.W. and P.J. Parks. 1996. "Reforestation Cost-sharing Programs." *Land Economics* 72(2): 248–260.

Helfand, G.E. and B.W. House. 1995. "Regulating Nonpoint-source Pollution Under Heterogeneous Conditions." *American Journal of Agricultural Economics* 77(4): 1024–1032.

Herriges, J.A., R. Govindasamy, and J.F. Shogren. 1994. "Budget-balancing Incentive Mechanisms." *Journal of Environmental Economics and Management* 27(3): 275–285.

Just, R.E. and N. Bockstael, eds. 1991. *Commodity and Resource Policies in Agricultural Systems.* Berlin, Germany: Springer-Verlag.

Kumm, K. 1990. "Incentive Policies in Sweden to Reduce Agricultural Water Pollution," in J.B. Braden and S.B. Lovejoy, eds., *Water Quality and Agriculture: International Perspectives.* Boulder, CO: Lynne Rienner Publishers.

Laffont, J. 1994."Regulation of Pollution with Asymmetric Information," in C. Dosi and T. Tomasi, eds., *Nonpoint Source Pollution Regulation: Issues and Analysis.* Boston, MA: Kluwer Academic Publishers.

Lohr, L. and T. Park. 1995. "Utility-consistent Discrete-continuous Choices in Soil Conservation." *Land Economics* 71(4): 474–490.

Norton, N.A., T.T. Phipps, and J.J. Fletcher. 1994. "Role of Voluntary Programs in Agricultural Nonpoint Pollution Policy." *Contemporary Economic Policy* 12(1): 113–121.

Reichelderfer, K.H. 1990. "National Agro-environmental Incentive Programs: The U.S. Experience," in J.B. Braden and S.B. Lovejoy, eds., *Water Quality and Agriculture: International Perspectives.* Boulder, CO: Lynne Rienner Publishers.

Ribaudo, M.O. 1998. "Lessons Learned about the Performance of USDA Agricultural Nonpoint Source Pollution Programs." *Journal of Soil and Water Conservation* 53(1): 4–10.

Segerson, K. 1988. "Uncertainty and Incentives for Nonpoint Pollution Control." *Journal of Environmental Economics and Management* 15(1): 87–98.

_____. 1990a. "Incentive Policies for Control of Agricultural Water Pollution," in J.B. Braden and S.B. Lovejoy, eds., *Agriculture and Water Quality: International Perspectives.* Boulder, CO: Lynne Rienner Publishers.

_____. 1990b. "Liability for Groundwater Contamination from Pesticides." *Journal of Environmental Economics and Management* 19(3): 227–243.

_____. 1995. "Liability and Penalty Structures in Policy Design," in D. W. Bromley, ed., *The Handbook of Environmental Economics.* Cambridge, MA: Basil Blackwell.

_____. 1998. "Voluntary vs. Mandatory Approaches to Nonpoint-pollution Control: Complements or Substitutes?" Working Paper, Department of Economics, University of Connecticut, Storrs, CT (November).

Segerson, K. and T.J. Miceli. 1998. "Voluntary Environmental Agreements: Good or Bad News for Environmental Protection." *Journal of Environmental Economics and Management* 36(2): 109–130.

Shortle, J.S. and D.G. Abler. 1994. "Incentives for Nonpoint Pollution Control," in C. Dosi and T. Tomasi, eds., *Nonpoint-source Pollution Regulation: Issues and Analysis.* Boston, MA: Kluwer Academic Publishers.

Shortle, J. and J. Dunn. 1986. "The Relative Efficiency of Agricultural Source Water Pollution Control Policies." *American Journal of Agricultural Economics* 68(3): 668–677.

Smith R.B.W. and T.D. Tomasi. 1995. "Transaction Costs and Agricultural Nonpoint-source Water Pollution Control Policies." *Journal of Agricultural and Resource Economics* 20(2): 277–290.

Stranlund, J.K. 1995. "Public Mechanisms to Support Compliance to an Environmental Norm." *Journal of Environmental Economics and Management* 28(2): 205–222.

Tomasi, T., K. Segerson, and J. Braden. 1994. "Issues in the Design of Incentive Schemes for Nonpoint Source Pollution Control," in C. Dosi and T. Tomasi, eds., *Nonpoint Source Pollution Regulation: Issues and Analysis.* Boston, MA: Kluwer Academic Publishers.

Weitzman, M.L. 1974. "Prices vs. Quantities." *Review of Economic Studies* 41(4): 477–491.

Wiebe, K., A. Tegene, and B. Kuhn. 1996. *Partial Interests in Land: Policy Tools for Resource Use and Conservation.* Agricultural Economic Report No. 744, U.S. Department of Agriculture, Washington, DC.

Wu, J. and B.A. Babcock. 1995. "Optimal Design of a Voluntary Green Payment Program under Asymmetric Information." *Journal of Agricultural and Resource Economics* 20(2): 316–327.

_____. 1996a. "Purchase of Environmental Goods From Agriculture." *American Journal of Agricultural Economics* 78(4): 935–945.

_____. Forthcoming. "The Relative Efficiency of Voluntary vs. Mandatory Environmental Regulations." *Journal of Environmental Economics and Management.*

_____. 1996b. "Spatial Heterogeneity and the Choice of Instruments to Control Nonpoint Pollution." Center for Agricultural and Rural Development, Working Paper No. 96-WP 164, Iowa State University, Ames, IA (September).

Wu, J., M.L. Teague, H.P. Mapp, and D.J. Bernardo. 1995. "An Empirical Analysis of the Relative Efficiency of Policy Instruments to Reduce Nitrate Water Pollution in the U.S. Southern High Plains." *Canadian Journal of Agricultural Economics* 43(3): 403–420.

Xepapadeas, A.P. 1991. "Environmental Policy under Imperfect Information: Incentives and Moral Hazard." *Journal of Environmental Economics and Management* 20(2): 113–126.

Xepapadeas, A.P. 1992. "Environmental Policy Design and Dynamic Nonpoint-source Pollution." *Journal of Environmental Economics and Management* 23(1): 22–39.

Xepapadeas, A.P. 1995. "Observability and Choice of Instrument Mix in the Control of Externalities." *Journal of Public Economics* 56(3): 485–498.

7 SUSTAINABILITY, TECHNOLOGY AND INCENTIVES

Madhu Khanna, Katti Millock and David Zilberman

University of Illinois, Urbana, IL
University of Copenhagen, Denmark
University of California, Berkeley, CA

While there are several definitions of sustainability, its premise is recognition of the exhaustible nature of environmental quality and recognition that the lack of well-defined property rights on environmental quality creates incentives for its overexploitation. The flow of polluting residues, which inevitably accompany production processes, could be larger than the assimilative capacity of the environment and, thus, could hinder the attainment of sustainability. Pollution is an externality associated with human activities that, in the absence of government regulation or well-defined property rights on environmental quality, individuals have no incentive to control. This implies that there is a role for environmental policy, which creates incentives to control pollution and induces a shift toward production processes that reduce the generation of polluting residuals. For sustainable development to be consistent with the growing demand for food and manufactured goods, it needs to be accompanied by technological development that increases input productivity while it reduces the generation of pollution per-unit of input or per-unit of output. Numerous technologies with these features have been developed in the past, but their adoption has been limited because of a lack of incentives. In the future, government may play an important role for inducing the adoption of efficient technologies and for developing improved technologies.

INTRODUCTION

Sustainability has become a broadly supported guiding principle for the management of agriculture and natural resources. The beauty and main limitation of this concept is that it is vaguely defined and may be interpreted quite differently by various groups (Batie, 1989; Ruttan, 1994). Nevertheless, it implies that resource development systems must be designed to recognize the vulnerability and limitations of natural systems. It must also balance this with the desire for high economic returns. This perspective suggests that economic research is a valuable tool in further defining the concept of sustainability. It also suggests that economists could, and should, play an important role in providing the foundation for establishing sustainable resource policies. The key element in the pursuit of sustainable agriculture is technological change. The forms of such change include innovation and the discovery of new technologies; changes in the existing technology mix (that is, shifting from technologies that are more harmful to the environment to those that are environmentally friendly); and the application of recent technologies in a manner that is more consistent with environmental and natural resources constraints.

Major breakthroughs have occurred in the economics of technological change. Fifty years ago it was quite common to assume that technological change was neutral and was mostly affected by random events, such as inspiration and innovative genius. In the 1970s and 1980s, seminal works (Hayami and Ruttan, 1985; Binswanger and Ruttan, 1978) provided strong empirical support for the induced innovation hypothesis. According to this hypothesis, innovation is an economic activity that is influenced by economic incentives. Thus, the direction of innovation can be explained by scarcity and by changes in relative prices. Similarly, Griliches (1957) spawned extensive literature (Feder et al., 1985) that both conceptually and empirically related technology adoption to economic incentives.

This chapter provides a perspective on how economics can play an important role in creating incentives for technological change that are consistent with the pursuit of sustainability. Segerson's framework (this volume) emphasizes the economic and policy considerations required obtaining flexible incentives. In this chapter, the first section introduces a modeling framework that incorporates agronomic and biological features, and explicitly aims to characterize technological change and sustainability. In spite of its potential, it seems that economic research, especially in agriculture, has not played a major role in policymaking or in guiding the formulation of least-cost strategies to achieve sustainability.

In the second section, we argue that this situation can be remedied by improving economic literacy among scientists and by incorporating biological and physical relationships in the framework of economic decision making. The third section begins by discussing the need for policy intervention at different stages of pollution and damage generation in order to address the multidimensional sources of environmental damage. We then focus, in the third section, on the role of economic incentives that induce the adoption of technologies that control pollution by reducing the amount of unutilized input residues. This framework is useful for addressing problems of environmental contamination by chemicals, water logging and by-catch of fish (Khanna and Zilberman, 1997), and is based on a model introduced by Caswell et al. (1990). The model assumes that precision technologies (PTs), so-called because of their capacity to increase the precision with which inputs are applied to the production process, can improve the efficiency with which inputs are used and can reduce pollution. Because these technologies may also increase the economic returns earned by producers, to some extent their adoption may occur voluntarily—even in the absence of government regulations (Casey and Lynne, this volume). Such voluntary adoption rates for these technologies, however, may not be adequate to achieve sustainability. Government regulations would then be needed to provide additional incentives for their adoption.

Under the policy mechanisms used to provide these incentives, regulators often find it necessary to have detailed information about production characteristics and to have the ability to monitor the polluting residuals that they discharge. Unfortunately, regulators often lack such detailed information. In the fourth section, we review the problem of incomplete information and some of the other obstacles that prevent the establishment of first-best pollution control policies. Acknowledging the difficulties raised in the fourth section, the fifth section analyzes alternative policies for environmental regulation and argues for practical approaches to problems of heterogeneity and incomplete information. In particular, it suggests new incentive-based policies that could induce the adoption of residue monitoring equipment. In the sixth section, we discuss the role that the public sector has in encouraging research and the emergence of institutions (such as input use consultants) that can effectively disseminate information about new technologies. We also argue that sustainability should be viewed as a nonrenewable resource management problem—one that requires dynamically consistent policies. These policies must provide incentives for innovation and adoption of precision and residue monitoring technologies.

EDUCATION AND COOPERATION

In the past 20 years, environmental economics has been one of the fastest growing areas of research. Policy prescriptions of economists have been adopted to address major environmental problems. Market trading of air pollution rights and the introduction of trading in water and water rights are three of the important policy innovations that can be attributed to economic research. Economists have been less successful when addressing problems of agricultural waste disposal and agricultural residues from pesticides and chemicals. Economists may have halted the reliance on inefficient policies, such as broad bans on pesticide use, but their contribution to an actual solution has been insignificant.

One plausible reason for this result may be the lack of overall economic literacy among scientists and the public. The public may not politically support the use of economic incentives that address problems associated with human health and environmental quality. Furthermore, many natural scientists have minimal exposure to economics. Some academic programs in biological and environmental sciences have no economic course requirements. Several science majors have reported that introductory economics classes were uninspiring. These classes, however, may have emphasized technical concepts rather than the usefulness and the relevance of economics for policymaking. Furthermore, others view economics as inherently biased against the environment. Exposure to general courses in environmental economics and policy can alleviate many of these prejudices. Economists are challenged to develop such courses and to integrate them in the curricula of biological sciences and other related disciplines. Thus, the stage is set for such efforts, and we should pursue it.

Some of the limitations of our own economic research need to be recognized. Most of the economic research in policy and environmental issues is inward-oriented. It rarely incorporates concepts and information developed by other disciplines. Economists are often as ignorant about biological and physical sciences as biologists are about economics. The training of environmental economists should include background education in the basic sciences, such as physics, chemistry and biology. This type of information should be incorporated into our economic models to make them more useful and relevant.

Lancaster (1968) introduced the notion of family production functions and the demand for product quality characteristics. He argued that overemphasis of the generality of results makes classical consumer theory limited in its application. In the pursuit of generalities, environmental economists tend to avoid specific assumptions about the structure of production and the technological linkages between production and pollution. This tends to limit the range of practical results and policy prescriptions to which environmental economics research can lead. With a more specific structure, Lancaster (1968) was able to obtain much richer results. The same may be true with a more precise modeling of environmental economics. This chapter will rely on basic results from natural sciences, and from stylized observations of real-world phenomena to provide specific assumptions about production and the role of environmentally friendly technologies in reducing pollution. These assumptions will enable economists to draw some specific policy prescriptions and results.

PRODUCTION PROCESSES AND THE ENVIRONMENT

Risk assessment models estimate the damage of production activities, such as the use of agricultural chemicals and their effects on humans and other species. These damages are measured by the numbers (or percentages) of deceased, sick or injured members of the affected populations. Several processes link input use to environmental and health damages (figure 7.1). These processes include input application; residue generation, transport and disposal; exposure; and dose response (Bogen, 1985). The damages that are caused by the contamination of surface water and groundwater by pesticides, for example, depend on the volume and method of application of pesticides; the resulting pesticide residues and their movement in water; the extent of exposure of humans or other species to the water through drinking or washing activities; and the vulnerability of exposed populations to chemical residues.

The environmental and health risks of agricultural production can be reduced by interventions in the processes that link input use to risks. Risks can be reduced by the decreased amount of emissions at the source, by the abatement of residues, and by the reduction in the exposure and vulnerability of the populace to chemicals. For example, the outcomes of the transfer and the fate of many types of liquid wastes are determined by the design and functioning of drainage and sewage ca-

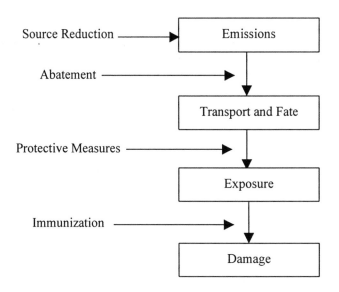

FIGURE 7.1 Policy Intervention at Various Stages of Damage Control

nals and of treatment facilities. Reduction of exposure from the use of agricultural chemicals may be achieved by wearing protective clothing during pesticide applications, by using water filters, and by subsidizing the use of bottled water. The introduction of immunization programs to reduce environmental vulnerability reduces health risks through the reduction of dose-response coefficients. Sound economic analysis must take into account all of these options in order to quantify their benefits and costs and to design policies that reduce risks (or increase safety) at least cost.

Source reduction is the most straightforward and complete pollution control strategy. Strategies, which aim to reduce human health risks from pollutants by reducing human exposure and vulnerability, often do not reduce risks for other species. The key parameters of exposure and dose-response processes are shrouded with uncertainties. These uncertainties carry over to policies that aim to reduce risk by controlling these processes, thus increasing the complexity of policymaking. Recent developments within information and communication technologies have led to the introduction of precision farming technologies. These technologies have a significant potential to reduce residues at their source. This chapter develops a framework to analyze policies that encourage the adoption of such emission reducing technologies.

Source Reduction Through the Adoption of Precision Technologies

Ayres and Kneese (1969) have emphasized that material balance relationships play a crucial role in explaining the environmental side effects of production processes. Not all inputs are completely utilized in production processes and the residuals, or unutilized inputs, are often pollutants. For example, only certain parts of the nutrients that are applied to cornfields are available to, or are absorbed by, the plants. The unutilized nutrients or residuals may leach into the groundwater and cause problems with nitrification or salinization. Similarly, when pesticides are applied aerially, as much as 75 percent may drift away from the fields to which they were applied. We define input use efficiency (IUE) as the percentage of applied input that is actually utilized in the production process. It is inversely related to the percentage of input that ends up as residue. IUEs are generally less than 100 percent. The extent to which residues occur depends upon both the application technology and the environment in which the application is made.

We distinguish between two types of input application technologies: traditional versus PTs. Generically, we define PTs as those technologies that enable the user to adjust the quantity of input used over space and time for a specific application. This would result in increased IUE when compared to the efficiency of traditional technologies. PTs include a broad range of production methods that meet the above criteria; they are not just technologies that have been commercially labeled as such. PTs are assumed to cost more in terms of extra capital or labor compared to traditional application technologies. The gain in IUE that is associated with a transition to PTs depends upon the characteristics of the physical environment, such as soil type, and the IUE of the traditional technology. Therefore, the gains in IUE, as a result of switching to PT, may vary significantly across locations and firms. For example, with traditional irrigation technology, the IUE may be 0.2 or 0.3 on steep land with sandy soils, compared to 0.9 for heavy soils on level land. Thus, the gain from the adoption of drip or sprinkler irrigation would be much greater on sloped land with sandy soils.

There are several important examples of PTs in agriculture. These include modern irrigation technologies, integrated pest management (IPM), and variable rate applicators of fertilizers and pesticides. Compared to more traditional gravitational irrigation technology, which has an IUE of 60 percent in California, sprinkler and drip irrigation IUEs can increase to 85 percent and 95 percent, respectively. Also, drip irrigation and sprinkler irrigation improves the IUE of fertilizers and pesticides. One may view IPM as a PT because it requires careful monitoring of pests before the pesticide is applied. This helps to reduce the frequency and to increase the effectiveness of pesticide applications. Similarly, commercially labeled PTs improve the IUE of fertilizers and pesticides through the use of soil maps and geographic positioning systems.

PTs are not limited to the application of water or agricultural chemicals. A major problem in fisheries is by-catch in which there is a significant difference between the volume of fish caught and the volume utilized. This contributes to the depletion of many fish populations. In this case, PT reduces the percentage of unusable fish caught. Similarly, in forestry management, clear-cutting produces nonusable and low-value materials and causes severe environmental side effects, such as soil erosion. The technology that targets a specific tree species and does not harvest the rest of the forest should reduce undesirable environmental side effects.

Further examples of PT in industrial cases are provided in Khanna and Zilberman (1997).

A Micro-level Model of the Adoption of Precision Technologies

One approach in analyzing the adoption of PTs is to assume that farmers pursue multiple objectives (Casey and Lynne, this volume). These objectives could include profits, worker safety, environmental quality, and social acceptability. The adoption of green technologies could be enhanced by a desire to increase worker safety, concern for environmental quality, or willingness to adhere to community standards. Other incentives for the adoption of green technologies could include price premiums for pesticide-free produce and agricultural products that are produced by environmentally friendly processes (Van Ravenswaay and Blend, this volume). These considerations suggest a potential for establishing voluntary arrangements that could lead to the adoption of PTs. They also suggest that the adoption of environmentally friendly technologies could be increased by raising the awareness of the environmental and safety implications of alternative technologies.

In this chapter, we take a more narrow approach and assume that farmers are motivated solely to maximize profits. The results are policies that hold even when farmers are not sympathetic to environmental and safety considerations. These policies may not be fully executed but should at least be designed as a backup to educational efforts and to voluntary agreements. We rely on Khanna and Zilberman's (1997)[1] analysis of PT choices[2] that are made by profit-maximizing and price-taking farmers. Production activities are assumed to take place in micro-units (for example, a field) using variable inputs, such as water, fertilizer, pesticides, and application technology.[3] Producers are assumed to have a discrete choice between traditional and precision technologies. Each micro-unit has several quality characteristics, such as fertility and a leakage coefficient. The leakage coefficient denotes the unused fraction of an applied input that ends up in residues, which may be of environmental concern. These characteristics vary across micro-units.

The terms pollution, residue and leakage are used interchangeably since they all denote the same phenomenon. The leakage coefficient for the traditional technology varies across different micro-units depending on soil characteristics. PT requires higher fixed cost per-acre, but it tends to reduce the leakage coefficient and, thus, improve IUE. We assume that PT impacts the leakage coefficient more in locations with low environmental quality and high leakage than it does in locations with good environmental quality and low leakage. Thus, PT can reduce environmental contamination by reducing leakage.

For each micro-unit, the profit-maximizing manager must determine technology and IUE levels. Without environmental regulation, when the explicit or implicit price of residues is zero, profit for the micro-unit equals revenue from production minus variable and fixed costs. A sequential process determines the optimal choice of technology for a micro-unit. First, the optimal variable input mix for each technology is determined and the profits associated with each technology are computed. Profits are compared across technologies and the most profitable technology is selected. This analysis can be used to show that the adoption of PTs can increase the productivity of some variable inputs because they tend to increase yield and to reduce variable input use at given prices.

Different types of PTs may vary to the extent that they are input saving, yield increasing or pollution reducing. The adoption of a PT generally is expected to reduce residues, per-unit of input and output, for a micro-unit. Some instances, in which traditional technology is operated under adverse environmental and economic conditions, the transition to PT could increase input use. Thus, it is possible that the adoption of PT could increase residues per acre when the increase in input use and the increase in output are very large.

When contemplating the adoption of PT, a micro-unit manager should compare the gain (which is associated with the possibility of higher crop yields) and the lower cost (which is a result of a reduction in variable input use) to the additional cost of the PT. Adoption of PT, therefore, is most likely to occur with high value crops that are grown in locations where input and output prices are higher. Locations with low environmental quality (in terms of high leakage coefficients) are also likely to be among the first adopters of PT. Thus, PT can create positive economic returns on lands with high leakage coefficients, thereby expanding the utilized land base.

Aggregate pollution, input demand and output supply are derived by aggregating over micro-units. Introduction of PT tends to increase agricultural supply at both the intensive (existing crop acreage) and extensive margins (new acreage brought under cultivation). The impact of introducing precision farming on aggregate variable input demand is not clear since adoption tends to reduce variable input demand at the intensive margin but increases demand at the extensive margin.

The increase in aggregate output induced by the adoption of precision farming tends to reduce output price, given a downward sloping demand curve. The output price effect could be relatively high if the demand for the final product were inelastic. A reduction in output price would have a secondary effect on variable input use, technology choice, and aggregate pollution. It would reduce incentives for technology adoption, and would reduce variable input use and pollution at the intensive and extensive margins.

In the absence of penalties for residue generation, producers would adopt PTs only if profits were greater than those earned using the traditional technology. When this is the case, profit-maximizing farmers would not attribute economic value to the pollution reducing attributes of a PT. Consequently, the imposition of a pollution tax would provide an incentive for the adoption of PTs just to reduce pollution generation. A pollution tax may also induce some land at the extensive margin to be taken out of production, while inducing other micro-units to adopt PT. Thus, a pollution tax would tend to reduce variable input use and output per-acre at the intensive margin, under both technologies, but it may actually increase output per-acre for micro-units that adopt PTs. The net impact of a pollution tax on aggregate pollution output and on output price would depend on the relative strengths of the intensive and extensive margin effects and on the extent to which the pollution tax would induce the adoption of PTs.

These observations are consistent with the results of Abler and Shortle (1999) and with empirical evidence on the adoption of irrigation technology (Caswell, 1991), soil testing (Babcock et al., 1997) and pest scouting (Carlson and Wetzstein, 1993). These studies confirm that financial incentives, such as high input prices, tend to lead to the adoption of technologies, which reduce variable input use and increase output. Inelastic demand for output may, however, set a boundary for the output expansion effect. The reduction in pollution intensity associated with PT is likely to reduce overall pollution even with a substantial expansion in output. Studies also confirm that adoption tends to occur mostly in locations with low IUE and high leakages. In these cases, environmental problems could be significantly reduced when PTs replace the traditional ones. Other studies, however, confirm that the adoption of center-pivot and low-volume irrigation systems in Nebraska and Kansas expanded irrigated land and led to a significant depletion in groundwater reservoirs. The introduction of technology that increases IUE could not only increase productivity, but also could generate new environmental problems by expanding production into areas not previously used.

Other Factors Influencing the Adoption of Precision Technology

The adoption of PT occurs selectively and is often triggered by extreme circumstances, as is shown by the results of other studies. Economic incentives for adoption may also depend on the institutional structure within which production occurs. For example, the adoption rate of drip irrigation has been below its potential, even though it has been around for about 30 years and can be quite effective in addressing water logging and drainage problems (Dinar and Zilberman, 1991). This situation occurs because water subsidies and restrictions on water trading have reduced incentives to adopt PTs. Therefore, adoption is mainly observed on high value crops and at locations with high water prices. With this and other PTs, however, it is evident that input prices will have to increase substantially to cause a significant increase in adoption, given the institutional structure. In the cases of drip and center-pivot irrigation, adoption has been triggered by dramatic events, which have significantly affected input availability and prices. In California, drip irrigation has received a significant boost because of the droughts of 1977 and 1987–1991. The adoption of center-pivot and low-volume center-pivot irrigation systems in the Midwest was triggered by increased output prices in the early 1970s, followed by increased oil prices in the late 1970s and early 1980s.

New PTs have to be perfected over time through the process of learning by doing. For example, drip irrigation did not reach its full potential until extension specialists, dealers, support staff, and farmers understood and perfected its application. Its introduction required the establishment of a support infrastructure for the product. More recently, with the development of computerized irrigation systems that use weather information, new types of professionals (such as irrigation consultants) have begun to play an important role. Some irrigation consultants are private, independent companies that mostly serve small- and medium-sized farms. Some of the mega-sized farms have their own in-house irrigation consultants. With the aid of these consultants, the weather information services and computerized irrigation systems, the efficiency of water use in some areas of the San Joaquin Valley, California has increased significantly. This has led to both higher crop yields and substantial savings in water applications (Parker and Zilberman, 1996).

These cases also demonstrate some of the adjustments that would have to occur in terms of training, product support, and the organization of agriculture. Most of the increase in automation and PT in California farms has been a result of the severe droughts of recent years. This is a vital point: *Incentives matter.* Improving productivity while addressing environmental problems through technological change may be feasible, but its adoption must be induced by appropriate incentives.

OBSTACLES TO FIRST-BEST POLLUTION CONTROL POLICIES

When the residue from agricultural production is variable and its direct damage is known, the implication of theory suggests that a pollution tax equal to the marginal damage will result in a first best solution. In this case, a tax will induce all farmers to reduce input use. It may induce some farmers with high leakage per-unit of output to exit, and may lead others who have high leakage coefficients per-unit of input to adopt PT. There are several constraints that limit the use of a first-best tax to control pollution (Segerson, this volume). Some of these constraints and alternative approaches to address them are discussed below.

Measuring the Damages Due to Pollution

It is very difficult to establish a monetary value for the marginal damage associated with pollution, such as contamination of groundwater or air. Therefore, policymakers normally take the Baumol and Oates' (1974) approach, which is to set regional target levels for aggregate pollution and to develop policies limiting pollution at that level. This approach is consistent with the view of sustainability as a policy paradigm, which aims to contain environmental quality within certain bounds by constraining the flow of pollution to be equal to the assimilative capacity of the environment.

Political Economy Considerations

Theoretical analysis suggests that four types of financial incentives can contain pollution at the same level: nonuniform input taxes; uniform pollution taxes; subsidies for pollution reduction; and tradable pollution permits. All of these may lead to efficient resource allocation, but they could have different distributional effects. Farmers would obviously prefer subsidies for pollution reduction and would oppose taxation. Since farmers are most affected by pollution reduction policies, they would use their political power to affect the policy choice.

When farmers utilize land of varying quality, or use technologies with different impacts on environmental quality, uniform pollution taxes can be replaced by nonuniform input taxes to achieve efficient resource allocation. In practice, differentiated input taxes may be difficult to implement for political reasons and for reasons of equity. Uniform input taxes on environmentally damaging inputs (such as pesticides, fertilizers and gasoline) may then be used and could provide some positive environmental benefits by allowing incentives to switch to input saving technologies. The use of input tax revenues, to subsidize the adoption of PTs or to subsidize research for the development of PTs, may make these taxes more palatable to producers. This scenario has been adopted in Scandinavian countries. Some water districts in California have raised the price of water to encourage its conservation. They have used some of the revenue from this increase in water price to subsidize the adoption of conservation technologies (Zilberman et al., 1997). Revenue from taxes on agricultural technologies has been used for research and development, but most of it has been used to find pesticide substitutes and to improve input application technologies. Gasoline tax revenue, on the other hand, usually has not been returned to the industry, but has been allocated to the general revenue.

Political economy strategies suggest that tradable permits would be more acceptable for pollution control than would input taxation. This scenario requires that the government establish an overall target, introduce permits that would be distributed to producers (generally in proportion to their past pollution levels) and allow permits to be traded. Alternatively, the government could set an upper bound on the regional use of inputs, such as pesticides, and could introduce initial allocations of pesticide use so that farmers would have the right to trade these permits. Yarkin et al. (1994) suggest that, if the aggregate use of methyl bromide were reduced by 50 percent but a transferable permit system was established, then about 80 percent of the benefits associated with methyl bromide in California agriculture could be preserved.

Transaction Costs

Missing markets are a major cause of inefficiency and can negatively affect the environment. Coase (1960) argued that the externality problem could be solved efficiently through transactions between the concerned parties. He also recognized that high transaction costs could prevent an efficient outcome. Historically, water has been allocated on a first-come, first-served basis. While such systems may have been sufficient in the past and even been efficient when water was abundant, they now prevent water scarcities from being reflected in water prices, and hinder water conservation. The transition to market mechanisms, however, is being hampered by the high costs of establishing property rights, monitoring, and enforcing. When considering institutional reform, transaction costs have to be taken into account. If the efficiency gain is smaller than the transaction costs, then reform may not be justified.

Over time, transaction costs could be reduced through technological innovation, so that the benefits of a transition to market mechanisms would increase. In the case of the Wetlands of California, the introduction of water markets has been aided by innovations in electronic communication (Olmstead et al., 1997). Much of the motivation for the introduction of water trading has come from decreased water availability and the desire to increase water availability for environmental purposes. The elimination of restrictions on water trading in California coincided with a reallocation of water for environmental purposes. The introduction of trading may lead to the adoption of modern irrigation technology and other PT practices. The efficiency gain that is associated with improved water efficiency could compensate producers for the reduction in aggregate water supply.

Uncertainty and Randomness

The environmental impact of agricultural inputs depends on the residues from crop production and the vulnerability of the environment. It could also be affected by random events. For example, the air quality effects of rice burning depend on wind conditions and temperature. The extent to which chemical residues contaminate a body of water depends on rain conditions. If financial incentives were designed to accommodate all these considerations, they would vary according to where, when and how chemicals were applied. Thus, they would become extremely nonuniform over space and time.

To compute the optimal pollution tax, one needs to know the marginal benefits and costs of pollution. These relationships cannot be known with certainty and must be estimated statistically. Even under the best of circumstances—when data are available and the appropriate statistical processes are applied—estimates may vary greatly from the true values of key parameters. A tax that is based on these estimates may be suboptimal. Similarly, the key relationships that affect pollution generation are subject to random shocks from weather changes, disease infestation and other factors. Policies are frequently established *ex ante*, before the true state of nature is known, and this may also lead to suboptimality.

A framework developed by Weitzman (1974) was used to compare the performance of pollution taxes and direct controls under uncertainty and randomness. Under these conditions, he demonstrated that neither policy instrument would always be superior. For example, Weitzman found that taxes, on average, performed better when both the marginal costs and benefits of pollution reduction were highly elastic with respect to pollution. Direct controls that explicitly specify the levels of pollution (rather than affecting them indirectly through incentives) were found to perform better, on average, when both marginal relationships were inelastic. Thus, with uncertainty and/or variability, there can be circumstances in which taxes are suboptimal and less preferred than direct controls.

Heterogeneity

While the implementation of a tax is less information intensive than are direct controls when firms are homogeneous, the theoretical efficiency gains from using taxes are pronounced when firms are heterogeneous. Optimal taxes, however, are also more difficult to determine, implement and enforce. When firms are heterogeneous in the location and timing of pollution, it is impractical to introduce trading or transferable permits because the price of permits must be adjusted by coefficients that reflect this heterogeneity.

The efficiency of input use in agricultural production varies across locations because of variations in weather, geography, soil type, and other factors. The environmental impact of residues also varies over space. For instance, drifting air pollution will be much more severe in cities than it will be in the country. Groundwater contamination in an aquifer that serves a large city has a greater impact on society than does a contaminated aquifer that is used primarily for irrigation.

With heterogeneity of environmental impact, the cost of a residue unit will vary among locations. Combining this with the heterogeneity of residue production that results from differences in application technology, the Khanna-Zilberman framework can be extended so that each location is characterized by two coefficients, one of which accounts for heterogeneity in residue production and the other of which accounts for differences in environmental vulnerability. The vulnerability coefficient would represent the vicinity to watersheds or to specifically sensitive environments. Even with full information about residue levels, the optimality of resource allocation will not be restored by a uniform residue tax or by the restriction of aggregate residue levels. In this case, the optimal policy is a residue tax that varies by the environmental vulnerability index. Such a policy, however, would be difficult to implement and to enforce.

When there is heterogeneity in both leakage and environmental variability, adoption patterns will depend on the distribution of land with respect to these variables. Under an optimal tax policy, the adoption of PT is more likely to occur in locations with high leakage and high environmental vulnerability coefficients. When there is heterogeneity in environmental vulnerability, optimal residue taxes should vary across locations. If residues are not observed but policymakers have information on locational characteristics and technology choices, then the optimal tax rate could vary accordingly. Such a variable tax policy may not be easy to implement. Other policies that approximate the optimal resource allocation may be considered. For instance, if policymakers have good information on the technology and past input levels used by farmers, then policymakers could design a technology differentiated tax on the variable input.

Multidimensionality

Pesticide use may create multiple, simultaneous side effects. For example, pesticide use may simultaneously affect food safety, worker safety, groundwater quality and the biological viability of other species in the ecosystem. Aerial spraying of pesticides may affect workers in the field; surface water quality; water ecosystems; land animal populations; equipment; real estate; and other food crops (which may lead to food contamination and to human health problems). Thus, having a uniform tax per-unit of pollution could be suboptimal. Each of these dimensions would require special attention. Policymakers have to know the mechanisms of the transmission of toxins and the processes of exposure as well as dose response in order to determine optimal policies. A tax must be adjusted to account for both locational differences and the dimensional consequences of pollution. It may be accompanied and even replaced by other forms of intervention to address issues of exposure. In some cases, direct mandates may be part of an efficient policy.

Observability and Asymmetric Information

Lack of knowledge about the behavior of individual producers is another obstacle in implementing first-best pollution taxation. Since many agricultural pollution problems are nonpoint-source polluters in nature (for example, groundwater contamination), policymakers cannot identify the contributions of individual producers. The control of point-source pollution (PSP) problems, in which the pollution of individual producers is known, is easier to address. In some cases, an ambient tax (Segerson, 1988) can efficiently address a nonpoint-source pollution (NSP) problem. Nevertheless, it can involve large and highly variable transfers in order to assure that the lower bound of regional water quality is not exceeded. This is a type of policy that can be difficult to implement when polluters are risk-averse.

The distinction between PSP and NSP problems, however, is often technology dependent. When appropriate monitoring technology is introduced, an NSP problem may become a PSP problem. Thus, environmental policy choices could be modified with the introduction of improved monitoring technologies. Public support for research and development activities that would lead to better monitoring technologies will be an important component of a long-run environmental management strategy.

In the literature of mechanism design, Groves (1973) developed a complex set of procedures that regulators can use to induce individuals to reveal their true level of pollution, allowing a tax to be adjusted on a case-by-case basis. However, mechanism designs can be very complex. Such mechanisms may not be easily applicable because policymakers prefer to have simple policy solutions to complex problems. Nevertheless, the emphasis on mechanism design is crucial for developing alternative policies that address environmental pollution and induce the adoption of PT.

DESIGNING POLICIES IN AN IMPERFECT WORLD

In the previous section, we reviewed a sample of major obstacles to the establishment of first-best pollution control policies and to the establishment of necessary policy adjustments. Solutions that overcome three of these obstacles—asymmetric information, heterogeneity and multidimensionality—are presented below.

The Problem of Incomplete Information

In assessing policies within the Khanna-Zilberman framework, a tax or a transferable right system that targets residues may not be feasible because of the difficulty of monitoring these residues. In the absence of complete information about the residues generated by a farmer, a technology or input dependent tax would be a second-best alternative to the residue tax. The tax on the variable input (for example, pesticides, water or fertilizers) would depend on application technology.

When a farmer applies a pesticide aerially, he or she would have to pay a higher tax than he or she would if he or she were to apply the same pesticide with low pressure-precise application equipment. Implementation of such a tax could require information, neither easily available nor verifiable, on input use for individual fields. Therefore, a uniform input or output tax is often used as a third-best policy.

An input tax is less efficient than a technology dependent tax but is more easily applied. It is collected from input sellers rather than from individual farmers. It is inefficient since it does not differentiate between different application technologies. A more efficient policy consists of a combination of an input sales tax (a mill tax on fertilizers or pesticides) and a per-acre tax, or subsidy, based on technology choice. Since the choice of irrigation technology or the use of consultants is observable, a tax or subsidy based on these activities can be easily implemented.

Output taxes could also be considered as means of pollution reduction and, if sufficiently high, they could have a significant impact on the generation of pollution. Much of this impact may be through the extensive margin, where the output tax leads to a reduction of output and pollution by forcing firms to downsize or to discontinue business. In cases in which farmers can choose between several application technologies, output taxation is an inefficient way to reduce pollution. Users of PTs could have higher yields with less pollution per-acre but would have to pay higher taxes per acre than users of traditional technologies that generate more pollution per-acre.

Rather than relying upon uniform input or output taxation, the regulator can directly address the problem of insufficient information about polluters in several ways. Traditionally, enforcement of pollution control has relied upon random monitoring combined with fines for noncompliance (Linder and McBride, 1984) in which the expected penalty equals the marginal damage cost. The regulator could save on inspection costs by imposing a self-reporting requirement on the producer (Malik, 1993). This requirement would impose less stringent fines on polluters who voluntarily report accidental pollution.

When the costs of monitoring and enforcement are high and government budgets are shrinking, the regulator could transfer monitoring costs to the polluters. This is especially appropriate for metering water and energy use, but it may also be applied to metering residues. Millock et al. (1997) proposed a mechanism aimed at addressing the problem of unobservability of individual actions. They developed a taxation scheme that gave incentives to individual polluters to invest in monitoring equipment in a cost-effective manner. Under these schemes, firms that install monitoring equipment are taxed at a level based on their actual pollution. Non-monitored firms pay a tax based on the average pollution of the group of firms. Under this policy, monitoring serves as a mechanism by which efficient polluters signal the regulator that they pollute relatively less. Such an incentive scheme explicitly recognizes the asymmetry of information as well as the cost of monitoring, and aims at balancing this cost with the gain from controlling the residues from an additional micro-unit.

Policies Addressing Heterogeneity

Babcock et al. (1997) argued that, in many cases, environmental vulnerability is concentrated in relatively small subsets of any given region. There are several ways in which policy may be designed to address the problem of heterogeneity.

Zoning and Direct Controls

One management strategy to address heterogeneity is to target strong environmental regulations (such as high taxation and direct control) at regions with the most pollution and to target less radical regulations (such as moderate input taxes) at regions with less pollution. Thus, under strategies of this type, land would be zoned and treated differentially. For example, the penalty for environmental contamination in riparian areas should be much higher than in less environmentally sensitive areas.

Zoning is one method that can be used to directly control environmental problems. Traditionally, direct controls have taken the form of best management practices for the more vulnerable lands. The problem with this form of policy is the close monitoring that is required for its enforcement. Thus, when the cost of monitoring is high, financial incentives that make the producer responsible for his or her own land use are preferable.

Resource Conservation Funds

A different approach to improving environmental quality is to establish resource conservation funds that pay producers for the avoidance of environmentally damaging activities. An example is the Conservation Reserve Program (CRP) in the United States. Such funds are established either to reduce the negative side effects associate with pollution or to preserve environmental amenities. For example, to preserve an endangered species or a certain ecosystem, it may be necessary to restrict or even to disallow agricultural production in certain regions.

The CRP is designed to address environmental problems of this second type, namely, to develop mechanisms that reduce, or eliminate, the agricultural activities in an environmentally valuable area. To some extent the CRP can be seen as a subsidy program that induces farmers to reduce their agricultural activities. Babcock et al. (1997) demonstrated that there is a significant degree of heterogeneity in areas targeted for the CRP. Relatively small amounts of land may contain most of the environmental amenities. If the purchasing schemes were designed efficiently (for example, by targeting the land that is most sensitive), then it could be possible to obtain 80 percent to 90 percent of environmental benefits with a substantially smaller capital outlay.

The experience gleaned from the development and implementation of the CRP and other purchasing programs is especially encouraging. It indicates that society is willing to pay for altering economic activities in locations with high environmental vulnerability, or value, and for expanding and refining these programs to increase the returns for the amount spent. Designing such programs effectively is a major challenge to agricultural economists. There are limits, however, to the effectiveness of purchasing programs and to the public's willingness to pay for altering production activities to meet environmental quality objectives. Policymakers need to combine *carrots* with *sticks* when addressing these problems. The principle that the polluter must pay (*sticks*) is appropriate for controlling externalities. Purchasing program funds (*carrots*), however, are better utilized for environmental conservation purposes.

To address pollution problems and to attain environmental policy objectives, the policymaker has a significant arsenal of regulatory instruments, including input taxes that are dependent or independent of technology choices; transferable permit mechanisms for input use; subsidization for land diversion; and the reduction of certain input uses. Financial incentives are powerful and essential policy tools for addressing environmental problems in agriculture and for promoting sustainability. As we have shown above, however, the design of such incentives is not straightforward. It may need to be complemented by other forms of policy intervention, such as information, education and direct control.

Policies That Address Multidimensionality

Recall that policies to improve environmental health are not limited to the reduction of residues. Residues cannot be eliminated in livestock systems. In some crop systems, it may be more cost effective to tolerate some residues but to reduce their impact by developing mechanisms that reduce exposure, and even susceptibility, to toxicity. This suggests that the use of financial incentives need not be limited to the taxation of chemical use, the subsidization of the adoption of PT or to other means that reduce residues. Financial incentives may be needed to improve conveyance facilities of waste products so that environmentally harmful leakages during transport and disposal may be limited. These incentives could include subsidies for the construction of safer conveyance facilities and subsidies, taxation, and liability for the control of the damages associated with waste disposal. One important policy challenge is to provide incentives that will enable reuse and recycling of animal wastes (Norris and Thurow, this volume).

Financial incentives could also target the reduction of exposure and vulnerability to waste products. This may include the funding of research and development for the purpose of improving protective clothing, monitoring chemical residues and designing more effective filter systems. Once better equipment is available, financial incentives can induce the adoption of these technologies. For example, liability rules or taxation could lead to reduced exposure to pollution through the use of protective clothing, filters and appropriate work practices. Other policies reduce exposure to pollution, such as the construction of two water systems—one for drinking and washing, and the other for irrigation. The development of an efficient medical emergency system to treat exposure to toxins is another component of an environmental health policy that targets dose-response processes. Thus, policies to reduce environmental and human health risks should be derived from a framework that considers intervention in all stages of the risk generation processes.

PUBLIC SECTOR ACTIVITIES AND NEW INSTITUTIONS

The role of government and the public sector in attaining sustainability should not be limited to instituting taxes, providing subsidies and introducing regulations. Given the enormous lack of knowledge that hampers the effectiveness of environmental policies, the public sector must be provided with research and information. Individuals may be unaware of environmental problems that result from their activities, so education is the key to information. Education leads individuals to internalize environmental costs and to modify their behavior. The importance of the social norms and values, which result from the public's education regarding environmental pollution, is apparent when one observes the success of recycling schemes and the cleanliness factor that it enhances in the common areas of different regions.

Government has an important role when facilitating collective action to address environmental problems. Regional cooperation and coordination is needed for the success of pest eradication programs (Carlson and Wetzstein, 1993), regional pest monitoring efforts, and the introduction of beneficial species for the regional biological control of pests. These components of environmental improvement strategies are likely to be under-supplied by the private sector; thus, they justify public sector efforts.

The public should be able to identify and monitor environmental problems and to design and implement solutions. This perspective pinpoints an important role for university agricultural extension programs, natural resource conservation agencies and similar organizations that operate at a regional level. Their role is to detect problems and to facilitate solutions to these problems. Some of these organizations (environmental groups) may be established through the voluntary collective action of concerned citizens. Because of the free-rider problem, however, there is likely to be under-provision of environmental monitoring and of problem solving. This problem is the reason government intervention and institutions are needed.

The public sector's role is also to provide an infrastructure for the development of technological and institutional solutions to these environmental problems. In the long run, however, the technological implementation (such as recycling, waste management and disposal) of an effective strategy for attaining sustainability may be better carried out by the private sector. These private industries could require financial and technical support in the initial stages. In some cases, the details of technical or institutional solutions are quite clear, but private entities do not initiate the implementation of the strategy. For instance, it took some time before private sector activities were developed in the areas of waste management and recycling. Thus, government officials in environmental agencies and agricultural extension programs should also play the role of agents for change. They must encourage initiatives, reduce transaction costs and provide better incentives and support.

Economists tend to ignore the role of the public sector when encouraging entrepreneurship and when initiating solutions that are ultimately performed by the private sector. Public institutions, such as agricultural extension programs, have played a major role in providing the foundations for a stronger farming sector and successful private enterprise. Also, agricultural extension programs have played a major role in establishing institutions for collective action, which include the Farm Bureau, farm cooperatives, and pest management districts. Agricultural extension specialists, for example, were crucial to the introduction of drip irrigation in California (Caswell et al., 1990). Similarly, agricultural extension specialists have introduced biological control and IPM activities to the private sector.

A recent study on technology transfer (Parker and Zilberman, 1996) led to the finding that most of the leading firms in the biotechnology area (for example, Genentech, Chiron, Calgene and Amgene) were established by researchers from different universities or were based on university discoveries. In some cases, university offices of technology transfer provided the support that enabled these enterprises to be established.[4] In this way, public sector activities can be important in initiating a strong private sector. The two are not independent, and sometimes are symbiotic. Consequently, one challenge for policy is to encourage positive dependencies and to discourage many of the abuses that may occur in some symbiotic relationships. Therefore, the public sector's support of green industries may be justified. Public research and extension, in many cases, provide these types of firms with research, development and marketing efforts that otherwise may not be affordable. A specific example is the satellite system first developed by the military for surveillance purposes. These systems are now being adapted by the private sector for the geo-referencing of crop yield and for the soil quality information necessary for precision farming.

The public sector has an important role to play in fostering changes in technology and management that address environmental problems. In particular, public sector activities may lay the groundwork for establishing new industries that can provide better solutions to environmental problems. Consequently, we need to develop economic criteria for the efficient management of these public sector activities. We also need to develop better measures of performance for public sector activities that initiate enterprise and change. We must develop rules on how and when to build the capacities for change in the public sector and how and when to reduce, or even discontinue, them.

The major contribution of the public sector to private sector activities is a credible and efficient government apparatus that protects and safeguards property rights, lays down a stable legislative framework and disseminates information about emerging environmental problems in a timely fashion so that private sector initiatives can be developed. The government can be proactive but, above all, it must be efficient and consistent.

Certified consultants should play an increasing and major role in addressing and resolving agro-environmental problems. Tax incentives are limited in their capacity and potential to modify the behavior of producers. This is because of the extreme heterogeneity and complexity of agricultural systems and because of political and economic reasons. Reliance on prescribed command-and-control policies is inefficient and often results in very complex and diverse sets of optimal policies for the same reasons. Thus, when it comes to production activities that give rise to negative environmental side effects, it is important that decision-makers have enough flexibility and awareness to address and efficiently resolve both private and public concerns.

Highly trained individuals should be assigned to make decisions on environmentally sensitive activities. This is part of a new policy that is emerging in California and other states in which certified consultants make decisions on restricted chemical use. Ideally, such consultants have constraints and different incentives to guide their solutions of pest problems or other problems with environmental side effects. Their choices should be documented and subject to scrutiny; they can make choices that maximize the well being of their clients.

SUSTAINABILITY AND BEYOND

This chapter argued that sustainable development could be achieved by using appropriate incentive-based policies. The multidimensionality and complexity of environmental systems require the use of multiple policy tools. These include incentives for the adoption of PTs, resource conservation funds, flexible regulations and institutions for collective action.

Sustainability implies a pursuit of economic growth while maintaining environmental quality. Environmental quality can also be considered a luxury good. If this is the case, then as society gets wealthier, the demand for this good will increase. Therefore, the objective of sustainability may not be ambitious enough and society should strive to continue improving environmental quality rather than to simply maintain the status quo. The framework that we use to analyze the adoption of PTs allows for the introduction of incentives, which encourage producers to shift toward more environmentally friendly technologies and which control pollution from existing production processes.

The global arena is where the environmental situation is most worrisome and where the notion of sustainability and prevention of deterioration is even more critical. This analysis has emphasized the domestic context, but the same notions of incentives, collective action and efficient intervention apply to global problems as well. The manner in which incentives and technological change can contribute to the sustainability of the global environment is a major area for further research.

ENDNOTES

1. The model is a generalization of previous models that were introduced for the adoption of irrigation technology by Caswell and Zilberman (1986) and by Dinar and Zilberman (1991).
2. In some contexts (such as forestry and fisheries) it may be useful to use the term harvesting technology.
3. In many cases there is variability in input use efficiency (IUE) even within a micro-unit. An example is a field, in which there are patches of land that have high leakage coefficients and others that have low leakage coefficients. As a result, two dimensions of environmental quality of concern may be average leakage and variability of leakage. In these cases, PT reduces both the variability of leakage coefficients and the average leakage coefficient. The analysis gets much more complex in this case, but the results essentially do not change.
4. Technology transfer specialists were crucial to matching inventors with venture capitalists and providing the financial and institutional know-how that led to the establishment of new ventures.

REFERENCES

Abler, D.G. and J.S. Shortle. 1999. *Handbook of Environmental Economics.* London, UK: Edward Elgar Publishers.

Ayres, R.U. and A.V. Kneese. 1969. "Production, Consumption and Externalities." *American Economic Review* 59(3): 282–297.

Babcock, B.A., P.G. Lakshminarayan, J. Wu, and D. Zilberman. 1997. "Targeting Tools for the Purchase of Environmental Amenities." *Land Economics* 73(3): 325–339.

Batie, S.S. 1989. "Sustainable Development: Challenges to the Profession of Agricultural Economics." *American Journal of Agricultural Economics* 71(5): 1083–1101.

Baumol, W.J. and W.E. Oates. 1974. *The Theory of Environmental Policy.* Englewood Cliffs, NJ: Prentice Hall.

Binswanger, H.P. and V.W. Ruttan. 1978. *Inducted Innovations: Technology, Institution, and Development.* Baltimore, MD: The Johns Hopkins University Press.

Bogen, K. 1985. "Uncertainty in Environmental Health Risk Assessment: A Framework for Analysis and Application to a Risk Assessment Involving Chronic Carcinogen Exposure." Unpublished Ph.D. dissertation, Department of Agricultural and Resource Economics, University of California, Berkeley, CA.

Carlson, G.A. and M.E. Wetzstein. 1993. "Pesticides and Pest Management," in G.A. Carlson and D. Zilberman, eds., *Agricultural and Environmental Resource Economics.* Oxford, UK: Oxford University Press.

Caswell, M. 1991. "Irrigation Technology Adoption Decisions: Empirical Evidence," in A. Dinar and D. Zilberman, eds., *The Economics and Management of Water and Drainage in Agriculture.* Boston, MA: Kluwer Academic Publishers.

Caswell, M., E. Lichtenberg, and D. Zilberman. 1990. "The Effects of Pricing Policies on Water Conservation and Drainage." *American Journal of Agricultural Economics* 72: 883–890.

Caswell, M. and D. Zilberman. 1986. "The Effects of Well Depth and Land Quality on the Choice of Irrigation Technology." *American Journal of Agricultural Economics* 68: 798–811.

Coase, R.H. 1960. "The Problem of Social Cost." *Journal of Law and Economics* 3: 1–44.

Dinar, A. and D. Zilberman. 1991. "Effects of Input Quality and Environmental Conditions on the Selection of Irrigation Technologies," in A. Dinar and D. Zilberman, eds., *The Economics and Management of Water and Drainage in Agriculture.* Boston, MA: Kluwer Academic Publishers.

Feder, G., R.E. Just, and D. Zilberman. 1985. "Adoption of Agricultural Innovations in Developing Countries: A Survey." *Economic Development and Cultural Change* 30: 59–76.

Griliches, Z. 1957. "Hybrid Corn: An Exploration in the Economics of Technological Change." *Econometrica* 25: 501–522.

Groves, T. 1973. "Incentives in Teams." *Econometrica* 41: 617–631.

Hayami, Y. and V.M. Ruttan. 1985. *Agricultural Development: An International Perspective.* Baltimore, MD: The Johns Hopkins University Press.

Khanna, M. and D. Zilberman. 1997. "Incentives, Precision Technology and Environmental Quality." *Ecological Economics* 23(1): 25–43.

Lancaster, K.L. 1968. *Consumer Demand: A New Approach.* New York, NY: Columbia University Press.

Lichtenberg, E. and D. Zilberman. 1988. "Efficient Regulation of Environmental Health Risks." *Quarterly Journal of Economics* 103: 167–178.

Linder, S. and M. McBride. 1984. "Enforcement Costs and Regulatory Reform: The Agency and Firm Response." *Journal of Environmental Economics and Management* 11: 327–346.

Malik, A. 1993. "Self-reporting and the Design of Policies for Regulating Stochastic Pollution." *Journal of Environmental Economics and Management* 24: 241–257.

Marra, M. and D. Zilberman. 1993. "Agricultural Externalities," G.A. Carlson, D. Zilberman, and J.A. Miranowski, in eds., *Agricultural and Environmental Resource Economics.* Oxford, UK: Oxford University Press.

Millock, K., D. Sunding, and D. Zilberman. 1997. "An Information-revealing Incentive Mechanism for Nonpoint-source Pollution." Unpublished manuscript. Department of Agricultural and Resource Economics, University of California, Berkeley, CA.

Olmstead, J., D. Sunding, D. Parker, R. Howitt, and D. Zilberman. 1997. "Water Marketing in the 90s: Entering the Electronic Age." *Choices* 3rd Quarter: 24–28.

Parker, D. and D. Zilberman. 1996. "The Use of Information Services: The Case of CIMIS." *Agribusiness* 12: 209–218.

Ruttan, V. 1994. "Constraints on the Design of Sustainable Systems of Agricultural Production." *Ecological Economics* 10: 209–219.

Segerson, K. 1988. "Uncertainty and Incentives for Nonpoint Pollution Control." *Journal of Environmental Economics and Management* 15: 87–98.

Weitzman, M.L. 1974. "Prices vs. Quantities." *Review of Economic Studies* 41: 477–491.

Yarkin, C., D. Sunding, D. Zilberman, and J. Siebert. 1994. "Methyl Bromide Regulation . . . All Crops Should Not Be Treated Equally." *California Agriculture* 48(3): 10–15.

Zilberman, D., C. Yarkin, and A. Heiman. 1997. "Agricultural Biotechnology: Economic and International Implications." Paper presented at the XXIII International Conference of Agricultural Economists, *Food Security, Diversification, and Resource Management: Refocusing the Role of Agriculture?* Sacramento, CA.

8 Using Ecolabeling to Encourage the Adoption of Innovative Environmental Technologies in Agriculture

Eileen O. van Ravenswaay and Jeffrey R. Blend
Michigan State University, East Lansing, MI
Montana Department of Revenue, Helena, MT

The potential of ecolabeling to create economic incentives for the adoption of environmental technologies in agriculture is examined in this chapter. Ecolabeling programs were described. A theoretical framework was developed and used to derive the necessary economic conditions for ecolabeling programs to generate adoption incentives. The extent to which these conditions could be met in agriculture was investigated. The investigation used survey data on consumer demand and information on producer costs in reference to several new ecolabeling programs in agriculture. Since little empirical research has been completed on this subject, no definitive conclusions were made about the prospects of ecolabeling, but key research needs were identified.

INTRODUCTION

Ecolabeling is an institutional mechanism for establishing a market for environmental protection during the production or consumption of products.[1] The ecolabeler establishes the terms of trade between the buyer and seller by defining environmental protection standards that are to be met during the production or consumption of the product. The concept is similar to setting standards for product performance or safety.

At first glance, ecolabeling may seem like an unlikely candidate for creating market incentives for environmental protection. Economists typically assume that environmental damage associated with production or consumption is an externality. In other words, the damage is to some third party, not to the producers or consumers in the market. But what if environmental damage from the production of a good affected the consumers of that good? In this case, it is no longer appropriate to analyze the problem as a producer externality. The consumer now faces a trade-off because his or her consumption is the ultimate cause of the environmental damage he or she suffers. Without an ecolabel, the consumer has only one way that he or she can reduce the damage, namely, to consume less. With an ecolabel, he or she has the option of buying from a seller who has reduced the damage for him or her.

While there may be gains from trade in this situation, there are high information costs associated with defining the good and an asymmetric information problem associated with enforcing the terms of trade. An ecolabeling agent could address these problems, but the costs may be prohibitive.

This chapter examines the potential of ecolabeling as a means to encourage the adoption of environmental technologies in agriculture. First, ecolabeling programs are explained with reference to what they generally involve, and the legal constraints on such programs in the United States are given. Next, the necessary conditions for an ecolabeling program to generate revenues sufficient to encourage technology adoption are analyzed. Empirical information is examined to assess the extent to which these conditions could be met in agriculture. Evidence on potential consumer demand for ecolabeled foods is examined first. Then costs of establishing and operating ecolabeling programs in agriculture are examined. Since there is little empirical work on this subject, the chapter identifies a number of areas in which research is needed.

ECOLABELING PROGRAMS

More than 20 countries and the European Community have adopted public ecolabeling programs to encourage the development of manufacturing processes and products with less environmental impact (EPA, 1993a, 1993b, 1993c, 1994). In the United States, there are several private non-profit ecolabeling programs (for example, Green Seal and Scientific Certification Systems) and voluntary environmental labeling programs that are supported by the government (for example, the EPA Energy Star Program). These programs have been facilitated by the development of environmental marketing rules at the state, federal and international levels (Grodsky, 1993; Kuhre, 1995 and 1997; Lamprecht, 1996; van Ravenswaay, 1996).

An ecolabel identifies environmentally preferable products based on an environmental impact assessment of a product compared to other products in the same category. An important feature of this impact assessment is that it is not limited to the environmental impacts from use and/or disposal of the product. It also includes impacts from the production of the product. A third party, either public or private, conducts the impact assessment.

Ecolabeling is only one form of environmental labeling that is seen in markets today. Two other common forms of such labeling are government-mandated labels and self-declaration labels (Kuhre, 1997). Examples of government-mandated environmental labels are fuel-efficiency ratings that are required on new automobiles, energy-use guides that are required on household appliances and environmental hazard warnings that are required on pesticides and products containing chlorofluorocarbons (CFCs) or toxic substances. Examples of self-declaration labels are manufacturer claims about recyclability, recycled content, solid waste reduction, biodegradability and non-use of certain chemicals (for example, phosphates).

Two key features differentiate ecolabeling from these other forms of environmental labeling. Unlike government-mandated labels, ecolabels are voluntary. Unlike self-declaration labels, ecolabels involve standard setting and enforcement by a third party.

Ecolabels are much like a seal of approval. They are awarded by a public or private non-profit organization that (1) establishes environmental standards for product categories and (2) certifies that products meet those standards. Thus, an ecolabel is like a seal of approval because it is a signal of high standards as well as a signal that products meet standards.

An ecolabeling organization performs three key tasks: standard setting, certification and marketing. Standard setting determines the environmental standards a product must meet to qualify for the ecolabel. Certification determines whether a given product meets those standards. Marketing develops customer awareness of, and trust in, the claim.

Green Seal, for example, is a private ecolabeling program that operates in the United States. It develops environmental standards for product categories (for example, paper, fluorescent lamps, household cleaners and paint) that pertain to the product's characteristics (for example, energy efficiency) and how it is to be made (for example, the type of de-inking and bleaching process that may be used for recycling paper).[2] The task of certifying whether a product meets those standards is contracted out to Underwriters Laboratories. If a product is certified to meet its standards, Green Seal licenses its mark to the product manufacturer, subject to various contractual terms, such as periodic monitoring (EPA/PPT 1993b). Green Seal also identifies potential customers for products with the seal.

Ecolabeling programs vary substantially in terms of the comprehensiveness of their environmental standards.[3] Some ecolabels concern themselves with a single environmental impact within a single stage of the life cycle of a product (figure 8.1). For example, the Flipper seal of approval on tuna is concerned only with the impact of tuna fishing on dolphins, and the EPA Energy Star program focuses only on energy conservation in the use of computer equipment. In contrast, ecolabeling programs like Green Seal and those of many European countries consider multiple environmental impacts throughout the stages of the life cycle of a product.

Ecolabeling standards are sometimes based on a method known as life-cycle assessment (LCA), which is defined as involving four sets of tasks (EPA/PPT 1993c). The first task is to define what constitutes the life cycle of a product. This includes the extraction of raw materials, manufacturing, distribution, product use and disposal. The second step involves an inventory of environmentally significant inputs (for example, energy and water) and outputs (for example, emissions

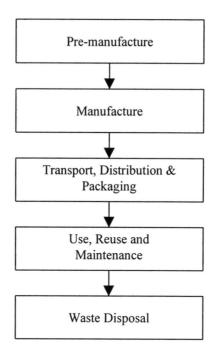

FIGURE 8.1 Product Life Cycle

to air, to water and of solid waste) throughout the various life-cycle stages. The third step is to assess the impacts of environmental inputs and outputs on ecosystems, human health and natural resource stocks. Of all these steps, the impact of environmental inputs and outputs is the most controversial because there is still great scientific uncertainty about the fate and effects of various pollutants. The final step is to evaluate options for reducing environmental impacts throughout the product's life cycle.

In most ecolabeling programs, private or public, the standard setting process is very lengthy and usually involves some variation of the following steps. First, a product category is identified by the ecolabeling organization—typically through proposals from industry or from environmental groups. The next step is to develop a description of some or all the stages of a product's life cycle and the kinds of environmental impacts associated with each stage. In practice, it is impossible to examine all impacts. Most programs try to identify impacts that differ the most across the various companies' products. Standards are then proposed for reducing these environmental impacts. These standards are made available for public review and comment. The standards are revised and then finalized to reflect public comment. A scientific review panel and an appeals process may also be part of the standard setting process. Finally, periodic review could be included to ensure that standards reflect technological progress.

To see the implications for agriculture, suppose that a major food processor wanted to obtain a Green Seal for its product. Green Seal would look at all of the life-cycle stages of the product, not just at the production practices of the food processor. To qualify for obtaining the Green Seal, the food processor would need to ensure that the raw agricultural commodities used in this product would have been produced in an environmentally friendly manner. This would include an assessment of the energy used to produce, pack and ship the commodities, and an assessment of the impact the commodities would have on natural resources (for example, agronomic practices on water quality, soil quality and biodiversity).

Ecolabeling could become important in competing in foreign and domestic markets. The International Standards Organization (ISO) has proposed, and will soon adopt, international standards for environmental labeling known as ISO 14020, 14021, 14022 and 14023. These labeling standards are part of the broader set of standards placed on environmental management systems and on environmental audits known as ISO 14000 (Kuhre, 1995 and 1997; Lamprecht, 1996). The Global Ecolabeling Network, a voluntary organization of national and multinational Ecolabel Licensing Organizations, is trying to establish an ecological criteria databank that could be used by members when setting standards.[4] The United Nations Task Force on Environmental Labeling is facilitating the discussion of principles of equivalency in ecolabeling environmental criteria and of potential international trade issues (such as mutual recognition of ecolabeling schemes).

NECESSARY CONDITIONS FOR CREATING ADOPTION INCENTIVES

The economic conditions necessary for ecolabeling to create adequate incentives for firms to adopt innovative environmental technologies are described in this section. To simplify the discussion, it is initially assumed that there is one type of ecolabel and it affects the environmental impacts, but not the safety or nutritional characteristics of food.

In standard economic theory, uncompensated environmental damage is usually treated as a negative production or consumption externality. That is, the environmental damage is assumed borne by a party that is unable to seek compensation from the market's participants and that is external to the product market. Consequently, the damage is not accounted for by consumers or producers in the product market and is not reflected in the equilibrium price and quantity.

The premise behind ecolabels is that some of the uncompensated disutility from environmental damage, which is associated with the production or consumption of the product, is experienced by the market's consumers. In this study, this uncompensated disutility is called an internality.

Suppose, for example, that production or consumption of a product resulted in wastes that harmed the environment. If the consumer were to believe that he or she suffered from this environmental harm and was not compensated, he or she would experience an internality in the form of disutility from consumption. Thus, he or she would face a trade-off between the marginal utility that he or she would derive from additional consumption and the marginal disutility that he or she would derive from additional uncompensated environmental damage.

The internality premise could be plausible today because there is a much greater appreciation of the interdependency among the elements of an ecosystem. Environmental science has shown us that what we once regarded as separate, independent elements of the ecosphere are, in fact, elements that are highly interdependent. Under these conditions, a change in one element could have many indirect effects and would not be accounted for in the current set of property rights that govern goods traded in markets. Thus, consumers are learning that money is not the only sacrifice that they make to acquire goods.

If it is assumed that all other product qualities remain unchanged and that all production causes some type of uncompensated environmental damage, then the consumer's problem can be expressed more formally as

(1)   Maximize $U(X,Q(X,E))$
         s.t. $PX = M,$

in which U is a quasi-concave utility function, X is the quantity of goods purchased, Q is environmental quality, E is an exogenous amount of environmental damage, P is the price of X, and M is income.[5] The internality is captured by the effect of X on Q. The effect of Q on utility is strictly positive ($\partial U/\partial Q > 0$). The effects of X and E on Q are negative ($\partial Q/\partial X < 0$; $\partial Q/\partial E < 0$). Thus, the marginal utility of X may be positive or negative depending on the relative magnitude of the direct ($\partial U/\partial X > 0$) and indirect ($\partial U/\partial Q \cdot \partial Q/\partial X < 0$) effect of X on utility.

This model can be used to capture the effects of internalities caused during either production or consumption. Since we are interested in the potential effect on the producer's adoption of environmental technologies, in this chapter we only consider the producer as the source of the internality.[6]

Suppose that some producers adopt an innovative environmental technology that does not change any of the performance characteristics of X (that is, there are no changes in product safety or quality), but it reduces the amount of environmental damage that is created per-unit of output. Suppose an ecolabel was developed to advertise this environmental improvement to consumers. Let X' be the quantity purchased of the ecolabeled version of X. Assume, also, that the firm truthfully advertises the relationship between X' and Q compared to that between X and Q and that the consumer is aware of and fully understands it. Assume that this environmental technology raises the marginal costs of production. Because marginal costs are higher, X' is sold at a higher price P'. The consumer's problem becomes

(2)   Maximize $U(X,X',Q(X,X',E))$
         s.t. $PX + P'X' = M.$

Since, by assumption, $\partial U/\partial X = \partial U/\partial X'$, then the first-order conditions imply that when

(3) $P' - P > \{\partial U/\partial Q(\partial Q/\partial X' - \partial Q/\partial X)\}/\lambda$,

in which λ is the Lagrangian multiplier, the consumer will not purchase any X'.[7] Thus, the necessary condition that must be present for ecolabeling to create an adoption incentive for firms is that the difference in marginal costs of the new method of production does not exceed the marginal value of the environmental improvement to the consumer of the last unit sold.

This condition is illustrated in figure 8.2. The demand for X is represented by curve D, and D′ represents the demand for X′. MC and MC′ represent the marginal costs of supplying X and X′. In figure 8.2, the difference between the marginal costs is less than the difference in demand. Consequently, both consumer and producer surpluses are greater with X′. If all firms do not face identical costs, and not all consumers value environmental improvement the same, then a market with both kinds of products would likely result.

Since consumers do not have omniscience and it is costly and difficult for them to observe whether a producer has truly improved environmental quality, the model, as developed so far, is too simple. Some account must be given to consumer trust in the ecolabel claim. Since trust determines the consumer's expectation of the relationship between X, X′ and Q, the trust variable can be incorporated by weighting the Q production function by a probability function Prob(Q). Thus, (2) becomes

(4)   Maximize $U(X, X', (Q(X, X', E)Prob(Q))$
          s.t. $PX + P'X' = M$,

in which the producer's claim is represented by the Q function and the trust in the claim is represented by the probability that weights function Prob(Q).

Perceived truthfulness would depend, in part, on producers' reputation for truthfulness as well as on the perceived effectiveness of anti-deception laws to ensure truthful labeling. In other words, Prob(Q) is conditional on reputation (R), and perceived effectiveness of anti-deception laws (A), or

(5)   Prob(Q;R,A).

There are various actions that producers could take to increase R. For example, they could seek national standards (such as those being developed for organic products). Alternatively, they could use a third-party labeler/certifier who is widely known for being accurate and truthful. The accuracy and truthfulness of the labeler/certifier would be insured because his or her profits would depend on this reputation (for example, Underwriters Laboratory) or because his or her revenues would come from protecting the environment or consumers (for example, Consumers' Union). This investment could increase marginal costs of production but also could increase the marginal benefits to consumers. As long as the extra costs do not exceed the extra revenues, firms and consumers will be better off with the labeler/certifier.

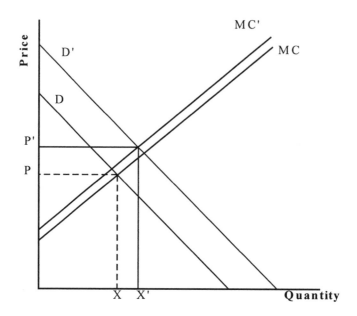

FIGURE 8.2 Demand and Supply of Conventional and Ecolabeled Goods

The model developed thus far is too simple in two other respects. The model is based on the assumption that there are only two brands of a product, namely, conventional and ecolabeled. In agricultural markets, the existence of the organic label as a potential substitute should be considered. Also, since production practices that affect environmental quality could also affect the safety and other quality attributes of the final product, these factors must be taken into account. Since a number of theoretical models of demand for food safety and quality already exist (van Ravenswaay and Hoehn, 1996), this theoretical issue is not pursued in this chapter.

DEMAND FOR ECOLABELED AGRICULTURAL PRODUCTS

Although much has been written on the growth of green marketing and environmental marketing (Kuhre, 1995 and 1997; Wasik, 1996; EPA/PPE, 1991; EPA/PPT, 1994; Peatti, 1995; Makower, 1993; Polonsky et al, 1995; Lamprecht, 1996; Cairncross, 1995), there is little scholarly research about the potential market demand for ecolabeled products. There is, however, a substantial amount of proprietary research on green consumers. Marketing research and public opinion survey companies have completed this research, of which a small portion is publicly available at this time.

The most relevant marketing study comes from the Food Marketing Institute (FMI, 1997). The Hartman Group (1996), a marketing research firm that is under contract with the Northwest Food Alliance, collected the data for this study.[8] Survey data were collected in two stages. First, an 8-page questionnaire was mailed in February 1996 to the principal grocery shopper in a nationally representative sample of 2,900 households. Sixty-five percent of the questionnaires were filled out and returned to yield a national household sample of 1,879 respondents. When it came to food shopping, statistical techniques were used to classify respondents into groups that represented consumers who did or did not care about environmental issues. A 12-page questionnaire was then mailed to the 903 respondents who were classified as caring about the environment. Seventy-nine percent of those questionnaires were filled out and returned to yield a green national household sample of 715 respondents.

The FMI study (1997) found that there is a large group for whom environmental friendliness was a tie breaker when choosing among brands within a food product category, but they will not pay more (table 8.1). They also found that while the majority of consumers were not likely to buy environmentally friendly products if they cost more, a small, but still significant, number of consumers indicated that they would likely pay a premium (table 8.1).

The amount of environmental concern and the degree to which this concern is likely to affect food shopping patterns is characterized by a cluster analysis of 30 questionnaire items that results in six consumer segments or groups (FMI, 1997; Hartman Group, 1996). These groups are *True Naturals, New Green Mainstream, Young Recyclers, Affluent Healers, Overwhelmed* and *Unconcerned.* The percentage of respondents in each consumer segment is shown in table 8.2. The *True Naturals* group is very environmentally knowledgeable and concerned. It is the only group of respondents for which environmental considerations are core food purchase criteria. The *New Green Mainstream* group is concerned but not very knowledgeable about environmental issues related to food. These consumers occasionally express their environmental concern through food purchase decisions, but only if there are no sacrifices in product quality (for example, taste, appearance, cleanliness and convenience). The *Young Recyclers* are environmentally concerned but unwilling to pay more for environmentally friendly products. The *Affluent Healers* are only somewhat environmentally concerned but are very concerned about nutrition and health and are willing to pay more for healthier foods.

TABLE 8.1 Purchase Intentions for Environmentally Enhanced Products

Purchase Intentions	Very Interested	Somewhat Interested
	percent	
Purchasing environmentally enhanced products	25	46
Purchasing environmentally enhanced products if price is 10 percent higher	8	38

Source: Food Marketing Institute (1997).

TABLE 8.2 FMI Consumer Segments

Group	Percent of Sample
True Naturals	7
New Green Mainstream	23
Young Recyclers	10
Affluent Healers	12
Overwhelmed	30
Unconcerned	18

Source: Food Marketing Institute (1997).

The *Overwhelmed* group (30 percent) is too concerned with personal survival to worry about environmental issues. The *Unconcerned* group (18 percent) does not believe the environment is in danger.

Some differences in the purchase intentions of these groups are shown in table 8.3. The *True Naturals* group is much more likely than the other groups to express interest in purchasing environmentally enhanced products, and they indicate the purchase of some kind of environmentally friendly organic products within the preceding month. In contrast, 90 percent of *Unconcerned* group of respondents say that they are not interested in purchasing environmentally enhanced products.

The findings of the FMI study are corroborated by the results of another national poll. Roper Starch maintains a syndicated survey, called Green Gauge, that is based on annual national samples of 2,000 households since 1990.[9] The Green Gauge tracks consumer attitudes and behaviors related to environmental issues. The results are proprietary, and only a small percentage of these statistics is publicly available.

The Green Gauge respondents are separated into five categories (Stisser, 1994; List, 1993). The percentage of respondents within each consumer segment is shown in table 8.4. The cluster results are somewhat similar to those of the FMI study. For example, 55 percent of respondents to this survey fall in one of the three environmentally concerned categories (table 8.4) as compared to the 52 percent found in one of the four environmentally concerned categories in the FMI study (table 8.2).

Although the FMI study suggests that less than 10 percent of consumers would pay a premium for environmentally friendly food, there is evidence that this estimate may be low. For example, in a survey of Colorado households, Sparling et al. (1992) found that about one-half of the consumers would pay a small premium of up to 8 percent for organic food—.25 percent of consumers would buy organic produce at a 24 percent premium; and fewer than 3 percent of consumers would pay a premium of 64 percent. The average price premium for organic food in the markets in which consumers were sampled was above 60 percent. Thus, Sparling et al. (1992) concluded that the organic premium is too high to reveal the true market potential for organic products.

TABLE 8.3 Purchase Intentions by Consumer Segment

	True Naturals	*New Green Main-stream*	*Young Recy-clers*	*Affluent Healers*	*Over-whelmed*	*Uncon-cerned*
				percent		
Very interested in purchasing environmentally enhanced products	67	43	25	19	12	10
Very interested in purchasing environmentally enhanced products if price is 10 percent higher	38	16	6	4	2	3
Purchased an environmentally friendly product in past month	72	49	40	30	26	29
Purchased an organic product in past month	42	15	8	5	4	6

Source: Food Marketing Institute, 1997.

Similarly, a review of studies of consumer willingness-to-pay for reduced pesticide residues suggests that at least 10 percent and perhaps as many as 40 percent of consumers are willing to pay a 10 percent premium (van Ravenswaay, 1995). Of course, this finding reflects consumers' food safety concerns as well as environmental concerns. The FMI study suggests that there is a significant group of *Affluent Healers* who focus only on the food safety aspect.

A drawback of the studies discussed so far is that they pertain to situations where consumers are aware of the ecolabeled alternative. Awareness, however, takes time and resources to develop. A major advantage of the organic label is increased consumer awareness. More than 90 percent of consumers are familiar with the organic label (van Ravenswaay, 1995). Green Seal, however, which is a relatively new ecolabel and has standards for products in 28 product categories, is recognized by only 14 percent of U.S. consumers (FMI, 1997).

A comparison of ecolabeling programs in other countries indicates that consumer awareness of an ecolabeling program takes many years to develop. For

TABLE 8.4 Green Gauge Consumer Segments

	1990	1993
	percent	
True-blue Greens	11	14
Greenback Greens	11	6
Sprouts	26	35
Grousers	24	13
Basic Browns	28	32

Sources: Stisser (1994) and List (1993).

example, the oldest ecolabeling program is Germany's Blue Angel seal that was established in 1977. According to a 1988 survey, the Blue Angel seal is recognized by 79 percent of German households (EPA/PPT, 1993b). Canada's Environmental Choice Program was founded in 1988. A 1992 survey found that 42 percent of the consumers recognize the logo (EPA/PPT, 1993b). Japan's EcoMark program was established in 1989. A 1990 survey found that only 22 percent of the public was aware of their program (EPA/PPT, 1993b).

Another drawback of these studies is the vague description of the ecolabeled good and its price. In the case in which product quality is unchanged, our theoretical framework suggests that there are two sources of value that consumers could obtain from an ecolabeled good: (1) environmental improvement (that is, increased Q); and (2) the assurance that the promised improvement will actually be made (that is, Prob Q). Without knowledge of the product type and quality, the amount and type of environmental improvement, the assurance of delivery, and the product price, it is difficult for respondents to accurately predict their own actions. Moreover, different respondents would likely make different assumptions about these features. For example, there is consumer confusion about what the organic label implies about environmental attributes and other qualities (Park and Lohr, 1996; van Ravenswaay, 1995). The FMI study suggests that the vast majority of consumers know little about the relationship between farming and environmental quality and how it differs for different crops. Additional research is needed to evaluate how ecolabel features may affect purchases.

COSTS OF AGRICULTURAL ECOLABELING

We now turn to the empirical issue of the likely costs that are associated with an agricultural ecolabel. Because there are few agricultural ecolabeling programs, there is little empirical information about how ecolabeling costs would vary. Consequently, some of the new agricultural ecolabeling programs that are now being developed will be discussed in this section along with the major categories of costs.

Using standards created by the New York integrated pest management (IPM) Program, ecolabels that are used on agricultural products are being, or have been, developed by Stemilt's Responsible Choice Program®, the Core Values Program, the Massachusetts IPM Partners With Nature Program, California Clean Growers and Wegman's Food Stores.[10]

Stemilt's Responsible Choice Program® requires all growers to follow European Integrated Fruit Production (IFP) guidelines. Growers may not exceed the point goal for each fruit crop. Points are given for each pesticide used on a particular crop. They are based on 8 attributes, which include pesticide efficacy, leaching potential, pre-harvest interval, soil half-life and biological disruption.

The Core Values Program requires growers to follow the Northeastern Stewardship Alliance Guidelines. These Guidelines require up-to-date training of farm managers in all aspects of IFP. Included in these requirements are attendance at regular training, updating and review meetings; reduction of the applications of herbicides (by the use of alleyways or travel lines between tree rows); tree pruning; and the use of chemicals that are based on the lowest ecological disruption.

Under the Massachusetts Integrated Pest Management Partners with Nature Program, crop-specific guidelines are specified for every aspect of production that is addressed under integrated pest management (IPM). Points are awarded for each set of guidelines. In order to qualify for ecolabeling, farmers must attain 70 percent of the possible points for a particular crop. Guideline production categories for apples include soil and nutrient management; cultural practices; pesticide application and records; insect management; disease management; weed management; vertebrate management; weather monitoring; and crop monitoring (Hollingsworth et al., 1996).

California Clean Growers (CCG) require farmers to follow general guidelines that incorporate using ecologically sound practices; strengthening farm soils through natural enrichment programs; promoting and developing wildlife refuges; using natural biological pest controls; establishing safe working conditions; and producing products with superior taste and nutrition.

New York IPM guidelines, established by Cornell University faculty, are currently being used by Wegman's Food Stores. Growers are required to meet 80 percent of the points required by the Northeastern Stewardship Alliance Guidelines for each crop. A private consultant inspects grower records and determines whether the criteria have been met. A licensing agreement has been signed between Wegman's Food Stores and the Cornell Research Foundation for the use of the Cornell IPM Logo on Wegman's products. Canned sweet corn in 1997 was the first product to carry the Cornell IPM label.

The relevant costs of ecolabeling to producers include the labeling of fees, the satisfaction of record keeping requirements and, in some cases, higher input costs and the risk of reduced yield (Grant et al., 1990; Agnello et al., 1994). Additional producer costs that may occur (but are not considered here) include the transaction costs involved with changing suppliers (for example, search costs and costs of a new contract) and the lost productivity from equipment that can no longer be used under label standards.

Agricultural ecolabeling programs are so recent that they are either waiving the fees or setting arbitrary fees that are not necessarily based on actual costs. The fee under the Partners with Nature Program, for example, is $20 for the first crop and $15 for each successive crop. In order to stimulate added participation, some programs (such as CCG) are not yet charging fees to producers. Green Seal and other national programs of nonagricultural items and organic certifiers base their fees on a percentage of firm revenue. In California, for example, organic growers pay yearly registration fees between $25 and $2,000, depending on the gross sales (Klonsky and Tourte, 1994).

The major costs to the labeler include the research and development of environmental standards and grower training programs; the collection and analysis of certification information on each producer for each submitted crop; and marketing and consumer education.

The costs of setting ecolabel standards depend on how comprehensive the environmental performance standards are and how precisely they are measured. The fact that existing programs in the United States are based on input or process standards (for example, best management practices) rather than performance standards (that is, environmental impact) suggests that the transaction costs of developing and monitoring performance standards are prohibitive.

Researchers are currently working on measurement tools that could, with further refinement, be used by labelers to create performance standards. These standards consist of monitoring systems, simulation models and indexing systems (Riha et al., 1997). Examples include the Environmental Impact Quotient (EIQ) Index being developed at Cornell (Kovach et al., 1992), the nutrient yardstick of the Netherlands, and others identified in Roberts and Swinton (1996). The nutrient yardstick is currently being used in the Agro-Milieukeur ecolabel of the Netherlands (Poppe, 1992). In the case of the EIQ, the environmental impacts of particular pesticides on farm workers, consumers and farm ecology are ranked and weighted into one index. Environmental impacts of various growing practices that might be measured by these tools include the concentration of nitrates and other chemicals in the soil and groundwater; the amount of beneficial soil organisms on a plot; the toxicity to animals from particular chemicals; the half-life of pesticides used; the amount of plant and animal biodiversity on a farm along with population levels of each species; and the total usage of water.

A comprehensive ecolabeling standard would take into account as many of the impacts measured by these tools as possible, including energy use, concentrations of air pollutants (for example, carbon dioxide and methane), concentrations of surface water pollutants, amounts of solid and toxic waste, and the environmental impacts from shipping products for retail. A less comprehensive standard would only focus on a few of these impacts.

A precise measure of environmental improvement would involve the direct sampling of environmental media and ecosystem elements that include water, soil, air, wildlife and plants. A less precise measure would involve using a simulation technique, such as fate models, that would obtain an estimate of the effect of different farming systems on environmental indicators. The least precise measure would dictate only the practices to be used.

Since none of the domestic agricultural labels use monitoring or simulation techniques, it is hard to say what the costs of using a more precise measure would be. There are several reasons why we should expect these costs to be quite high:

- It would be complicated and difficult to compare alternative cropping systems, because multiple performance criteria would have to be developed.
- Different technologies could have differing time lags of effectiveness.
- It would be difficult to deal with multiple crops that are all interrelated on a farm.
- The question of how much weight should be given to the undesirable side effects of production would have to be resolved (Roberts and Swinton, 1996.
- Field monitoring and laboratory testing of environmental indicators would be costly.

The certification of growers requires the management of huge data sets; the coordination of tests with laboratories; the monitoring of applicants' products and practices; the continual update of standards and records; the collection of individual producer data on crops, farm parameters, and current and past growing practices. All of these requirements necessitate grower record keeping efforts. Also, in order to comply with anti-deception statutes, such data must be maintained in a secure manner.

Marketing costs involve consumer research and general public education. The Core Values Program is currently conducting a pilot study of two different consumer education strategies. The first provides consumers at participating stores with brochures, grower profiles and a large poster advertising the labeled good. A second plan involves consumer interaction with actual Core growers through staffed tastings at which brochures, recipes and samples are distributed.

The marketing costs for the installation of Wegman's Food Stores' IPM Labeling Program have included training videos for employees based on the products that they are selling and the promotion of these products; short in-store videos for consumers; brochures; signs; radio time; ads; and cable-TV time. The brochure describes Wegman's commitment to safer and more environmentally friendly foods and defines IPM. Bill Pool of Wegman's Food Stores estimates that it will take them about 3–4 years to educate their customers.

Grower costs of record keeping can be substantial. For example, Grant et al. (1990) estimated the average grower time requirements to comply with New York IPM standards for adequate record keeping. Data administration is estimated at five hours per week at data entry rates of six fields per hour for pest activity and the recording of pesticide application; five fields per hour to create spray summaries; and two and one-half fields per hour to create threshold graphs for pest control. Inspectors from the labeling organization and government are estimated to visit an average of about two farms per day. In 1990, the salary of an average New York inspector with an IPM background was $31,200.

In terms of inputs and outputs, the costs of alternative practices would depend on the nature of the program. In some cases, the costs seem comparable, and in others, they appear to be much higher. The study results of the costs of growing practices that are associated with environmental improvements are summarized in table 8.5.

TABLE 8.5 Input Costs and Yield Losses from Adopting Environmental Technologies

Study	Input Costs of Technology	Yield Loss
Gut and Brunner compare costs between conventionally grown apples using the organophosphate pesticide, asynphos methyl, and apples grown using pheromone mating disruption (PMD).	Under PMD, input costs are $55–120 more per acre. Other costs include time and effort for greater information processing and recording.	Greater losses under PMD in years of high pest pressure.
Williamson et al. compare the economic and pest control results of apples grown using conventional versus PMD pest control in the Yakima Valley in Washington state.	Under PMD, input costs are $188.14 more per acre due to higher labor requirements and the high cost of pheromone emitters.	In 1992, a year of severe pest pressure, outcome under PMD was less favorable.
A paper by Agnello et al. (1994) evaluates alternative pest monitoring programs in New York state.	Long-run savings might occur from using IPM practices in apple orchards. Short-run material costs about the same with increased expenditures of time and effort.	
Swinton and Scorsone study the short-term impacts on Michigan's apple, tart cherry and blueberry production from the loss of several pesticides.	Gross margins over pest control costs using the next best alternative are expected to fall by 16–21 percent	
Klonsky and Livingston compare four types of farming systems: 1) conventional four-year rotation; 2) a conventional two-year rotation; 3) a low input system; and 4) an organic system.	Neither organic nor low input systems achieved equal profits to either of the conventional systems on a whole farm basis, even though they did for some crops. Fertility and weed management was the biggest challenge for them.	
Sellen et al. (1994) studied several types of organic vegetables.	Premiums of 41 percent to 92 percent are needed for organic to be as profitable as non-organic.	

The use of ecolabels could also result in certain types of cost savings. Producers, for example, may save money from reduced input usage (that is, less pesticide use under IPM). There also exists the potential of reduced legal expenses for farmers who comply with governmental regulations, and of their reduced liability insurance costs. If a producer farms in such a way that regulations are met, he or she does not have to worry about penalties for noncompliance. Ecolabels may also reduce a producer's marketing costs, or they offer him or her access to new markets. Finally, environmental improvements may result in higher land values and in improved relationships with insurers, lenders, investors, workers and the surrounding community.

CONCLUSION

Ecolabels are essentially voluntary environmental seals of approval on products. They certify that a product meets higher environmental standards than other products in the same category. If consumers were to value this environmental claim enough to cover its costs, ecolabels could give such products a competitive edge. Thus, ecolabels could be useful for encouraging the development and adoption of innovative agro-environmental technologies.

For consumers to value the ecolabel, negative externalities from production must affect consumers in the product market. In this case, consumers face an internality or a trade-off between the benefits of consumption (net of price) and the environmental costs of consumption.

The conditions that are necessary for ecolabeling to create incentives for producers to adopt innovative environmental technologies are that (1) consumers value ecolabeled products more than they value conventional products and that (2) the difference in value is equal to, or greater than, the difference in the marginal costs of producing the two types of products.

Gains in market share or revenue from ecolabeling are very uncertain since little is known about the potential demand for specific ecolabeled products. Available marketing studies suggest that there is a substantial market niche for ecolabeled products if the premium is less than that of organic food. Indeed, a niche market for ecolabeled products, which is appropriately targeted and promoted, could more easily assure that higher production costs are compensated by higher product prices. Moreover, ecolabel use is potentially a useful way to differentiate among food products and to gain market share within a product category. Existing research, however, is based on very general survey questions for which respondents may not be able to accurately forecast their behavior. More specific survey data is needed to learn how consumers would react to more or less comprehensive and precise environmental claims that are made on different types of food products.

Ecolabeling programs in agriculture are too new to establish costs precisely. The major categories of identified costs include the formation and monitoring of standards; the costs of marketing; the costs of record keeping; the potential of yield reductions; and the potential of additional input costs; and the potential for cost savings, such as those through reduced pesticide expenditures. The net costs depend on the comprehensiveness of environmental standards and on the precision of tests used to certify that producers meet the standards.

A comprehensive ecolabel standard would involve a life-cycle assessment of a broad range of environmental impacts, which include energy use, water use, waste generation, soil quality and biodiversity. Many of these impacts are not considered in current agro-ecolabeling programs, which tend to focus mainly on reducing pesticide use. By expanding the list of the environmental impacts to be considered, new opportunities for competing in domestic and foreign markets may be created.

The present technology for measuring the environmental impacts of production does not provide much precision; so most standards are expressed in terms of best management practices rather than in terms of environmental impacts. This makes it more difficult to explain to consumers the value of the ecolabel. Research on improving environmental impact measurement could lower labeling fees and stimulate consumer demand for ecolabeled products.

ENDNOTES

1. The USDA/ERS, Cooperative Agreement No. 43-3AEL-6-80060, and the Michigan Agricultural Experiment Station provided research for this chapter.
2. Further information on Green Seal is available at their web-site (http://www.greenseal.org).
3. A detailed description of ecolabeling programs is contained in a series of four reports commissioned by the EPA/PPT (1993a, 1993b, 1993c and 1994).
4. Information about the Global Ecolabeling Network and its ecolabeling members can be obtained at the Global Ecolabeling Network web-site (http://www.interchg.ubc.ca/ecolabel/gen.html).
5. Since both food and environmental quality are necessary for life, subsistence levels of both may be incorporated by using the Stone-Geary form of the utility function.
6. Many of the consumer externalities associated with a product involve the use of complements, such as waste disposal services, energy use and water use. This can be incorporated into the specification of the production function for environmental quality, Q. A main result is that the producer will reformulate the product if the difference in marginal costs is less than, or equal to, the difference in marginal willingness to pay.
7. If X and Q are additively separable and corner solutions between X and X' arise, the first-order conditions for a maximum are:
    $\partial U/\partial X + \partial U/\partial Q \cdot \partial Q/\partial X - \lambda P \leq 0$, and if $\partial U/\partial X + \partial U/\partial Q \cdot \partial Q/\partial X - \lambda P < 0$, then X=0;
    $\partial U/\partial X' + \partial U/\partial Q \cdot \partial Q/\partial X' - \lambda P' \leq 0$, and if $\partial U/\partial X' + \partial U/\partial Q \cdot \partial Q/\partial X' - \lambda P' < 0$, then X'=0;
    M-PX-P'X' = 0.
8. The Northwest Food Alliance, mentioned above, is in the process of establishing an ecolabeling program in agriculture.
9. The Green Gauge web-site (http://www.roper.inter.net/research/syndicated/green.htm) is the basis for this description. Results for each year are available from Starch Roper for about $15,000. The survey results presented here were obtained from marketing magazine articles (Stisser, 1994; List, 1993).
10. Much of the information in this section was obtained in discussions with the following individuals: Molly Anderson, Tufts University; Paul Buxman, California Clean Growers; David Granatstein, Washington State University; Larry Gut, Michigan State University; Curtis Petzoldt, Geneva Agricultural Experiment Station, Cornell University; Bill Pool, of Wegman's Food Stores; William Coli, University of Massachusetts, Amherst; and Joe Kovach, Geneva Agricultural Experiment Station, Cornell University.

REFERENCES

Agnello, A.M., J. Kovach, J.P. Nyrop, W.H. Reissig, D.I. Breth, and W.F. Wilcox. 1994. "Extension and Evaluation of a Simplified Monitoring Program in New York Apples." *American Entomologist* 40(1): 37–49.

Cairncross, F. 1995. *Green, Inc.* Washington, DC: Island Press.

EPA/PPE (U.S. Environmental Protection Agency, Office of Policy, Planning and Evaluation). 1991. *Assessing the Environmental Consumer Market.* EPA21P-1003, U.S. Environmental Protection Agency, Report Number EPA21P-1003, Washington, DC (April).

EPA/PPT (U.S. Environmental Protection Agency, Office of Pollution Prevention and Toxics). 1993a. *Evaluation of Environmental Marketing Terms in the United States.* EPA21P-1003, U.S. Environmental Protection Agency, Washington, DC (February).

_____. 1993b. *Status Report on the Use of Environmental Labels Worldwide*, pp. 44, 50, 56–57, and 72–76. EPA 742-R-9-93-001, U.S. Environmental Protection Agency, Washington, DC (September).

_____. 1993c. *The Use of Life Cycle Assessment in Environmental Labeling.* EPA 742-R-9, U.S. Environmental Protection Agency, Washington, DC (September).

_____. 1994. *Determinants of Effectiveness for Environmental Certification and Labeling Programs.* EPA742-R-94-001, U.S. Environmental Protection Agency, Washington, DC (April).

FMI (Food Marketing Institute). 1997. *The Greening of Consumers: A Food Retailer's Guide.* Washington, DC: Food Marketing Institute.

FTC (U.S. Federal Trade Commission). 1992. "Guides for the Use of Environmental Marketing Claims." *U.S. Federal Register* 57(157): 36363–36369 (13 August).

Grant, J., J. Tette, C. Petzoldt, and J. Kovach. 1990. *Feasibility of an IPM-grower Recognition Program in New York State.* State Integrated Pest Management Program, IPM Report No. 3, Cornell University, Ithaca, NY (November).

Grodsky, J.A. 1993. "Certified Green: The Law and Future of Environmental Labeling." *Yale Journal on Regulation* 10:147–227.

Gut, L.J. and J.F. Brunner. 1996. *Implementing Codling Moth Mating Disruption in Washington Pome Fruit Orchards.* Tree Fruit Research and Extension Center Information Series No. 1, WSU Cooperative Extension Office, College of Agriculture and Home Economics, Washington State University, Pullman, WA.

Hartman Group, The. 1996. *Food and the Environment: A Consumer's Perspective.* Bellevue, WA: The Hartman Group.

Hollingsworth, C.S., W.M. Coli, and R.V. Hazzard, 1996. *Integrated Pest Management Massachusetts Guidelines: Commodity Specific Definitions.* Extension IPM Program, Bulletin No. SP136, University of Massachusetts, Amherst, MA.

Klonsky, K. and P. Livingston. 1994. "Alternative Systems Aim to Reduce Inputs, Maintain Profits." *California Agriculture* 48(5): 34–42.

Klonsky, K. and L. Tourte. 1994. *State Registration and Organic Certification: A Guide for California Growers.* University of California, Cooperative Extension Department of Agricultural Economics, University of California, Davis, CA.

Kovach J., C. Petzoldt, J. Degni, and J. Tette. 1992. *A Method to Measure the Environmental Impact of Pesticides.* Agricultural Experiment Station, Bulletin No. 139, Cornell University, Ithaca, NY.

Kuhre, W.L. 1995. *ISO 14001 Certification: Environmental Management Systems.* Upper Saddle River, NJ: Prentice Hall.

_____. 1997. *ISO 14020s: Environmental Labeling Marketing.* Upper Saddle River, NJ: Prentice Hall.

Lamprecht, J.L. 1996. *ISO 14000: Issues and Implementation Guidelines for Responsible Environmental Management.* New York, NY: American Management Association.

List, S.K. 1993. "The Green Seal of Eco-approval." *American Demographics* (January) 15(1): 9–10.

Makower, J., J. Elkington, and J. Hailes. 1993. *The Green Consumer.* New York, NY: Penguin Books.

Park, T.A. and L. Lohr. 1996. "Supply and Demand Factors for Organic Produce." *American Journal of Agricultural Economics* 78(3): 647–655.

Peattie, K. 1995. *Environmental Marketing Management.* London, UK: Pitman Publishing.

Polonsky, M.J. and A.T. Mintu-Wimsatt, eds. 1995. *Environmental Marketing.* New York, NY: Haworth Press.

Poppe, K.J. 1992. "Accounting and the Environment," in G. Schiefer, ed., *Integrated Systems in Agricultural Information.* Bonn, Germany: ILB

Riha, S., L. Levitan, and J. Hutson. 1997. "Environmental Impact Assessment: The Quest for a Hollistic Picture," in S. Lynch, C. Greene, and C. Kramer-LeBlanc, eds., *Proceedings of the Third National IPM Symposium*, pp. 40–58. Natural Resources and Environment Division, Miscellaneous Publication number 1542, USDA/ERS, Washington, DC. (May).

Roberts, W.S. and S.M. Swinton. 1996. "Economic Methods for Comparing Alternative Crop Production Systems: A Review of the Literature." *American Journal of Alternative Agriculture* 11(1): 10–17.

Sellen, D., J.H. Tolman, D. Glenn, R. McLeod, A. Weersink, and E.K. Yiridoe. 1994. "A Comparison of Financial Returns During Early Transition From Conventional to Organic Vegetable Production." Department of Agricultural Economics and Business, Working Paper No. WP94/12, University of Guelph, Guelph, Ontario, Canada.

Stisser, P. 1994. "A Deeper Shade of Green." *American Demographics* 16(3): 24–29.

Swinton, S. and E. Scorsone. 1997. *Short-term Costs and Returns to Michigan Apple, Blueberry, and Tart Cherry Enterprises with Reduced Pesticide Availability*. Research Report No. 551, Michigan Agricultural Experiment Station, East Lansing, MI. (April).

van Ravenswaay, E.O. 1995. *Public Perceptions of Agrichemicals*. Task Force, Report No. 123, Council on Agricultural Science and Technology, Ames, IA (January).

_____. 1996. *Emerging Demands on Our Food and Agricultural System: Developments in Eco-labeling*. Department of Agricultural Economics, Staff Paper No. 96-88, Michigan State University, East Lansing, MI (20 September).

van Ravenswaay, E.O. and J.P. Hoehn. 1996. "The Theoretical Benefits of Food Safety Policies: A Total Economic Value Framework." *American Journal of Agricultural Economics* 78(5): 1291–1296.

Wasik, J.F. 1996. *Green Marketing and Management*. Cambridge, MA: Blackwell Publishers.

Williamson, E.R., R.J. Folwell, A. Knight, and J.F. Howell. 1996. "Economics of Employing Pheromones for Mating Disruption of the Codling Moth." *Crop Protection* 15(5): 473–477 (August).

9 ENVIRONMENTAL EXTERNALITIES AND INTERNATIONAL TRADE: THE CASE OF METHYL BROMIDE

M.S. Deepak, Thomas H. Spreen and John J. VanSickle
U.S. Department of Agriculture
Economic Research Service, Washington, DC
University of Florida, Gainesville, FL

Methyl bromide is a broad-spectrum pesticide that serves simultaneously as an insecticide, nematicide, herbicide and fungicide. In Florida, methyl bromide has been used both as a pre-plant soil fumigant and as a post-harvest fumigant to control a wide array of pests for many of the fruit and vegetable crops produced in the state. Based on determinations made by the Montreal Protocol on Substances that Deplete the Ozone Layer (Montreal Protocol), the U.S. Environmental Protection Agency (EPA) was required to list methyl bromide as a Class I Substance. The 1990 Amendments to the U.S. Clean Air Act stipulate that the production and importation of Class I Substances are phased out by the year 2001. Consequently, the United States has a shorter time span for the phaseout of methyl bromide than have all of the other developed and developing countries.

A brief theoretical description of the problem provided the conceptual framework of this chapter. To compare the industry-specific impacts of various domestic measures to regulate methyl bromide, an empirical model of the U.S. winter vegetable and fruit industry was developed. Several flexible policy instruments, including marketable quotas and a Pigovian tax, were proposed as alternatives to an outright ban. Results from the application of the empirical model were then presented in the base case scenario and in three regulatory scenarios; producer impacts were discussed; and the four scenarios were compared. A brief overview of the mathematical underpinnings of the empirical model is presented in this chapter's appendix, which includes a brief discussion of the advantages and disadvantages of this type of analysis.

INTRODUCTION

Methyl bromide is a broad-spectrum pesticide, which serves simultaneously as an insecticide, nematicide, herbicide and fungicide. In Florida, methyl bromide has been used both as a pre-plant soil fumigant and as a post-harvest fumigant to control a wide array of pests for many of the fruit and vegetable crops that are produced in the state.[1]

As a pre-plant fumigant, methyl bromide has been used in conjunction with plastic mulch to provide effective control of weeds, nematodes and other soil-borne pests that inhibit the production of tomatoes, bell peppers, eggplant, cucumbers, squash, strawberries and watermelons. As a post-harvest fumigant, methyl bromide has been used for the control of fruit flies that may be present on fresh citrus that is shipped out of Florida. Methyl bromide, by virtue of its multi-pronged efficacy, has been critical to the competitiveness in both national and international markets of Florida's fruit and vegetable production, and has been used for a number of years.

In 1991, total world consumption of methyl bromide was 71,260 metric tonnes, with the following geographical distribution: North America (43 percent); Europe (24 percent); Asia (24 percent); Africa (4 percent); South America (3 percent); and other regions (2 percent). In 1991, total U.S. consumption was 25,490 metric tonnes, which constituted nearly 36 percent of total world consumption. Seventy-nine percent of U.S. consumption of methyl bromide is used for soil fumigation. California and Florida together accounted for about 75 percent of total soil fumigation use of methyl bromide in 1991 (Hathaway and Giudice, 1996).

In April 1992, a working group, which represented the parties of the Montreal Protocol on Substances that Deplete the Ozone Layer (Montreal Protocol), concluded that the man-made emissions of methyl bromide were decidedly deleterious to the ozone layer of the stratosphere. Since the use of methyl bromide is heavily concentrated within developed countries, in 1995 the parties of the Montreal Protocol specified that the use of methyl bromide in those nations be frozen at its 1991 levels. They also specified that the use of methyl bromide be phased out 25 percent by 2001, 50 percent by 2005 and 100 percent by 2010. Developing nations were allowed an additional 10 years to eliminate the use of methyl bromide. Consumption of methyl bromide, however, will be frozen in 2001 based on the average of its 1996–98 consumption. In 1998, the ninth Meeting of the Montreal Protocol amended the phaseout schedule for developed countries with a 25 percent reduction in 1999, 50 percent reduction in 2001, 70 percent reduction in 2003 and 100 percent reduction in 2005. The schedule for developing countries was modified with a 20 percent reduction in 2005 and a 100 percent reduction in 2015.

Based on these determinations by the Montreal Protocol Working Group, the U.S. Environmental Protection Agency (EPA) was required to list methyl bromide as a Class I Substance, as stipulated in the 1990 Amendments to the U.S. Clean Air Act. These amendments stipulated that the production and importation of Class I Substances, which now includes methyl bromide, are to be completely phased out by the year 2005. Consequently, the United States has a shorter time span for the phaseout of methyl bromide than have all of the other developed and developing countries. Given the predominance of the United States in the total worldwide use of methyl bromide, the environmental thrust of this unilateral initiative is laudable. However, it does not ensure the compliance of other user countries (either current or potential) in the control of a truly global externality.

The outright ban of methyl bromide by the EPA represents an inflexible technology-based form of environmental regulation. It not only imposes significant technological constraints on certain agricultural industries in the United States, but it also places them at a significant economic disadvantage to competing industries in other developed and developing countries. The objective of this chapter is to propose and evaluate more flexible policy alternatives for the reduction, or elimination, of methyl bromide in the United States.

A brief theoretical description of the problem is provided in a static framework in the second section that places alternative unilateral policy options in perspective. An empirical model of the U.S. winter vegetable and fruit industry is described in the third section to illustrate the industry-specific impacts of various U.S. regulation measures on the use of methyl bromide. Flexible policy instrument alternatives to an outright ban of methyl bromide are proposed that include marketable quotas and a Pigovian tax. Results from the application of the empirical model to scenarios imposed by government regulation are presented in the fourth section. In the fifth section, a discussion of the measurement of producer impacts as a result of the application of the empirical model to the scenarios is presented. The various scenarios are compared in the sixth section. The appendix of the chapter contains a brief overview of the mathematics of the empirical model and a discussion of its advantages and disadvantages for this type of analysis.

THEORETICAL BACKGROUND

Any proposed solution to the methyl bromide problem must address two important dimensions. First, sustained protection of the ozone layer can be achieved only through multilateral action. Any unilateral move toward regulation may be effective in the short run, but it could prove self-defeating in the future if it were to induce the adoption of a methyl-bromide-based technology across international borders. Second, domestic displacement costs must be minimized—the solution must be as Pareto improving as possible.

For analytical purposes, we divide the world into three representative countries: a user country, a competitor to the user country, and the rest of the world. The user country, A, is one that uses methyl bromide as an input in the production process of some industry. This production process generates an environmental externality that affects the consumer sector everywhere. A competitor to the user country, B, is a country that produces the same goods without using methyl bromide, but it contains consumers who are affected by the externality that is generated by the user country. In the rest of the world, R, the consumer sector is again a victim of the externality that is generated by the user country. It is a country, however, that does not directly compete with either of the other two countries in the markets for the goods that are under study in this chapter.

The utility functions of the three countries are denoted as $U_A(C_{A1}, C_{A2}, H)$, $U_B(C_{B1}, C_{B2}, H)$ and $U_R(C_{R1}, C_{R2}, H)$, respectively. C_{ij} denotes the consumption level of good j in country i. Good 1 represents a good that is produced using methyl bromide in country A and without using methyl bromide in countries B and R. Good 2 is a composite good that denotes all other goods, and its production does not involve the use of methyl bromide in any of the three countries. H is a non-priced public health good, the global availability of which is reflected in the omission of a subscript. Health is affected by the amount of the externality generating input used in the production of good 1 in country A, which is denoted by G_{A1}.[2]

All three goods are assumed normal in countries A, B and R so that all marginal utilities are positive. Good 1 is traded freely between countries A and B, and the combined market is assumed to clear at a common price. Trade between the rest of the world, R, and either country A or country B is ignored in order to simplify the analysis. Thus, countries A and B are related through both the market for good 1 and the external effect on health that is caused by the use of the externality generating input for the production technology of country A. On the other hand, country R is related to country A only through the effect of the externality on the health of its citizens.

As a welfare maximizing decision-maker, the government of country A wishes to set the optimal level of use of input G, subject to competition with country B in the market for good 1. Two versions of this decision problem are examined below.

Case 1: A Selfish Approach

The government ignores countries B and R completely and chooses a selfish level of use of input G to maximize the collective welfare of country A—namely, the domestic welfare function U^A—subject to the condition that the total consumption of good 1 in countries A and B is equal to the total amount of good 1 that is supplied by countries A and B.

Assuming that the current level of use of input G in industry 1 for country A is socially excessive, the government must contend with the following domestic adjustments when contemplating the reduced use of input G: (1) a social loss in the market for good 1 as the price of good 1 increases and consumer surplus shrinks; (2) a social benefit from the increased availability of the health good H; and (3) a social loss in the market for good 2 if the demand for good 1 is price inelastic, since consumers are forced to spend less on good 2.

Imports of good 1 play a crucial role in equilibrating the effects of these adjustments. The free flow of imports cushions the effect on the final price of good 1 as domestic producers cut back on their supply of good 1.

Case 2: An Altruistic Approach

The government of country A now recognizes the external effects that it inflicts on countries B and R through its use of input G. It optimizes domestic welfare, subject to both the market-clearing condition for good 1 and the additional constraints in which the collective welfare levels in countries B and R are, at least, equal to some given levels, U^{B*} and U^{R*}, respectively.

In comparison to the solely domestic adjustments required of a selfish government, an altruistic government internalizes the various effects that its use of input G has on countries B and R as well. Since countries A and B share the market for good 1, a decrease in the level of use of input G in country A will have a direct negative effect on the consumers of good 1 in country B, a negative ripple effect in the market for good 2 in country B (if the demand for good 1 in country B is price inelastic) and a positive effect on the consumption of the free health good H. Country R will benefit from country A's reduced use of input G only through the health externality.

It is of some significance that, even by adopting a selfish approach, the government of country A ameliorates the severity of the externality everywhere, given the global nature of the non-priced health good H.

AN EMPIRICAL MODEL

The theoretical model outlined above presents the scope of the problem of methyl bromide and of environmental externalities. The theoretical model does not permit alternative means by which to regulate the use of input G, nor does it outline the impacts of input G on the producers of good 1 in country A who are the generators of the externality.[3] In view of the diffuse and immeasurable nature of the pollution that is caused by methyl bromide, the empirical model that is presented in this chapter does not purport to measure the global health benefits that would emanate from a reduction in the use of methyl bromide in the United States. To address the positive aspects of methyl bromide use, an industry-specific (partial equilibrium) quadratic programming model is developed. This model quantifies the effects of various policy instruments that could be employed to regulate methyl bromide in the North American winter fresh vegetable market. A conceptual overview of the model is provided in the appendix of this chapter. With reference to terms of the theoretical model, countries A and B are likened to the United States and Mexico, respectively. Good 1 is identified as a bundle of winter fresh vegetables and fruit crops that are currently grown in the United States with and without methyl bromide, and bundles of winter fresh vegetables and of fruit crops that are grown in Mexico without methyl bromide [4].

The particular fruit and vegetable crops we include in the model are tomatoes, bell peppers, cucumbers, squash, eggplant, strawberries and watermelons. Not all of these crops use methyl bromide as a soil fumigant in all circumstances; nevertheless, all are included in the model because they are part of a double-cropping system that requires the use of methyl bromide. Farming activities that do not require the use of methyl bromide are omitted from the model, so the results could exaggerate the welfare impacts on agricultural producers.

Florida is included in the model as the dominant U.S. producer of all the crops influenced by the use of methyl bromide and as a producer that experiences significant foreign competition from Mexico. Texas is included as an American producer of bell peppers in the winter market, even though it does not use methyl bromide. California also uses methyl bromide and is included as a major supplier of fresh strawberries for the winter market.

This model is developed to mathematically illustrate the production of the above crops from California, Florida, Texas and Mexico during the winter months. The commodities produced in these regions are shipped to New York, New York; Atlanta, Georgia; Chicago, Illinois; or Los Angeles, California. They are wholesale markets that represent the northeast, southeast, midwest and west regions of the United States that have a demand for these crops (Scott, 1991).

All production systems are modeled as constant marginal cost or Leontief technologies. Production costs by crop are determined for each of the producing regions based on budgets that were developed for the 1990–91 season. The constrained optimization model was solved using the General Algebraic Modeling System (GAMS) software. For a complete mathematical statement of the model, see Spreen et al. (1995) or Deepak et al. (1996).

After solving the model for a baseline scenario that represents current specifications within the industry, three alternative regulatory scenarios are analyzed. The first scenario is a total ban on methyl bromide, in which budgets and yields are altered using the next-best alternative available, to reflect the production of crops without the use of methyl bromide.

The second scenario is one in which a quota limits the total potential use of methyl bromide in each production region in the United States to 50 percent of its current consumption. This quota is implemented either by allowing individual producers to market the 50 percent of the unused quotas to other producers in the country, or by imposing a quota on methyl bromide suppliers that allows them to absorb rents received as the market allocates methyl bromide to its highest and best use.

The third scenario is the application of regional quotas that limits the use of methyl bromide in each production region to 50 percent of its current regional consumption. Thus, producers in each area are given a quota equal to one-half of its historical use level. They can use the quota either to produce crops with the highest and best return, or to market their quota only to other producers within their region. Hence, they cannot market the quota in regions other than their own.

In each of the three quota scenarios, individual producers are free either to reduce total acreage that is treated with methyl bromide or to reduce the rate of application of methyl bromide on the currently treated acreage. Since this facility allows a farm-level choice of abatement technology, both quota instruments may be classified as flexible incentives because they indirectly specify an environmental end, but they do not specify the means by which to attain that end (Batie and Ervin, this volume). Furthermore, the tradability of individual allotments ensures compliance with the prescribed limits at the national level (in the case of the total quota) and at the regional level (in the case of area quotas) while giving individual operators added flexibility with their permits (Batie and Ervin, this volume; Segerson, this volume).

The results corresponding to each of these scenarios are summarized in tables 9.1 through 9.4, and are then discussed in sequence. From the perspective of the theoretical model, an outright ban is a truly altruistic measure because the marginal benefits of the health good, H, in countries A, B and R are seen as outweighing the marginal losses that accrue to the consumers of good 1 in countries A and B. Each quota option of the second and third scenario, however, is a more moderate measure of the impact of methyl bromide use relative to the selfish and altruistic case extremes.

EMPIRICAL RESULTS

Ban on Methyl Bromide

The solution of the base model reflects the current industry practices of methyl bromide use on winter fresh fruit and vegetable crops in the United States and Mexico. The base model is then adjusted by changing the production costs and yields of the various cropping systems that use methyl bromide to represent production without methyl bromide. (Producer practices, costs and yields are changed according to the recommendations of production scientists and economic analysts who are familiar with the industry.) A comparison of the two solutions provides a quantitative assessment of the impact that a total ban on methyl bromide would have on producers and consumers.

A total ban will have sizeable impacts on U.S. fresh fruit and vegetable producers and on the final consumers of these products in the four representative markets (table 9.1). Tomato production in Florida decreases by more than 60 percent, and the total supply to the terminal markets is reduced by 7 percent, even with an offsetting growth of 83 percent in the supply of tomatoes from Mexico. The eggplant market exhibits a similar pattern as Florida ceases its production and total supply decreases by nearly 16 percent—in spite of an increase of 123 percent of Mexican eggplant production. In the case of bell peppers, both Mexico and Texas increase production significantly but cannot prevent a decrease of about 9 percent in overall supply, attributable to a cutback of 63 percent of production in Florida.

In the cucumber market, the increase in Mexican production is much smaller than the decrease in Florida production. As a result, total market supply decreases significantly by 21 percent. Squash production in Florida actually increases when the use of methyl bromide is eliminated, but a more than proportionate decline in the supply from Mexico results in smaller total output. Strawberry production decreases by 52 percent overall because both major suppliers, California and Florida, use methyl bromide and will suffer significant yield losses because of the ban. Watermelon production also has a marked decrease of 40 percent because no alternative supplier has yet been identified to substitute for production that would be lost because of the ban of methyl bromide in Florida during the month of May.

The reason for the substitution of production in Florida with that in Texas and Mexico is that a unilateral ban on methyl bromide will not affect the production systems in the latter areas for the crops considered in this study. Bell peppers are currently grown in Texas without methyl bromide, and Mexican producers use it on only a limited number of acres. Even with the limited use of methyl bromide, Mexico is not immediately affected by the proposed ban because, in 1995, the Montreal Protocol allowed developing countries to use methyl bromide until the year 2015 before forcing them to switch to alternative production practices.

Projected free on board (f.o.b.) revenues—both by crop and for the total—for all the major suppliers in the model are presented in table 9.2 for the different scenarios. With the exception of squash, the total ban of methyl bromide has negative revenue impacts on Florida that range from 32 percent for cucumbers and watermelons to 100 percent for eggplant, with an overall decrease of 54 percent. The total f.o.b. revenues for Texas and Mexico are expected to increase by 141 percent and 65 percent, respectively; those for California will decrease by 32 percent.

TABLE 9.1 Production of Selected Vegetable and Fruit Crops Planted in the United States and Mexico, by Crop and Scenario

Crop	Base Run (with MB)	Ban on MB	Percent Change	Total Quota on MB (50%)	Percent Change	Area Quotas on MB (50%)	Percent Change
				thousand hundredweight			
Tomatoes							
Florida	20051	7742	-61.4	9832	-51.0	11706	-41.6
Mexico	12115	22132	82.7	21104	74.2	19288	59.2
Total	32166	29873	-7.1	30936	-3.8	30994	-3.6
Bell Peppers							
Florida	4823	1764	-63.4	2337	-51.5	2426	-49.7
Mexico	2604	4021	54.4	3685	41.5	3389	30.2
Texas	657	1597	143.1	1397	112.7	1613	145.6
Total	8084	7381	-8.7	7420	-8.2	7429	-8.1
Cucumbers							
Florida	4523	2426	-46.4	3994	-11.7	3994	-11.7
Mexico	4150	4458	7.4	4164	0.3	4131	-0.5
Total	8674	6884	-20.6	8158	-5.9	8125	-6.3
Squash							
Florida	1068	1119	4.8	1112	4.1	1113	4.2
Mexico	1381	1250	-9.4	1285	-6.9	1278	-7.4
Total	2448	2370	-3.2	2397	-2.1	2391	-2.3
Eggplant							
Florida	1286	0	-100.0	849	-34.0	879	-31.6
Mexico	778	1736	123.2	1049	34.9	1031	32.5
Total	2064	1736	-15.9	1898	-8.0	1910	-7.5
Strawberries							
Florida	1236	387	-68.7	763	-38.3	806	-34.8
California	2282	1290	-43.5	2351	3.0	2282	0.0
Total	3518	1677	-52.3	3114	-11.5	3088	-12.2
Watermelons							
Florida	4484	2690	-40.0	2897	-35.4	2889	-35.6

Source: Authors' estimates.

TABLE 9.2 Projected f.o.b. Revenues, in Millions of Dollars, by Production Area and Scenario

| | Crop | | | | | | | |
	Tomatoes	*Peppers*	*Eggplant*	*Cucumbers*	*Squash*	*Watermelons*	*Strawberries*	*Total*
	million dollars							
Florida								
Base run (with MB)[a]	690	145	25	77	32	60	115	1144
Ban on MB	289	66	0	52	33	41	43	524
Total quota on MB (50%)	349	83	18	74	34	43	74	675
Area quotas on MB (50%)	414	87	19	74	34	43	79	749
Texas								
Base run (with MB)		17						17
Ban on MB		41						41
Total quota on MB (50%)		36						36
Area quotas on MB (50%)		42						42
Mexico								
Base run (with MB)	376	67	14	73	34			564
Ban on MB	688	104	32	78	31			933
Total quota on MB (50%)	656	95	19	73	32			875
Area quotas on MB (50%)	600	88	19	72	32			810
California								
Base run (with MB)							187	187
Ban on MB							128	128
Total quota on MB (50%)							201	201
Area quotas on MB (50%)							196	196

[a] *MB = methyl bromide*

Source: Authors' estimates.

Losses in consumer surplus at the wholesale level—both by crop and for the total—are reported for the various scenarios relative to the base case scenario (table 9.3). Tomatoes and strawberries will suffer the greatest setbacks under the ban, with consumer surplus losses of about $40 million and $46 million, respectively. Overall, the ban will cost consumers $121 million because of switching to lower productivity or to higher marginal cost production systems.

Total U.S. Quota of 50 percent

As the proposed ban on methyl bromide represents a governmental command-and-control measure, a quota that limits the use of methyl bromide to 50 percent of current total U.S. consumption is examined as a flexible alternative. Ostensibly, the initial allocation of the quota is based on historical crop production or on methyl bromide consumption levels, and a market evolves to allow either a one-time sale or a temporary leasing of grower allocations on a national scale.

As a means of realizing the nationwide targeted reduction of 50 percent in the use of methyl bromide, farmers in production regions are free to reduce total acreage that is treated with methyl bromide or to reduce the rate of application of methyl bromide on currently treated acreage. The latter option would involve the use of thicker plastic mulch and, therefore, would have a higher application cost. This cost is partially offset by the reduced material cost. Yield levels, however, remain the same as they were under a full rate of application. Interestingly, this option is not exercised as a means to reduced methyl bromide use. Instead, farmers in the production regions choose to use a full rate of application of methyl bromide on fewer acres or they choose not to use methyl bromide at all.

TABLE 9.3 Losses in Consumer Surplus from Bans or Quotas on the Use of Methyl Bromide, by Crop and Scenario (in Millions of Dollars)

Crop	Total Ban on Methyl Bromide	50% Total quota on Methyl Bromide	50% Area quotas on Methyl Bromide
		million dollars	
Tomatoes	39.57	26.11	26.12
Bell peppers	10.29	10.22	10.09
Cucumbers	15.03	5.56	5.77
Squash	0.53	0.47	0.49
Eggplant	2.49	1.75	1.64
Watermelons	6.71	6.11	6.13
Strawberries	46.16	12.25	13.17
Total	120.78	62.46	63.41

Source: Authors' estimates.

The general pattern of impact of the 50 percent reduction of methyl bromide on total U.S. crop production is similar to that under a total ban on methyl bromide. There are varying degrees of reduction in the total market supply of all seven crops that range from 2 percent for squash to 35 percent for watermelons.

The f.o.b. revenue impact of the 50 percent quota on Florida is a decline of 41 percent overall, with revenues falling from a current level of $1.144 billion in the base case scenario to $675 million in the total quota scenario of a 50 percent reduction of methyl bromide (table 9.2). Crop-specific f.o.b. revenue declines vary from 4 percent for cucumbers to 49 percent for tomatoes; squash f.o.b. revenue, however, increases 6 percent. The total revenues of Texas, Mexico and California increase by 112 percent, 55 percent and 7 percent, respectively.

At the wholesale market level, consumer surplus losses range from $470,000 in the squash market to more than $26 million in the tomato market. Bell peppers and strawberries reflect significant declines of $10 million and $12 million, respectively. For all seven crops together, the total loss of consumer surplus is more than $62 million (table 9.3).

50-percent Area Quotas

To the extent that the total or national quota system examined above is twice removed from actual producers and is, therefore, subject to high transaction or information costs, an area quota system is modeled as a more decentralized flexible alternative. Under this scenario, each production region in the United States is limited to 50 percent of its current consumption of methyl bromide. Growers may trade allocations within their respective regions but not across regions. In this respect, area quotas may be marketable at the grower level within a region but are not marketable between regions.

As with the total quota system, growers could either reduce treated acreage by 50 percent or they could cut the rate of the application of methyl bromide on currently treated acreage by 50 percent. In contrast to the total quota system scenario, some production regions did exercise the latter option under the 50 percent area quota system.

Production impacts in Florida under an area quota are, in general, similar to those realized under a national quota. The exceptions are tomatoes and strawberries. Although Florida's production of these crops declines relative to the base run, the production declines are smaller compared to the national quota.

Total f.o.b. revenue for Florida declines by $395 million (35 percent) relative to total f.o.b. revenue in the base case scenario (table 9.2). This occurs as a result of revenue losses of between 4 percent and 40 percent for the various crops. Total f.o.b. revenue increases 147 percent in Texas, increases 44 percent in Mexico and increases 5 percent in California.

Consumer surplus losses in the wholesale market for the different crops are smallest in the case of squash, at $490,000, and largest in the case of tomatoes, at $26 million (table 9.3). For the other crops, consumer surplus losses in the wholesale markets vary from more than $1 million for eggplants to more than $13 million for strawberries. Under the area quota system, the cumulative loss in consumer surplus is more than $63 million.

PRODUCER IMPACTS

In the previous discussion, the impact on producers was measured through changes in f.o.b. revenues in the three competing production regions. One might ask why changes in producer surplus figures are not reported. As noted in the appendix of this chapter, the model uses an implicit supply function to represent the supply side of the market. Under the specification used, no direct measure of producer surplus is available. Given the extensive projected impacts for producers in both Florida and Mexico, producers in both regions will bear significant adjustments under either a total ban or a quota system should either scenario be imposed on them.

In the empirical model, the cost of production for all crops includes both fixed and variable costs. Under the restricted use of methyl bromide, it is assumed that Florida's fixed assets will be allocated for other uses. These uses include land, machinery, and packinghouse facilities that are utilized for the production of fresh fruit and vegetables in the base case scenario. If the markets for these factors were efficient, the returns to those assets that exited the industry would be lower under restricted methyl bromide use than they were in the base case scenario.

Precise quantification of the returns to those assets that are no longer utilized in winter fresh fruit and vegetable production is beyond the scope of this study. In Florida, however, the crop alternatives available to winter fresh fruit and vegetable producers are limited, so crude estimates are possible. While citrus production may compete for land and capital in southwest Florida, the primary alternative to winter fresh fruit and vegetable production in that region, and in the west central region, is the raising of beef cattle. The land rent paid by beef cattle operations in Florida is generally less than $20 per acre. In the East Coast production regions of Palm Beach and Dade counties, the primary alternative to winter fresh fruit and vegetable production is urban development. In these regions, the return to land for non-agricultural use is likely much higher than it is for agricultural use. It is worth noting, however, there has been considerable public support in both of these counties for agricultural land preservation measures. In fact, in Palm Beach County, a restricted development zone has been established that encompasses much of the winter fresh vegetable production area.

Given these observations and the limitations imposed by the empirical model, a crude attempt to measure the impact on asset owners is presented. Smith and Taylor (various years) suggest that for the base case scenario of the model, the cost-of-production figures are divided into variable and fixed cost categories within the budget. Items included under fixed costs include land, machinery and equipment. In the base case scenario of the model, the value of all fixed assets required to support $1.14 billion of winter fresh fruit and vegetable production in Florida is $121 million. Under an outright ban of methyl bromide, this figure becomes $59 million. Thus, $62 million dollars of land, machinery and other equipment are no longer used in winter fresh fruit and vegetable production. This figure is an overestimate of the impact on asset owners in the Florida winter fresh fruit and vegetable industry because a large portion of these assets can find other productive uses. Under the assumption of efficient factor markets, however, the return to those assets will likely be significantly diminished. For example, land rent for vegetable production in West Central Florida is approximately $200 per acre, while its likely alternative use, beef cattle, earns approximately $20 per acre.

Following this same approach, the value of fixed assets utilized under a national methyl bromide quota is $67 million, a decrease of $54 million from the base case scenario. The value of fixed assets used when area quotas are imposed is $76 million, a decrease of $45 million when compared to the base case scenario.

COMPARISON OF POLICY INSTRUMENTS

In terms of both production and f.o.b. revenues (tables 9.1 and 9.2) and with the possible exception of squash, Florida growers are better off under either quota system scenario than they are under an outright ban on methyl bromide. Squash growers produce slightly more but earn slightly less f.o.b. revenue under a ban, relative to the two quota system scenarios. In direct contrast, vegetable growers in Mexico gains the most from the ban for all vegetables except for squash. For bell pepper growers in Texas, the ban on methyl bromide is preferable to the nation-wide quota scenario, but it is marginally inferior to the area quota scenario. As in Florida, California's strawberry industry is worse off under a ban than it is under either of the flexible policies, since it is a major user of methyl bromide.

Florida is better off with the area quota system scenario in the tomato, bell pepper, eggplant and strawberry markets—both in terms of production and of f.o.b. revenues—while cucumber, squash and watermelon growers are indifferent to both the area and the total quota scenarios. California, however, is somewhat better off under the total quota system than it is under the area quota system. As the figures in table 9.3 indicate, consumer surplus losses are somewhat smaller overall with a total quota scenario than they are with the area quota scenario.

In the empirical model, the objective function is the sum of consumer and producer surplus aggregated over the seven commodities and the four terminal markets. The value of the objective function for the various scenarios presented in this chapter is as follows: (a) base case scenario, $616 million; (b) outright ban, $495 million; (c) total quota, $596 million; and (d) area quotas, $592 million. Based on these results, any attempt at regulating the use of methyl bromide would involve a substantial loss of welfare at the terminal wholesale market level.

In both the base case scenario and under an outright ban, there is no measure of producer surplus in any of the markets. This is because of the constant marginal cost, or Leontief technology that is assumed in the model (see the chapter appendix for further discussion). Hence, the difference of $121 million—between the objective function values in the base case scenario and in the total ban—provides only a measure of the loss to consumers from the ban of methyl bromide (table 9.3).

Under the quota system scenarios, however, producers will realize rents for the quotas so that the objective function in these cases represents consumer surplus plus the rents that accrue to producers from the quotas. Under the total quota-system scenario, consumers lose about $62 million while producers gain about $42 million, giving a total welfare loss of $20 million. Under the area quota system, consumers lose about $63 million and producers gain about $39 million so that the total loss is $24 million. The value of fixed assets that must exit winter fresh fruit and vegetable production in Florida is $54 million (in the case of the national quota system scenario) and $45 million (under the regional quota system scenario). In terms of the final impact, it is unclear whether a total quota or a regional quota is preferable, however, either flexible approach is better than an outright ban.

The superiority of either flexible alternative to the command-and-control measure of an outright ban has social costs. A total ban would eliminate the external effect of methyl bromide on the ozone layer, whereas the flexible measures would only mitigate it. Given the intractable nature of the externality, any measure of the trade-off involved must necessarily be a value judgment.

The inability to isolate the damage caused by methyl bromide, in addition to the problems associated with a worldwide valuation of that damage, would clearly rule out a Pigovian tax as a policy option. An input tax on methyl bromide would also likely prove to be impractical. The shadow prices of methyl bromide are determined to be $4.14 per pound to achieve a 50 percent reduction in the level of methyl bromide use and $50.89 per pound to achieve complete elimination. Relative to a market price of $2.23 per pound, these shadow prices would be equivalent to tax rates of 186 percent for a 50 percent reduction and 2,282 percent for complete elimination. These figures attest to the wide-ranging effectiveness of methyl bromide as an input in the production of the winter vegetable and fruit crops that are included in this study.

Shadow prices for methyl bromide by production region are also determined from the regional constraints in the area quota system scenario of the empirical model. They are as follows: (a) Dade County, $0.29 per pound; (b) Palm Beach County, $3.82 per pound; (c) West Central Florida, $4.88 per pound; (d) Southwest Florida, $3.54 per pound; (d) California, $5.03 per pound and (e) Plant City (for Florida strawberries), $4.17 per pound. If production regions were allowed to trade their methyl bromide allocations with one another, the final distribution of regional use levels would be characterized by a common shadow price to which the regional shadow prices would converge. Therefore, regions with individual shadow prices below the common shadow price of $4.14 per pound (for a 50 percent reduction) may be expected to trade all or part of their allocations to those with individual shadow prices above the common price. This is verified by the results of regional methyl bromide use levels that correspond to the area quota (with no interregional trade) and to the total quota (with interregional trade) scenarios (table 9.4).

It is possible to identify regions with low (high) shadow prices—relative to the equilibrium value of $4.14 per pound—as those most (least) vulnerable to competition from regions that do not currently use methyl bromide. For example, tomatoes that are grown in Dade and Palm Beach counties and eggplant that is grown in Palm Beach County are susceptible to keen competition from Mexico. Bell peppers that are grown in Palm Beach County face competition from both Texas and Mexico. In the production of spring tomatoes, West Central Florida has a comparative advantage over Southwest Florida. This advantage, however, is offset somewhat by Southwest Florida's watermelon production. California and Florida, which are both strawberry growing regions in the model, use methyl bromide and are not currently threatened by imports.

TABLE 9.4 Methyl Bromide Consumption Levels by Production Area and Scenario (Million Pounds)

Area	Base Run	Area Quotas (50%)	Total Quota (50%)
	million pounds		
Dade County	1.86	0.93	0.00
Palm Beach County	3.52	1.76	1.63
West Central Florida	5.74	2.87	3.35
Southwest Florida	6.11	3.06	2.47
California	1.90	0.95	1.96
Plant City, Florida	1.29	0.64	0.79
Total	20.43	10.21	10.21

Source: Authors' estimates.

Although the final distribution of regional use levels of methyl bromide under the total quota system scenario is reflective of an equilibrium state, it must be emphasized that no explicit interregional trading mechanism for methyl bromide quotas is incorporated in the model. The possible accrual of income flows in the market for quotas—either through a one-time sale or through temporary leasing—could well change final trading patterns, and is, therefore, a challenging and a worthwhile extension that could be addressed in future work.

CONCLUSION

Methyl bromide is a critical soil fumigant in the production of an assortment of winter fresh vegetables and fruits in Florida and of strawberries in California. The Montreal Protocol identified methyl bromide as a significant depletor of the ozone layer in the stratosphere. It recently modified its recommended phaseout schedule in developed nations to a 25 percent reduction by 1999; 50 percent by 2001; and 100 percent by 2005. Developing nations, however, would be allowed an additional 10 years in which to achieve complete elimination of methyl bromide use.

Recent modifications to the U.S. 1990 Clean Air Act Amendments require that the production and importation of methyl bromide be completely banned by the year 2005. This study conducts an empirical analysis of the impact that this ban will have on the North American winter fresh vegetable and fruit industry and compares the ban with more flexible alternatives such as (1) a national quota of 50 percent of current usage and (2) regional quotas of 50 percent imposed at the level of the various production regions in the United States. Mexico and Texas are included as competing suppliers in the spatial equilibrium model used in the analysis although neither of them currently uses methyl bromide.

The results suggest that both flexible quota alternatives are preferable to an outright ban in terms of their respective impacts on domestic production and equilibrium price levels. Compared to the total quota approach, a regional quota approach has smaller impacts on output and f.o.b. revenues of growers in Florida while California is better off under a national quota. In terms of the total impact at the industry level, the national quota system scenario is, at least, marginally better than the area quota alternative.

The intractable nature of the global externality associated with methyl bromide precludes a Pigovian tax as a policy option. In the case of an input tax on methyl bromide as a possible regulatory instrument, tax rates as high as 186 percent and 2,282 percent would be required for a 50 percent reduction and for complete elimination, respectively.

Since the Montreal Protocol only calls for a phaseout of 25 percent by the year 2005 in developed nations, an outright ban in the United States by that year will clearly put domestic users at a distinct disadvantage relative to other large user regions. Besides the implied partial free riding by other user countries, there is also the possibility of current non-users' adoption of methyl bromide technology. These potential users could be either developing country signatories to the Montreal Protocol who have an additional 10 years to eliminate methyl bromide or non-signatories who did not honor the Montreal Protocol. These dynamic considerations would both accentuate the one-sidedness of an outright ban and dilute its effectiveness in protecting the ozone layer.

ENDNOTES

1. The authors acknowledge financial support provided by NAPIAP, USDA. The manuscript also benefited from comments provided by Andrew Schmitz, Frank Casey and Scott Swinton.
2. See Deepak, M.S., T.H. Spreen, and J.J. Van Sickle (1997) for a mathematical comparison of the two cases.
3. For an entirely producer-oriented analytical approach, see Conrad (1993).
4. Vegetable producers in Mexico use methyl bromide, but its use is not critical. Mexico is allowed to continue using methyl bromide until 2015 under the current agreement signed by the Montreal Protocol.

REFERENCES

Conrad, K. 1993. "Taxes and Subsidies for Pollution Intensive Industries as Trade Policy." *Journal of Environmental Economics and Management* 25: 121–135.

Deepak, M.S., T.H. Spreen, and J.J. VanSickle. 1996. "An Analysis of the Impact of a Ban of Methyl Bromide on the U.S. Winter Fresh Vegetable Market." *Journal of Agricultural and Applied Economics* 28(2): 433–443.

Hathaway, J. and J. Giudice. 1996. *Methyl Bromide Replacement Strategies.* Arizona Department of Environmental Quality, Waste Programs Division, Division Support and Pollution Prevention Section (June).

McCarl, B.A. and T.H. Spreen. 1980. "Price Endogenous Mathematical Programming Models as a Tool for Sector Analysis." *American Journal of Agricultural Economics* 62: 87–102.

Samuelson, P.A. 1952. "Spatial Price Equilibrium and Linear Programming." *American Economic Review* 42: 283–303.

Scott, S.W. 1991. "International Competition and Demand in the United States Fresh Winter Vegetable Industry." Unpublished Master's Thesis, Food and Resource Economics Department, University of Florida, Gainesville, FL.

Smith, S.A. and T.G. Taylor. Various Years. *Production Costs for Selected Vegetables in Florida.* Economic Information Report, Food and Resource Economics Department, Institute of Food and Agricultural Sciences, University of Florida, Gainesville, FL.

Spreen, T.H., J.J. VanSickle, A.E. Moseley, M.S. Deepak, and L. Mathers. 1995. *Use of Methyl Bromide and the Economic Impact of Its Proposed Ban on the Florida Fresh Fruit and Vegetable In-*

dustry. Agricultural Experiment Station, Technical Bulletin No. 898, Food and Resource Economics Department, University of Florida, Gainesville, FL (November).

Takayama. T. and G.G. Judge. 1971. *Spatial and Temporal Price and Allocation Models.* Amsterdam, Netherlands: North-Holland Publishing Co.

APPENDIX

The mathematical model used to depict the North American fresh winter vegetable market is a quadratic programming model. In this appendix, a simplified version of the model is presented and discussed. For more discussion regarding the empirical specification of the model, see Spreen et al. (1995) or Deepak et al. (1996).

To formulate the model, consider a market for a single commodity. Suppose that there are I supply (net exporting regions) and J demand or (net importing regions). Let X_i be the quantity produced in region i with i = 1, . . ., I, and Q_j be the quantity demanded in region j and j = 1, . . ., J. The inverse demand equation in region j is given by $P_j = a_j - b_j Q_j$, where P_j is price in region j. Let X_{ij} denote the quantity shipped from region i to region j. The per-unit transportation cost of shipment from i to j is t_{ij}. Because of the nature of the underlying production technology for the commodity, a supply function for X_i is not available. Suppose, however, that per-acre yields and production costs are known. Suppose that L_i denotes the acres planted in region i, d_i denotes yield per acre and e_i is production cost per acre.

A mathematical formulation of the competitive market equilibrium established among the supply and demand regions is:

$$Max \sum_{j=1}^{J}(a_j Q_j - 1/2 b_j Q_j^2) - \sum_{j=1}^{J}\sum_{i=1}^{I} t_{ij} X_{ij} - \sum_{i=1}^{I} e_i L_i$$

$$s.t. \quad \sum_{j=1}^{J} X_{ij} \le X_i \qquad i = 1, ..., I$$

$$\sum_{i=1}^{I} X_{ij} \ge Q_j \qquad j = 1, ..., J$$

$$d_i L_i = X_i \qquad i = 1, ..., I$$

$$X_i, L_i, Q_j, X_{ij} \ge 0.$$

Given this framework, a mathematical programming model can be written that will provide the competitive equilibrium level of production in each of the export regions, price and consumption in each import region and the quantity shipped from each export region to each import region.

This formulation is a price endogenous mathematical programming problem. It is a variation of the spatial equilibrium model first presented by Samuelson (1952) and later expanded by Takayama and Judge (1971). In the representation presented here, an implicit supply formulation is used as discussed by McCarl and Spreen (1980). The model is an implicit supply formulation because, instead of specifying $X_i = f_i(P_i)$ as the supply function in region i, a Leontief-type (fixed proportions) production function is used to depict the relationship between an input (acres planted) and output. A production budget is estimated in which the other inputs required to meet a specified level of per-acre yield are calculated.

This approach offers two significant advantages and one major disadvantage in the analysis of pesticide bans on agricultural producers and consumers. The prime disadvantage is that, since no upward sloping industry supply is available, direct calculation of producer's surplus is not possible. Thus, a widely accepted measure of producer welfare cannot be calculated.

A major advantage is that the impact of a pesticide ban can more directly be imposed on the model. In the case of methyl bromide, its most viable alternative was identified. Agricultural scientists were consulted, and they provided estimates of the yields and pre-harvest production costs that were associated with the alternative. In this way, a direct approach can be employed to reflect the impact of the ban on producers. A second advantage is that a more disaggregated model can be developed. In the case of the North American fresh winter vegetable market, both regional and temporal disaggregation are important issues in the specification of the model—for example, Dade County, Florida, markets in the December, January and February period. The west central region of Florida concentrates its marketings in November and again in April and May. Mathematical programming facilitates temporal disaggregation of production and allows the model to better capture the market windows served by individual production regions.

Given the limitations imposed by model specification on the measurement of producer welfare effects, two different measures of producer impacts are presented in the chapter. One measure is the change in f.o.b. revenues. This approach deals with the aggregate value of the industry at the shipping point. Large changes in f.o.b. revenues reflect significant expansion or contraction of the industry.

The second approach deals with the impact of pesticide bans on asset owners. The production budgets used in the analysis include both fixed and variable costs of production. Therefore, charges for land, machinery and management were included in the pre-harvest costs. The estimated pre-harvest cost per acre used in the mathematical programming model was divided into variable costs (for example, transplants, fertilizer and plastic mulch) and those fixed costs, that include land and machinery that have been allocated. The figures reported in the text represent the aggregate value of those fixed assets, which exited the industry under the total ban of methyl bromide or under a quota on its use. This estimate clearly overstates the impact on asset owners because most these assets will be allocated to other uses. Florida, however, represents a peculiar case because of the large gap in returns to high value crops that include vegetables for the fresh market and returns to the next best alternative—beef cattle production. In this case, the impact on asset owners is likely to be high as the opportunity cost of the next best alternative is low.

10 ALCOHOL FUEL TAX POLICY: SUGAR, CORN AND THE ENVIRONMENT

Andrew Schmitz and Leo Polopolus
University of Florida, Gainesville, FL

The United States has several tax and investment incentives for ethanol production. These include the federal alcohol mixture credit, federal excise tax exemption and tariff protection on imported ethanol fuel. The purpose of this chapter is to analyze the impacts of the federal, ethyl tertiary butyl ether (ETBE) tax credit upon the production of sugarcane and sugar beets in the United States. This new federal regulation provides tax incentives for producing ethanol from sugar crops when it is combined with isobutylene to produce ETBE. It is shown that the ETBE tax credit will have no effect upon the production of sugar crops in the United States for the following reasons: (1) the ETBE tax credit is merely a minor extension of a long list of other tax incentives and subsidies put in place to promote ethanol production from sugar crops; (2) the ETBE tax credit, as well as all the other related tax incentives and subsidies, has not had direct stimulus upon the production of ethanol from domestically produced sugarcane or sugar beets; (3) the current U.S. sugar program inhibits the production of ethanol from sugarcane and sugar beets because it provides both a price support and a price lid at levels too attractive to divert sugar crops from the production of high value sugar to the production of relatively low value ethanol; and (4) the ETBE tax credit is not expected to increase the prices of corn, corn sweeteners and/or sugar. Moreover, there will be no increase in sugarcane or sugar beets acreage as a direct result of the impacts of the ETBE tax credit on the production of corn-based ethanol.

INTRODUCTION

Many of the chapters in this volume are concerned with flexible incentives in which farmers have to contend with environmental externalities. The focus of this chapter is the use of subsidies that function as flexible incentives (Segerson, this volume) to encourage the adoption of a technology, which results in an improvement in air quality, but may also contribute to environmental problems in the sector producing the inputs used in the technology. That is, the creation of a new technology, through the use of flexible incentives, may create both positive and negative effects. The latter should be dealt with using the flexible incentive approaches discussed in this volume.

This chapter focuses on ethanol production, how it is affected by a tax credit (subsidy) and how ethanol production affects the derived demand for agricultural products, namely corn, which is the main ingredient for ethanol production. According to McNew and Gardner (1996):

> The Clean Air Act Amendments of 1990 require cleaner motor fuels to be used seasonally in geographic areas, which do not meet [The Environmental Protection Agency's] EPA's clean air standards. The two major pollution problems are carbon monoxide (CO) and ozone. CO emissions can be remedied by adding oxygen to gasoline, and ethanol is recognized by EPA as an oxygenate for this purpose. The ozone problem is the more complicated and requires reformulated gasoline. As an oxygenate, ethanol can be used in making reformulated gasoline. However, the environmental benefits of ethanol in these blends are not so clear cut. The problem is the greater volatility (tendency to evaporate) of ethanol as compared to methanol blends. For implementation of the Clean Air Act Amendments, EPA mandated that 30 percent of reformulated fuel must be ethanol based. The oil industry challenged this mandate in the courts as going beyond EPA's authority under the terms of the Clean Air Act, and won. So ethanol must compete, chiefly with substantially cheaper methanol-based blends, for the oxygenate market. Moreover, because of the volatility problem, EPA has mandated that ethanol not be used in reformulated gasoline for summer use in high temperature areas. (p 10)

The second policy is exemption of ethanol blends from the Federal excise tax on gasoline, or an equivalent tax credit for industrial users of ethanol (blenders) who cannot make use of the excise tax exemption. This policy makes ethanol-based fuels competitive in both the oxygenate- and standard-fuel markets. Under current law, which expires in the year 2000, the exemption is worth $0.54 per gallon of ethanol used. This subsidy, without the Clean Air Act mandate, is sufficient to keep ethanol competitive at current price levels of use, but it is not sufficient to generate the further expansion of ethanol fuel use that the 30 percent mandate was expected to provide.

There are many interesting aspects to the study of ethanol production. First, ethanol subsidies have both positive and negative environmental impacts. Second, numerous interested parties gain from the tax credits that encourage ethanol production. Third, in addition to the benefits of cleaner air that results from the use of ethanol versus alternative fuels, corn producers gain and so do ethanol manufacturers. Fourth, certain environmental groups oppose the production of ethanol, not because of its clean air properties, but because the increased production of the inputs needed for ethanol production results in significant environmental damage (see Batie and Ervin, this volume). Fifth, the current law expires in the year 2000 when it is up for renewal. On May 22, 1998, the ethanol tax credit program was extended through the year 2007. Under the terms of the new ethanol provision, the $0.054 per gallon tax incentive—a deduction for oil marketers from the $0.184 per gallon excise tax on gasoline—will be reduced to $0.053 per gallon in 2001, $0.052 per gallon in 2003 and $0.051 per gallon by the year 2005 (*Feedstuffs*, 1998).

Section 40(a)(1) of the U.S. Internal Revenue Service code grants tax credits of more than $0.50 per gallon to the producers of fuel alcohol (ethanol). The producers, in turn, use this fuel alcohol to produce a qualified mixture of alcohol and gasoline (gasohol). Under federal regulations, this more than $0.50 per gallon tax credit is not available to alternative fuels, such as methanol and methyl tertiary butyl ether (MTBE), which are derived from fossil fuels. Federal law also provides a $0.06 per gallon exemption from the $0.09 per gallon federal excise tax on gasoline if the taxable product is blended in a mixture with at least 10 percent alcohol to make gasohol. As defined here, alcohol includes only the alcohol that is derived from a source other than petroleum, natural gas or coal. A central issue is whether or not a tax credit for producing ethyl tertiary butyl ether (ETBE) could indirectly cause harmful environmental impacts in geographic regions where ethanol is produced from sugar crops or from corn. (ETBE is produced by combining ethanol and isobutylene. Ethanol can be produced from sugarcane, sugar beets, corn and certain other agricultural crops.)

There is a continuing debate in Washington, D.C. as to whether or not the tax credit should be extended. Also, various environmental groups have challenged the use of the tax credit for ethanol production, but the courts have ruled in favor of the U.S. government. In the early 1990s, the Florida Audubon Society and other interested parties sued the U.S. Secretary of the Treasury and the Commissioner of the U.S. Internal Revenue Service (*Florida Audubon Society et al. v. Nicholas F. Brady et al.*, 1991). The Florida Audubon Society instigated the lawsuit because the government failed to comply with the National Environmental Policy Act (NEPA) when promulgating the final rule of the federal alcohol fuel tax credit policy. The plaintiffs contended that the tax credit policy for ETBE would so increase sugar crop production in the United States that environmental pollution would be greatly increased. In contrast, the defendants argued that the ETBE tax credit policy would have no effect on the production of sugar crops in the United States. In 1997, the federal district court ruled against the plaintiffs on the grounds that their theory—the ETBE tax credit policy would injure the environment—was too indirect and too speculative to likely occur. The Federal Court of Appeals [94 Fed 3d (1996) 658] upheld the decision of the lower court in favor of the defendants. It argued that the Florida Audubon Society et al. was unable to prove that environmental injury had occurred in South Florida because of the ETBE tax credit policy. The court was well aware of the fact that ethanol had not been produced from sugar crops that were produced in the United States, even with the more than $0.50 per gallon tax credit that had been in effect for several years. Market conditions would have to change dramatically before ethanol would become an economically feasible output from sugar.

The purpose of this chapter is threefold. First, it provides theoretical and empirical evidence as to why ETBE tax credits would not bring about increased sugarcane production. Second, it explores the impact of the ethanol tax credit policy on corn production and prices, and the effects of these higher prices on high fructose corn syrup (HFCS) production and sugar prices. Third, it discusses the impact of the tax credit policy on the environment in the context of alternative technologies used to produce fuel. To tell the story, this chapter relates the arguments that were presented by both the plaintiff and the defense in the *Florida Audubon Society et al. v. Nicholas F. Brady et al.* case.

THE FRAMEWORK

A flow chart of the potential use of sugarcane in gasohol production is provided in figure 10.1. A discussion of whether sugar will be economical for use as an input in ethanol production is included in the following sections. The extent of its use is, in large part, a function of the U.S. sugar policy.

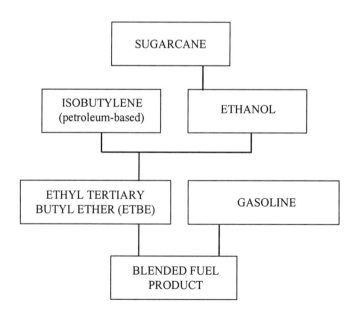

FIGURE 10.1 Flow Chart of Hypothetical Production of ETBE from Sugarcane

Sugar

The 1990 Farm Bill

The key features of the 1990 Farm Bill follow:

Market Stabilization Price (MSP). The MSP was established implicitly by the U.S. Secretary of Agriculture at a level high enough to permit domestic sugarcane and sugar beet production to clear commercial market channels rather than to be forfeited to the Commodity Credit Corporation (CCC). Clearing the market means that all domestically grown beet and cane sugar would be ultimately sold in private commercial market channels rather than purchased and handled by the U.S. government. For the 1990 fiscal year, the MSP was set at $0.2195 per pound for raw sugar. This essentially guaranteed that the gross revenue from raw sugar sales could exceed the gross revenue from equivalent volumes of sugarcane juice that were used to produce ethanol. As noted by Kiker and Gressel (1986), the U.S. sugar policy inhibited investments in ethanol distilleries by creating a high opportunity cost for sugarcane juice.

Loan Rates. The 1990 Farm Bill continued the previous legislation of a $0.18 per pound minimum loan rate for domestically produced raw sugar. While this program provided nonrecourse CCC loans for raw sugar, the 1.64 billion pounds placed under loan in 1988 were all redeemed (USDA/ERS, 1989). Thus, raw sugar prices moved above the MSP and permitted processors to retrieve their sugar, pay off their loans and sell it in the private sector. The sugar program was effectively administered to avoid the government accumulation of sugar stocks and to preserve the no-net-cost feature to the Federal Treasury—a statutory objective.

Tariffs. The United States installed a two-tiered tariff policy for sugar imports. The tariff rate for the first tier was $0.00625 per pound. This tariff rate was applicable to foreign imports that were subject to quotas up to 1.725 million tons of raw sugar for the 1990–91 sugar-crop year. For foreign imports above 1.725 million tons, quotas were suspended in lieu of second-tier tariffs of $0.16 per pound of raw sugar.

Import Quotas. Within certain constraints, the U.S. Secretary of Agriculture was permitted to adjust raw sugar quotas of foreign countries to fulfill the MSP objective of the sugar program. Thus, the Secretary had the authority to reduce foreign quotas over time to stabilize domestic raw sugar prices at the MSP level, and he could also increase quotas within the first tier of tariff rates. The 1990 Farm Bill imposed a minimum level of foreign sugar imports at 1.25 million tons of raw sugar.

Domestic Processing/Acreage Controls. The 1990 Farm Bill required that the U.S. Secretary of Agriculture impose marketing allotments (controls) upon sugarcane and sugar beet processors if the MSP were unable to be maintained with foreign sugar imports guaranteed at the minimum level of 1.25 million tons. The U.S. Secretary of Agriculture had the statutory authority to impose controls (proportionate shares) upon sugarcane and/or sugar beet growers to reduce the acreage planted, if the controls on the processors were not expected to sufficiently constrain domestic sugar production.

The 1996 Farm Bill

Although the Federal Agricultural Improvement and Reform (FAIR) Act of 1996 represented a major transformation in U.S. agricultural policy, the sugar program only experienced a few minor changes. The key components of the 1996 Farm Bill follow:

Loan Program. The national average of the raw sugarcane loan rate under the 1996 Farm Bill was fixed at $0.18 per pound. (In previous legislation it could be raised but it could not be lowered.) The national average of the sugar beet loan rate was also fixed at $0.229 per pound for refined sugar. (In previous legislation, this loan rate was determined relative to the sugarcane loan rate.)

Under the FAIR Act, loans were considered recourse when the tariff rate quota (TRQ) on sugar imports was at 1.5 million tons or less. A recourse loan means that the USDA can demand repayment of the loan at maturity regardless of the price of sugar. In contrast, nonrecourse loans require that the government accept the sugar when the loan matures in lieu of loan repayment in cash at the option of the processor. When the TRQ exceeded 1.5 million tons, loans became nonrecourse. Minimum payments were required only when the CCC offered nonrecourse loans. Whether loans were recourse or nonrecourse depended upon the amount of the quota (an administrative decision) and not on the actual amount of sugar that was imported (a market decision).

Without a minimum price guarantee and under the conditions of recourse loans, domestic sugar prices could become more volatile, particularly on the downside. This could also negatively affect the sugar producer's ability to borrow funds in order to finance their operations (Polopolus, 1996). Therefore, the degree of price risk in domestic sugar production was increased under the 1996 Farm Bill.

Sugarcane processors paid a penalty of one cent per pound for sugar that was forfeited to the government while sugar beet processors paid a penalty of $0.0107 per pound. Also, the interest rates that were established by the CCC each month for commodity loans (including sugar) were one percent higher in 1996 than they were under previous farm legislation. For example, the CCC interest rate for commodity loans that were taken out in August of 1996 was 5.875 percent for 1995 crops but it was 6.875 percent for loans on 1996 crops.

Marketing Controls and Assessments. The FAIR Act eliminated all previous authority of the domestic sugar and crystalline fructose marketing allotments or supply controls that were intended to guarantee foreign access of at least 1.5 million tons to the U.S. sugar market. As a result, there remained no incentive to expand the processing capacity simply to increase the historical base, and future domestic production/processing capacity was geared more closely to potential economic returns (Polopolus, 1996). Marketing assessments paid on all processed domestic sugar were increased by 25 percent from previous levels, or from 1.1 percent to 1.375 percent of the raw sugar loan rate for sugarcane processors. For sugar beet refiners, the assessment increased from 1.1794 percent to 1.47425 percent of their sugar loan rate. The no-net-cost provision was not modified.

Import Quotas. Import quota provisions are those discussed under the Loan Program (see the previous section). On imports above the TRQ, the tariff was $0.1717 per pound for raw cane sugar in 1996 and is scheduled to decline to $0.1536 per pound in the year 2000. The conceptual framework for analyzing the effect of the ethanol tax credit on sugar production is shown in figure 10.2. To fit the components of the U.S. sugar program, a support price, Ps, exists at which the United States imports Q2 – Q1 of sugar, where S is the supply of U.S. sugar and DD is total U.S. sugar demand. The world price of sugar is Pw. Exporters of sugar to the United States earn exporter quota rents of *abcd*. Under the 1996 Sugar Act, a minimum import quota was put in place that could be adjusted upward at the discretion of the Secretary of Agriculture.

Technologies are available for making ethanol from sugarcane and from sugar beets. In order to use these technologies, as is discussed later, the price of sugar would have to be significantly below the support price, Ps. The derived demand curve for sugar for use in ethanol production is D* (figure 10.2). The total demand curve for sugar now becomes DDo. Note that the potential to produce ethanol from sugar has no effect on sugar prices or on sugar production.

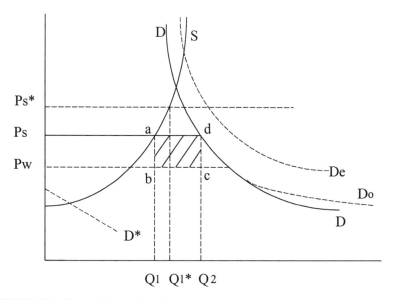

FIGURE 10.2 Sugar Market Fundamentals

As a result of ethanol tax credits, if the demand for sugar were to shift to the right over all ranges of the sugar prices, the price and the production of sugar for a given import quota of Q2 – Q1 would increase. For example, if demand shifted to De as a result of the ethanol tax credit policy, sugar prices would increase to Ps* and production would increase to Q1*. Because of the nature of technology and the relative price of sugar versus other crops that are suitable for ethanol production, sugar demand does not shift to De. Instead of shifting to De, the new demand is DDo, where demand shifts only at sugar prices well below Ps.

Corn

A simplified version of the U.S. corn program, which was in place prior to 1996, is shown in figure 10.3. The key components were target prices, loan rates, deficiency payments and acreage set-asides. Domestic supply, S, is adjusted for acreage set-asides. Domestic demand, Dd, is made up of feed demand plus other components that include the demand from corn sweetener and ethanol production. The target price is Pt, and production is Qt. The market price, however, is Pm given total demand D, which includes U.S. corn exports. Government deficiency payments total *PtabPm*. Unlike the case with sugar, users of corn pay prices below free market levels.

Under the 1996 Freedom to Farm Act, major changes took place. Target prices, acreage set-asides and deficiency payments were eliminated. In terms of market prices, as was the case prior to 1996, corn prices remained low relative to sugar prices. This was due, in part, because of the differential impact of U.S. farm legislation. It may well be that the new legislation will have little effect on market

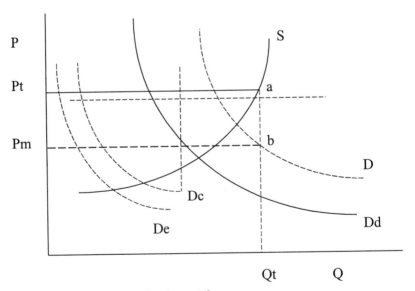

FIGURE 10.3 Corn Market Fundamentals

prices for corn and for corn production. If this were the case, ethanol manufacturers would continue to use corn, rather than sugar, in the production of ethanol because corn is less expensive than sugar. Because government farm programs kept the price of corn low relative to sugar, they were responsible for the sizable growth of the U.S. corn sweetener industry as well as for the growth in the production of ethanol, which is made from corn.

With the MSP set at almost $0.22 per pound for raw sugar, the opportunity cost of sugarcane juice used for ethanol production would be far beyond the price that a fuel ethanol distillery would be willing to pay. This price also provides substantial disincentives for investment in ethanol distillery capacity. Thus, the expansion of sugarcane or sugar beet acreage in response to federal regulations that allowed an ETBE tax credit would be inconceivable under the prevailing market and sugar program conditions.

EARLIER CONCLUSIONS

Polopolus and Schmitz (1991) reached the following conclusions:

- Because of a lack of economic feasibility for these products, the ETBE tax credit had no direct impact on ethanol or ETBE production from sugarcane.
- By way of increasing the prices of corn and HFCS, the ETBE tax credit had no indirect impact on ethanol or on ETBE production from sugarcane.
- The elasticity of cane sugar supply is highly inelastic. That is, sugar price increases do not lead to proportionate increases in sugar output.
- The ETBE tax credit had no effect on the sugar supply response because it was not a factor under current market and policy conditions.
- The ETBE tax credit was not expected to increase the price of corn, corn sweeteners or sugar.
- Water quality management programs that will be imposed by the South Florida Water Management District will likely adversely affect future sugarcane production in Florida. Losses of agricultural acreage in the Everglades Agricultural Area of Florida could exceed 190,000 acres. (pp. 11–12)

RECENT EVIDENCE

Ethanol Subsidies and U.S. Sugar Production

As argued by Polopolus and Schmitz (1991), the use of ethanol tax credits would not bring about an increase in the production of sugarcane or sugar beets. In terms of their earlier model, given the support prices for sugar, the demand for sugarcane and sugar beets that was used for making ethanol production was zero. The relative prices of sugar versus corn make sugarcane far more valuable as a feedstock for raw sugar production than for ethanol production. Hence, sugarcane and sugar beets have not been used in ethanol production. The basic methodology for determining the feasibility of producing ethanol from sugarcane involves a comparison of alternative revenue streams from one ton of Florida sugarcane in either ethanol or raw sugar product forms.

According to Gressel (1984), one average ton of Florida sugarcane (with a 10.5 percent sugar recovery rate) will produce alternatively:

17.86 gallons of ethanol   or   210 pounds raw sugar
and
6.5 gallons molasses

In contrast to ethanol that is produced from corn or from sugar beets, ethanol produced from sugarcane does not generate joint food or feed products (Hertzmark et al., 1980).

In a free ethanol market—a market without ethanol subsidies—the true value of ethanol would be equal to the price of unleaded gasoline. Given a spot price of this product at $0.66 per gallon (New York basis, oxygenated, May 20, 1997), the maximum revenue that would be generated from a ton of sugarcane used for ethanol production in a free market would be $11.79 per ton. This compares to $47.70 per ton (table 10.1), which was received from sugarcane in the form of raw sugar and molasses at the July 1997 futures price of $0.2163 per pound for raw sugar and $70 per ton for 6.5 gallons of molasses. On this basis, the production of ethanol from Florida sugarcane does not appear to be close to feasible or to competitiveness.

Even with the ETBE tax credit, it is uneconomical to produce ethanol from sugarcane. Adding the tax incentive effect of $0.60 per gallon brings the total price for ethanol to $1.26 per gallon. Although total revenue for ethanol derived from sugarcane now increases to approximately $22.50 per ton (table 10.1), this is still less than half the $47.70 per ton in revenue that could be derived from sugar and molasses.

Thus, based on a comparison of the revenue streams in table 10.1, there is no financial incentive for sugar mills in Florida to divert existing sugarcane production from current raw sugar production to the manufacture of ethanol. Moreover, there does not appear to be an economic rationale for the expansion of sugarcane acreage and production in Florida solely for the purpose of ethanol production. For the alternative total revenue streams to be equal, total subsidies for ethanol production that are received by sugar mills would have to be approximately $2 per gallon—more than three times the ETBE tax credit.

Even in the unlikely event that raw sugar prices would fall to the minimum government loan rate of $0.18 per pound, total revenue would still be $40.46 per ton of sugarcane. Thus, even with generous (in favor of the plaintiffs) assumptions regarding prices and ethanol subsidies, basic economics would strongly favor the production of raw sugar over ethanol from sugarcane.

TABLE 10.1   Revenue, in U.S. Dollars, from the Production of Ethanol or Raw Sugar Extracted from One Ton of Florida Sugarcane

	Ethanol	Raw Sugar and molasses
Production Alternatives		
Ethanol	17.86 gallons	
Raw Sugar		210 pounds raw sugar 6.5 gallons molasses
Ethanol Price		
Price of Unleaded Gasoline	$0.66 per gallon[a]	
Tax Incentive	$0.60 per gallon	
Total	$1.26 per gallon	
Total Revenue, Ethanol	$22.50	
Raw Sugar and Molasses Prices		
Raw Sugar		
(July 1997, Futures)		$0.2163 per pound[a]
Molasses		$70.00 per ton
Total Revenue, Raw Sugar		$47.70

[a]*Wall Street Journal, May 21, 1997.*

The Indirect Impact of an ETBE Tax Credit on Sugar Production

The Florida Audubon Society et al. (1991) argued that another way in which the ETBE tax credit could increase sugar production (and therefore ethanol production) would be by increasing the price of corn. This circular argument contends that the ETBE tax credit would increase the price of corn, which would then increase the price of HFCS. Because of an assumed cross price elasticity of demand between HFCS and sugar, the plaintiffs assumed that increasing the price of HFCS would increase the price of sugar. The implication of this argument is that the ETBE tax credit would indirectly cause ethanol to be produced from sugar crops because of the assumed close relationship between corn and sugar prices. The expanded demand for corn, as it is used in ethanol production (approximately 200 million bushels), would have a very minor impact on the price of corn (Polopolus and Schmitz, 1991). Supported by a number of studies, the long-term impact is estimated to be between $0.05 per bushel and $0.07 per bushel of corn, or only a 2.2 percent increase in corn prices.

Contrary to the plaintiffs' arguments, there is no strong, positive empirical relationship between corn prices and the prices of HFCS. As shown by the actual data, any causal relationship between corn prices and HFCS prices is difficult to predict. Part of this difficulty is explained by the fact that the HFCS market is segmented into two separate products—42-percent HFCS and 55-percent HFCS—each with different buyers and uses. Movements in corn prices, however, do not necessarily cause 42-percent HFCS and 55-percent HFCS to move in similar directions or with similar magnitudes (Polopolus and Schmitz, 1991).

The oligopolistic structure of the HFCS market further weakens the link between corn prices and HFCS prices. Research and development initiatives of the wet corn milling industry, along with the price support mechanisms of the U.S. sugar program, have given sellers of HFCS much larger gross profit margins than those that normally have been afforded to generic agricultural products. Assuming that corn costs represent only 30 percent of the gross revenue for HFCS (compared to the 65 percent sugarcane costs relative to raw sugar prices), HFCS manufa

cturers have unusually large margins for profit, advertising and research, and development expenditures. Thus, even large variations in corn prices can permit HFCS sellers to price their two products—42-percent HFCS and 55-percent HFCS—in patterns independent from the corn price movements. HFCS sellers can engage in these types of pricing strategies because of the oligopolistic nature of HFCS markets. For example, the four largest sellers of 42-percent HFCS account for 85.9 percent of the industry's processing capacity, while the four largest sellers of 55-percent HFCS account for 86.6 percent of the industry's processing capacity (Polopolus and Alvarez, 1991).

Corn markets, on the other hand, are driven almost completely by atomistic competition (where no individual corn seller has a perceptible influence on price), in which profit margins are negligible. That is not true with either of the two HFCS products, in which the few sellers are interdependent and actually influence market prices. Given the large, gross profit margins in the HFCS business, rival sellers may choose to sacrifice some profits when corn prices increase by holding the line on HFCS prices to gain market share or to achieve some other marketing objective.

The complexity of the corn and HFCS price relationship spills over to the cross price elasticity of demand relationships between sugar and HFCS. Polopolus and Alvarez (1991) concluded that increases in the price of 55-percent HFCS increased sugar prices while increases in the price of 42-percent HFCS decreased sugar prices. Again, the data conclusively fail to establish the existence of a simple empirical relationship between corn syrup and sugar prices.

By way of summary, higher corn prices do not necessarily translate to higher HFCS prices. Concomitantly, increases in HFCS prices do not necessarily cause sugar prices to escalate. At best, the linkages between corn prices and sugar prices are very weak. Thus, there is virtually no reliable way to predict that an increase in corn prices will make it economically feasible to produce ethanol from sugarcane. What is certain is that higher sugar prices, caused by whatever factors, provide added disincentives to the production of ethanol from sugar crops. (Any added sugar production that is caused by high sugar prices would be targeted for human food markets, and not for industrial fuel markets.)

One further interesting note concerns the relationship between corn and sugar prices: If corn prices were to cause sugar prices to escalate, this increase would be short-lived. In view of the theoretical discussion earlier, the Secretary of Agriculture would, in all likelihood, allow additional imports into the United States that would deter any major increase in sugar prices.

Supply Elasticities

The plaintiffs in the aforementioned case contended that there would be a large increase in sugar production due to the ethanol subsidy. They alleged that the supply schedule for U.S. sugar production is highly price elastic, which is not the case. They assumed the supply elasticity for the Florida sugar industry to be 4.23. If this were true, a 10 percent increase in the price of sugar would result in a 42.3 percent increase in Florida sugar output. The plaintiffs also assumed that one-third of the total increase in domestic sugar production would come from Florida sugarcane in response to the ETBE tax credit. The Florida sugar supply price elasticity of 4.23 is taken from Gemmill (1976) who described the sugar world of 1950 through 1975 and calculated sugar projections to 1985. This period was unique for the Florida sugarcane industry because of the takeover of Cuba by Fidel Castro in 1960. Cuba's immediate loss of sugar marketing rights in the United States led to an increase in Florida's sugarcane acreage in order to fill this void. Gemmill, however, cautioned readers about his supply elasticity estimate for Florida. He stated:

> The existence of (proportionate) shares in Louisiana and Florida greatly complicates the specification and estimation of time series models of supply . . . Preliminary attempts at time series estimation resulted in nonsignificant coefficients for all price variables. (p. 70)

When presenting a range of elasticities between zero (Puerto Rico) and 4.23 (Florida), Gemmill (1976) also stated: "one should be cautious in interpreting these elasticities." (p. 131)

For both sugarcane and sugar beets, the overriding economic consensus of opinion is that the elasticity of supply is price inelastic (table 10.2). The results of studies by Bhatti and Yanagida (1990), Roningen and Dixit (1989), Lopez (1989) and Advincula (1992) are given in table 10.2.

An inelastic supply elasticity of 0.23 for U.S. sugarcane means that, in the short-run, a 10 percent increase in sugar prices would lead to only a 2.3 percent increase in sugarcane production. For sugar beets, a 10 percent increase in beet sugar prices would cause only a 4.8 percent increase in sugar beet production. The recent research by Advincula et al. (1992) for Florida also validates the relative inelasticity of the supply response for sugarcane. There are several reasons why sugar supply elasticities are in the inelastic range:

- Most agricultural land is not suited for sugar beet or sugarcane production.
- Given the relative prices of other competing commodities, it is not lucrative for producers to expand acreages of sugarcane or sugar beets.

TABLE 10.2   Sugar Supply Price Elasticities, from 1989, 1990 and 1992 Studies

Author	Supply Elasticity
Bhatti/Yanagida	0.63 sugarcane supply (Pakistan)
Ronnigen/Dixit	0.5 U.S. sugar supply 0.17 EC sugar supply
Lopez	0.231 U.S. sugarcane, short-run supply 0.479 U.S. sugar beet, short-run supply 0.479 U.S. sugarcane, long-run supply
Advincula	0.51 Florida (expectational model) 0.35 Florida (Kalman filter tech niques)

- In the long run, the expansion of acreage could be a risky business since it is unclear whether the U.S. Sugar Program would remain in existence. If

- U.S. prices were to fall to the current world market price, there would be a dramatic reduction in domestic sugarcane and sugar beet acreage.

- Generally, to expand acreage and production, additional processing facilities must be constructed and they require massive investments. If a new processing plant were to be built, a substantial amount of sugar beet or sugarcane acreage would be needed to supply the raw product. Investors are reluctant to invest in additional processing capacity because of the uncertainties that surround the profitability of sugar production. This uncertainty stems, in part, from the continuous attack on the U.S. sugar industry by many economists who argue that U.S. sugar quotas should be greatly relaxed. This would result in a drop in the U.S. sugar price that would discourage investment in additional capacity.
- The 1990 Sugar Program essentially removed any incentive to invest in additional capacity. This is because under GATT, the U.S. Secretary of Agriculture is obligated to import 1.25 million tons of sugar annually. Under the 1996 Farm Bill, the safety net of nonrecourse loans for domestic sugar producers would disappear if foreign imports were 1.5 million tons or less. With·this constraint in place, any significant increase in production in the United States could trigger marketing allotments (or similar adjustments) in order to regulate the industry.

For the Florida sugarcane industry, in particular, there are several additional constraints to expanding sugarcane output. Soil subsidence is occurring on the muck lands that are best suited for growing sugarcane. This depletes the soil resource for sugar production. Water and other regulations inhibit the reclamation of additional wetlands for sugarcane production. Also, the expansion of sugarcane production to less-suitable sandy soils would result in higher per-unit costs of sugarcane production. Thus, factors—such as water, environmental cost and natural resources—are such that supply response is inelastic with respect to price.

Despite governmental support to sugar producers, worldwide sugar production has not increased substantially. There are several reasons for this:

- Sugarcane acreage is near the capacity afforded by climate and soil.
- For sugar beets, relatively high production costs allow other crop enterprises to compete for the same acreage. This is the case in California where high costs and recent disease problems actually caused a decline in sugar beet acreage. Other areas also incurred reductions in sugar beet production in 1988 and 1989.
- There is reluctance by processors to build new plants or to expand existing ones. Although plants in some locations are operating at capacity, it is apparently felt that the prices presently supporting sugar production are not reliable enough to justify the investment in new processing plants (Schmitz and Christian, 1990).
- Particularly, with regard to sugar beets, producers respond to relative product prices where multiple cropping is possible. They also respond to the degree of price and production risk that is associated with each crop. In terms of relative prices, grain markets have followed a price pattern similar to sugar. Grain prices soared throughout 1974, 1975 and in the early 1980s—just as sugar prices did. Hence, relative product prices have not changed a great deal, and in absolute terms, sugar production has not responded to high sugar prices (Schmitz et al., 1984). Given the relatively high price of sugar, sugar beet processors have never seriously considered producing ethanol. Because prices are not expected to vary greatly, the debate over the size of sugar beet and sugarcane supply price elasticities is academic. This is clear because, if everything else were equal, in order to obtain a supply response in sugar beets and sugarcane, their prices would have to increase. The likelihood of this is highly improbable. If an increase in the supply response were to occur, however, it likely would be a very little production response. Supply price elasticities for sugar beets and sugarcane are highly price inelastic.

More importantly, given that ethanol production from sugar is uneconomically feasible at the sugar support price, the supply elasticity may be of little concern within the relevant price range. For example, if the demand curve for sugar were to shift to D1 due to the ethanol subsidy and S2 was the supply schedule, the price of sugar would increase from P1 to P3 and its quantity would increase from Q1 to Q3 (figure 10.4). If S1 were the supply schedule, the price increase, P2, would be smaller, but the quantity increase, Q2, would be larger. We contend that the demand curve does not shift to D1, rather, the new demand curve moves to D* from D. Hence, ethanol subsidies have no effect on sugar production, regardless of whether sugar supply is S2 or S1.

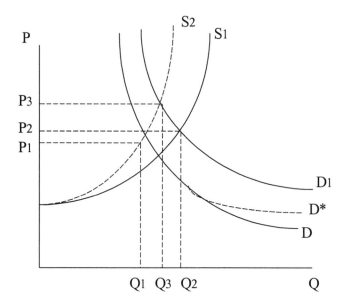

FIGURE 10.4 Ethanol Production and Demand Shifts Due to Ethanol Subsides

Ethanol Subsidies and U.S. Corn Production

Corn utilization in ethanol production from 1980 through 1995 is shown in figure 10.5. Utilization reached 600 million bushels in 1995. Food and fuel products produced from one bushel of corn are shown in figure 10.6. According to the National Corn Growers Association, one acre of corn (125 bushels) produces 313 gallons of ethanol, 1,362 pounds of 21-percent distillers' grain, 325 pounds of 60 percent gluten meal and 189 pounds of corn oil. Distillers' grain is a by-product of ethanol production from grain that can be used for protein in many animal feeds. Approximately 1.4 billion tons of distillers' grain are produced annually and the value derived from this market is critical to the economic viability of ethanol production.

Hauser and Braden (1982), using models of Hertzmark et al. (1980) and Womack et al. (1981), suggested that an increase in demand for corn which is to be used for ethanol, would have little impact on corn prices (table 10.3). Thus, the effect of ethanol subsidies on corn production and prices is not large enough to influence sugar production.

The price increases, which result from the expanded demand of corn for use in ethanol, were quite small—roughly an average of $0.06 per bushel for the period of the analysis (table 10.3). Given the average corn price for this period of $2.74 per bushel, the $0.06 per bushel increase represents a 2.2 percent increase in corn prices—a very small impact indeed. Using the above analysis, however, a 600

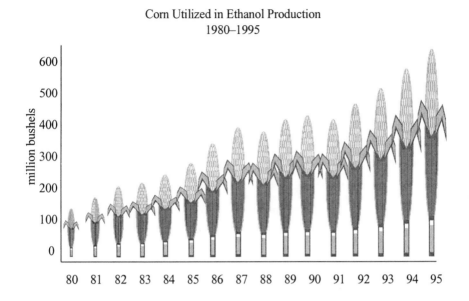

FIGURE 10.5 Corn Used for Ethanol Production, 1980 through 1995

Source: Renewable Fuels Association.

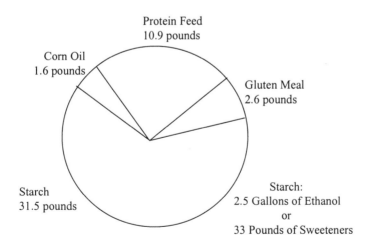

FIGURE 10.6 Food and Fuel Products from One Bushel of Corn

Source: National Corn Growers Association.

TABLE 10.3   Corn Prices and Ethanol Production: Estimates from previous studies, 1981 through 1982

Study	Assumed Increase in Production	Long-run Impact on Corn Prices
Hauser & Braeden model	173 million bushels	$0.05 per bushel to $0.068 per bushel
Hertzmark et al., model	173 million bushels	$0.051 per bushel
Womack et al., model	173 million bushels	$0.068 per bushel

million-bushel increase in corn production due to the ethanol subsidy would cause corn prices to rise by roughly $0.20 per bushel.

In a more recent study, Evans (1997) showed the effects of ethanol subsidies on corn prices to be higher—ranging from $0.40 per bushel to $0.50 per bushel. His results are presented in table 10.4. For 1995, Evans showed a price effect of $0.45 per bushel and a production response of 390 million bushels as a result of the ethanol, fuel subsidy program.

TABLE 10.4   Effects of Ethanol Subsidies on Corn Production and Prices, 1980 through 1995

Year	Price with Ethanol	Price without Ethanol	Difference	Product with Ethanol	Product without Ethanol	Difference
1980	2.94	2.88	0.05	6.33	6.20	0.13
1981	2.51	2.44	0.07	7.80	7.72	0.08
1982	2.47	2.34	0.13	7.99	7.93	0.06
1983	2.93	2.80	0.14	5.59	5.52	0.07
1984	2.57	2.43	0.14	7.59	7.47	0.12
1985	2.07	1.85	0.22	8.37	8.24	0.14
1986	2.01	1.80	0.21	8.02	7.84	0.18
1987	2.23	1.96	0.28	7.81	7.60	0.22
1988	2.78	2.49	0.29	5.59	5.41	0.18
1989	2.62	2.33	0.28	8.08	7.80	0.28
1990	2.07	1.77	0.30	8.37	8.08	0.28
1991	2.16	1.84	0.32	7.32	7.05	0.27
1992	2.10	1.76	0.35	8.67	8.32	0.34
1993	2.19	1.79	0.39	6.64	6.36	0.28
1994	2.59	2.17	0.42	9.17	8.74	0.43
1995	2.85	2.41	0.45	7.79	7.40	0.39

Source: Evans, 1997

The numbers of McNew and Gardner (1996) are much smaller than those provided by Evans (1997). According to McNew and Gardner (1996):

> The use of ethanol would decline about 335 million gallons, losing a third of its current market. This is estimated to cause a net cost to ethanol manufacturers of $165 million annually. The reduced ethanol production would cause the demand for corn to fall by 130 million bushels. However, because of lower corn prices for feed and less production of corn gluten and other feed byproducts of ethanol production, aggregate use of corn falls by only 40 million bushels. The result is a fall in the price of corn by $0.04 per bushel. This is much smaller on a percentage basis than the fall on the price of ethanol because the great bulk of corn use is not in ethanol production. However, because the corn price decline applies to the entire 8.5 billion bushels of corn for production, and not just to the 400 million bushels used to make ethanol, the corn growers' net loss of $258 million is even larger than the ethanol producers' losses. Nonetheless, as in the sugar quota case, the percentage loss to ethanol manufacturers is substantially greater that the percentage loss to corn growers. (p. 11)

CONCLUSION

The production of ethanol fuels is largely the result of government policy. Ethanol fuels are cleaner burning than nonrenewable fuels. Also, there is no evidence that ethanol-distilling plants are sources of major air pollution. It is somewhat ironic that, in view of the above, several groups sued the U.S. government over the use of the ethanol tax credit. They argued that the increased production of ethanol fuels would further pollute the environment. These groups alleged that sugar and corn production would expand, such that the resulting increase in produce could be used as raw materials for ethanol production. Concerning sugar, an industry already under close environmental scrutiny and especially in Florida, it was concluded that ethanol production had no effect on U.S. sugar production. Hence, the plaintiffs' arguments against the sugar industry were not valid. Sugar, an input for ethanol production, is too costly relative to corn. This is due, in part, to the U.S. government farm programs. The federal sugar program is designed to maintain and protect the domestic sugar industry. It is not intended to initiate an ethanol industry, fueled by sugarcane or sugar beets. The implicit MSP is a key sugar policy instrument used by the Secretary of Agriculture to minimize the likelihood that domestically produced sugar would be purchased by the CCC.

Prudent administration of the MSP requires the U.S. Secretary of Agriculture to adjust foreign import quotas, impose marketing allotments (or other adjustments) on domestic sugarcane and/or sugar beet processors, and/or impose acreage controls on sugarcane and sugar beet growers. Domestic producers and processors of sugarcane and sugar beets would take huge financial risks if they were to expand sugar output in the face of such potential controls and restrictions.

Unlike sugar, the amount of corn used for ethanol production is significant. There is, however, considerable disagreement on the exact magnitude of the impact of the fuel tax credit on corn use. Regardless, it remains an open question as to whether the added corn production that is needed for ethanol production has major polluting effects. If it does have major polluting effects, practices should be changed using the guidelines suggested in this book (Batie and Ervin, this volume; Khanna et al., this volume; Segerson, this volume).

The example of an alternative flexible incentive technology that was discussed in this chapter was shown to generate positive environmental effects—since ethanol is a cleaner burning fuel than gasoline. At the same time, the production of ethanol may well generate negative effects due to the expansion of corn production—an input that is used in the alternative technology. The extent of environmental damage, which depends on the amount of corn that is used for ethanol production, remains an empirical question. The incidence of the positive and negative effects of the alternative technology is very different. As an example, the residents affected by cleaner burning fuels (for example, California) are affected positively while those in corn producing states (for example, Iowa) have to bear the negative externalities (if, indeed, they do occur) that are attributable to increased corn production as a result of increased ethanol production.

REFERENCES

Advincula, C.A. 1992. "Supply Response of the Florida Cane Sugar Industry and Related Policy Implications." Unpublished Master's thesis, Department of Food and Resource Economics, University of Florida, Gainesville, FL (August).

Advincula, C.A., L.C. Polopolus, R.W. Ward, and J. Alvarez. 1992. "Supply Response of the Florida Cane Sugar Industry and Related Policy Implications." Department of Food and Resource Economics, Staff Paper SP 92-26, University of Florida, Gainesville, FL (November).

Alvarez, J. and L.C. Polopolus. 1990. *Sugar and the General Agreement on Tariffs and Trade.* Department of Food and Resource Economics, Sugar Policy Series No. 5, University of Florida, Gainesville, FL (November).

Andreas, M.L. 1990. "Testimony on Behalf of Archer Daniels Midland Company." Paper presented before the U.S. House Committee on Ways and Means, Public Hearing on Certain Tax and Trade Alcohol Fuel Initiatives, Washington, DC (1 February).

Bhatti, M.Y. and J.F. Yanagida. 1990. "Factors Affecting Farmers' Sugarcane Supply Response in Pakistan." *Pakistan Journal of Agricultural Social Sciences* 4: 67–72.

Evans, M.K. 1997. *The Economic Impact of the Demand for Ethanol.* Study prepared for the Midwest Governors' Conference, Lombard, IL (February).

Feedstuffs Staff Editor. 1998. "Congress Passes Ethanol Extension of Tax Credit." *Feedstuffs, the Weekly Newspaper for Agribusinesses* 70(22): 7.

Florida Audubon Society et al. Versus Nicholas F. Brady, Department of the Treasury et al. 1991. Civil Action No. 90–1185JSP, U.S. District Court, Washington, DC (June).

Gemmill, G.T. 1976. "The World Sugar Economy: An Econometric Analysis of Production and Policies." Ph.D. Dissertation, Department of Agricultural Economics, Michigan State University, East Lansing, MI.

Gressel, J.P. 1984. "United States Sugar Policy and Ethanol Production from Sugarcane in Florida." Unpublished Master's Thesis, Department of Food and Resource Economics, University of Florida, Gainesville, FL.

Hauser, R.J. and J.B. Braeden. 1982. "Economic Impacts of Corn Utilization in the Sweetener and Fuel Alcohol Industries." Paper presented at a Committee on Agriculture Hearing on the Surplus Agricultural Commodities Disposal Act, U.S. House of Representatives, Serial No. 97-BBBB: 72–73 U.S. Government Printing Office (15 September).

Hertzmark, D., S. Flaim, D. Ray, and G. Parvin. 1980. "Economic Feasibility of Agricultural Alcohol Production within a Biomass System." *American Journal of Agricultural Economics* 62(5): 965–971.

Hertzmark, D., D. Ray, and G. Parvin. 1980. *The Agricultural Sector Impacts of Making Ethanol from Grain*, p. 969. Golden Solar Energy Research Institute, Denver, CO (March).

Kiker, C. and J.P. Gressel. 1986. "Economic Appraisal of Ethanol from Sugarcane: Policy Effects." *International Journal of Energy Systems* 6: 90–95.

Lopez, R.A. 1989. "Political Economy of U.S. Sugar Policies in the 1980s." *American Journal of Agricultural Economics* 71(1): 20–31.

McNew, K. and B. Gardner. 1996. "A Sweet Deal or Just Gas for Corn Farmers?" *Economic View Points* 1(2): 9–11.

Polopolus, L.C. 1996. "World Sugar Markets and U.S. Sugar Policy." Food and Resource Economics Department, Staff Paper 96-7, University of Florida, Gainesville, FL (July).

Polopolus, L.C. and J. Alvarez. 1991. *Marketing Sugar and Other Sweeteners*. Amsterdam, Netherlands: Elsevier.

Polopolus, L. and A. Schmitz. 1991. "The Effect of the ETBE Tax Credit on the Production of Sugar Crops in the United States," in *Florida Audubon Society et al. Versus Nicholas Brady et al.*, U.S. District Court, Department of the Treasury, Washington, DC (25 June).

Ronnigen, V.O. and M.D. Praveen. 1989. *How Level Is the Playing Field? An Economic Analysis of Agricultural Policy Reforms in Industrial Market Economies*. Foreign Agricultural Economics, Report No. 239, ERS/USDA, Washington, DC (December).

Schmitz, A., R. Allen, and G.M. Leu. 1985. "The U.S. Sugar Program and Its Effects," in Gordon C. Rausser and Kenneth R. Farrell, eds., *Alternative Agricultural and Food Policies and the 1985 Farm Bill*. Giannini Foundation of Agricultural Economics, University of California at Berkeley, Berkeley, CA, and Resources for the Future, Washington, DC. San Leandro, CA: Blanco Publishers.

Schmitz, A. and D. Christian. 1990. "U.S. Sugar." Paper prepared for the U.S. State Department Conference on Sugar Markets in the 1990s, Washington, DC (May).

USDA/ERS (U.S. Department of Agriculture, Economics Research Service). 1989. *Feed Situation and Outlook*. USDA/ERS, Monthly Report, Washington, DC (February).

_____. 1990. *The 1990 Farm Act and the 1990 Budget Reconciliation Act*. USDA/ERS, Miscellaneous Publication No. 1489, Washington, DC (December).

_____. Various issues. *Sugar and Sweetener Outlook and Situation Report*. USDA/ERS, Monthly Report, Washington, DC.

_____. 1989. *Sugar and Sweetener Report*. USDA/ERS, Monthly Report, Washington, DC. (December).

Wall Street Journal. 1997. "Futures Prices and Cash Prices," p. C-12. New York, NY (21 May).

Womack, A., S. John, W. Meyers, J. Matthews, and R. Young. 1981. "Impact Multipliers of Crops: An Application for Corn, Soybeans and Wheat," in *Applied Commodity Price Analysis and Forecasting*. Ames, IA: Iowa State University Press (12–13 October).

11 PUBLIC VERSUS PRIVATE LAND OWNERSHIP TO PRESERVE WILDLIFE HABITAT

Fritz M. Roka and Martin B. Main
University of Florida, Immokalee, FL

Conversion and fragmentation of native habitats are considered the primary threats to endangered wildlife. While a purely market-based economy does not compensate private landowners for the public environmental services that their property provides the general public has expressed a willingness to pay for environmental services. This willingness to pay has been expressed through federal and state government agencies that buy and manage environmentally sensitive lands. Public ownership insures the provision of environmental services from a particular property. At the same time it provides the landowner with a measure of compensation for relinquishing his or her property rights. Annualized costs for acquisition and maintenance in southwest Florida were estimated to be $30 per acre. Limited public monies and a desire to improve flexibility of conservation programs have led to discussions of creating conservation leases for private landowners. Involving private landowners in a habitat conservation program has the opportunity to enhance the overall flexibility of achieving environmental goals. True flexibility of incentive payments to private landowners will depend upon the linkage of payments to habitat quality and the measurement of the level of environmental services provided by individual landowners.

INTRODUCTION

The majority of wildlife habitat in the United States is located on private lands, most of which support some form of compatible agricultural activity (Noss et al., 1997). Increasingly, however, habitat essential to threatened and endangered wildlife on private lands is rapidly being transformed into more profitable uses—a trend that is particularly evident in southwest Florida (Mulkey et al., 1997). In southwest Florida, state and federal agencies have aggressively pursued public ownership as a strategy to ensure habitat protection. Questions exist as to whether public ownership is the only feasible and desirable approach by which to protect and manage wildlife habitat, or whether more flexible and efficient mechanisms on private lands could exist through negotiated conservation leases.

This chapter explores the economic arguments behind public and private land ownership in the environmental policy context of conserving and managing wildlife habitat. It is argued that financial compensation to private landowners in return for conserving and managing wildlife habitat would provide a more flexible approach that complements conservation efforts on existing public lands. In this manner, the environmental goals associated with habitat protection could be attained at lower cost and could provide for the adoption of new technologies. However, as Batie and Ervin (this volume) point out, flexible incentives are means to an end and not ends in themselves. The long-term sustainability of habitat quality in south Florida will depend on a mixture of public and private land ownership. Costs of public ownership and management of land are presented as a starting point for comparing the economic efficiency of public versus private land ownership in order to achieve wildlife habitat protection.

LAND USE TRENDS IN SOUTHWEST FLORIDA

Since the 1500s when the Spanish occupied the Florida peninsula, the native habitats of southwest Florida have supported a vibrant cattle industry (Akerman, 1976). The conversion and fragmentation of this native habitat (comprised of pine flatwoods, cypress swamps and wetland prairies) are considered the primary threats to wildlife in southwest Florida. Since the turn of the century, the area supporting upland pine communities has declined by an estimated 88 percent (Mazzotti et al., 1992; Noss et al., 1995). More recently, increased economic incentives for row crops and citrus, along with a growing population, are responsible for a rapidly changing landscape. Between 1982 and 1992, total farm acreage in southwest Florida remained fairly constant between 44 percent and 47 percent of total land area (table 11.1). However, the mix of farmland uses has changed dramatically. During this period, open rangeland decreased by more than 190,000 acres. This was offset, in large part, by an increase in vegetable and citrus production by almost 150,000 acres. Perhaps the greatest threat to wildlife habitat in southwest Florida is its rapid human population growth, particularly within the coastal counties. This has created substantial economic incentives for land to be purchased and converted into urban developments (Mulkey et al., 1997).

Preserving native habitats for the protection of endangered species in south Florida is an issue currently receiving national attention. At the present time, thirteen mammalian, 25 avian, and 11 reptilian and amphibious species that depend on native habitats in southwest Florida are listed as endangered, threatened, or as species of special concern (Mazzotti, 1991). Of particular concern is the Florida panther (Puma concolor coryi), which is considered to be one of the most endangered species in the nation. It is estimated that only 30 to 80 Florida panthers survive today. The majority of these are believed to reside in southwest Florida

TABLE 11.1 Land Use Trends in Southwest Florida

	1982	*1992*	*Percent Change*
Land in Farms (percent of land area)	1,643,102 (47 %)	1,535,700 (44%)	-6
Harvested Cropland	163,067	312,601	+92
Pasture/Rangeland	1,034,980	844,013	-18
Number of Farms	1,152	1,580	+37
Average acres per Farm	1,426	972	-32

Source: Mulkey et al. (1997).

(Maehr and Cox, 1995). As part of the Florida Panther Habitat Preservation Plan, 926,300 acres of privately owned land in southwest Florida have been designated as priority habitat essential to the survival and recovery of the panther (Logan et al., 1994). The environmental policy objective of the Plan is to conserve priority habitat in private ownership and to increase ecological continuity with areas currently protected under public ownership. Since 1994, when priority habitat was designated for the Florida panther, the state or federal government has purchased only about 10 percent of these areas. It remains unclear how the protection and management of the remaining acreage of designated priority habitat will be ensured. It is also uncertain how these conservation efforts will affect the rights of private landowners and the long-term viability of the agricultural industry in southwest Florida.

Two conflicting views have emerged over how best to protect habitat on private land for panthers and other wildlife. One view supports the public purchase of privately owned ecologically sensitive lands by federal, state and local agencies. Another view argues for private ownership of land with financial incentives to conserve and manage wildlife habitat. This latter view argues that relying solely upon public land ownership to conserve and manage an additional 1 million acres in southwest Florida undermines existing agricultural industries and, given limited public treasuries, is economically unattainable.

THE ENVIRONMENTAL PROBLEM AND THE FIRST-BEST SOLUTION

Private landowners make land-use decisions by weighing their preferences with the potential economic returns of various land-use alternatives. When the expected stream of returns from one alternative is greater than the cost of converting from the current land-use system, private landowners can be expected to change their land-use patterns. Economic returns in a market economy accrue to those commodities with a well-defined set of property rights. Cattle, tomatoes, citrus and urban developments generate income to landowners because these commodities are rival and exclusive. With a complete set of property rights, market mechanisms can be used to make decisions as to how to best allocate resources to commodity production, which includes the conversion of native habitat into other uses.

Khanna et al. (this volume) cites the lack of well-defined property rights for environmental services as a reason for overexploitation of land. In this case, loss of native habitat to intensive agricultural and urban development stems from the fact that the private landowner views the environmental services of his or her property as public goods, making it unlikely that they will receive financial payments for these services. Hence, the landowner has little incentive to manage property in such a way as to insure long-term availability of wildlife habitat. Within the theoretical framework outlined by Khanna et al., a first-best solution toward the conservation of native habitats would be for landowners to receive an annual payment equal to the annual marginal benefit of the environmental services provided by the native habitat that they hold. If a landowner takes into account the marginal benefits of environmental services provided by their land holdings, the opportunity cost of converting native habitats to other uses would increase. The resulting land-use patterns would reflect an optimal mix of residential, agricultural, commercial and environmental services.

Several obstacles prevent the adoption of a first-best policy. First, the landscape and resulting services are heterogeneous. For example, cypress swamps provide a different mix of ecological services than pine flatwoods do. Second, multiple-dimensions to the environmental services are provided by native habitats. In addition to supporting wildlife, native habitats in south Florida provide areas for water conservation and retention, water quality enhancement, recreation, and, for some, spiritual enrichment. Third, while property owners may assert ownership rights over native habitat within their property boundaries, they cannot claim ownership of the wildlife that depends upon this habitat, or that crosses their property. Fourth, there is incomplete information and uncertainty with respect to evaluating the quality of wildlife habitat. Habitat quality depends on a number of ecosystem processes that are just beginning to be studied. Wildlife ecologists also believe that connectivity among habitats is an important feature for overall habitat quality (Meffe and Carroll, 1994; Noss et al., 1997). Connectivity allows animal movement within a whole geographic region and insures against the isolation of breeding populations. If agricultural production, urban development and roadways impede the movement of animals or other ecological processes (such as the movement of water and fire), then the habitat value of a property will be adversely affected by commercial development within neighboring land parcels.

Direct regulation over wildlife habitat has been attempted through various means, the most notable of which is the Endangered Species Act (ESA). Enacted in 1973, the ESA effectively transferred property rights from landowners to endangered wildlife (Goldstein, 1996). The presence of an endangered species on private land has become a financial liability because future development options have been restricted. The ESA does not offer any compensation to landowners who suffer losses from depreciated property values (Rohlf, 1989). Thus, the ESA has created a disincentive to private landowners to either disclose information on wildlife or to maintain high quality habitats that might attract endangered wildlife onto their property.

PUBLIC LAND OWNERSHIP

Land acquisition by public agencies to preserve wildlife habitat is an alternative policy to direct regulation. Fee-simple purchases transfer the complete bundle of property rights to the public interest with compensation to the prior landowners. Loomis (1993) argues for public ownership of wildland resources on the grounds of improved economic efficiency. A public land manager would recognize the nonmarket environmental values and, therefore, would supply an increased amount of public services from the wildlands. Offsetting these benefits would be the potential negative consequences to local and state economies caused by the loss of agricultural, mining, and timber businesses as a result of the transfer (Evans and McGuire, 1996).

The fee-simple purchase of private lands by government agencies or by non-profit organizations is a public policy that has been widely adopted in Florida to preserve the environmental amenities of native habitat (CARL, 1996). In southwest Florida, more than 655,000 acres, or just under 19 percent of the total area, are under federal or state ownership. Federal agencies, including the National Park Service and the U.S. Fish and Wildlife Service, account for more than 72 percent, or 474,913 acres of these public lands. Since 1974, there have been three state-sponsored land acquisition programs in Florida: the Environmentally Endangered Lands (EEL) bond program, the Conservation and Recreation Lands (CARL) program, and the Preservation-2000 program. As of 1995, these programs have successfully acquired nearly 900,000 acres of environmentally sensitive lands at a cumulative acquisition cost of $1.15 billion (CARL, 1996). When, and if, all land currently scheduled for acquisition is attained, federal and state land holdings will comprise more than 30 percent of southwest Florida.

When evaluating the economic costs and benefits of fee-simple purchases, it is important to recognize that the costs of public ownership do not stop with land acquisition. Ecosystems—whose natural processes have been disrupted by drainage, invasion of exotic plants and animals, intrusion of external diseases, poaching and other illegal and legal human activities—require active management (Meffe and Carroll, 1994). It is reasonable to assume that the acquisition of private lands would become counterproductive if these lands were to be acquired and then poorly managed due to limited resources. Innovative and flexible approaches to conserving and managing native habitat—for example, the development of economic incentives to private landowners for the provision of these services—may become an attractive alternative to fee-simple purchases. Documenting public expenditures on land acquisition and management can serve as a beginning basis for valuing management activities of private landowners that conserve and maintain environmental services on their properties.

Costs of Public Ownership in Southwest Florida

Data were collected on four publicly owned properties in southwest Florida to evaluate the costs of purchasing and managing public lands. The information for these case studies was obtained through reviews of annual budgets and of personal interviews with the respective public land managers. The four sites included Big Cypress National Preserve (BCNP), Florida Panther National Wildlife Refuge (FPNWR), Fakahatchee Strand State Preserve (FSSP), and Picayune Strand State Forest (PSSF). These sites, which are contiguous parcels in Collier County, are part of the greater Big Cypress watershed ecosystem and support similar habitats. This includes approximately one-half of the surviving population of Florida panthers (Maehr and Cox, 1995). The total area of these properties, both actual and projected, encompasses approximately 1 million acres (table 11.2). The general description, managing agency, mission statements and types of activities promoted within the site boundaries of each parcel are summarized in table 11.3.

Ecosystem preservation, restoration and management were common themes among the four land managers interviewed. More importantly, there is a common philosophy that the ecological health of the respective sites depends upon land-use practices outside the property boundaries. This common philosophy has translated into an exceptional degree of cooperation among the four managers interviewed, who are striving for consistent management practices over the broader ecosystem. For instance, FPNWR and BCNP contribute and absorb expenses associated with fire management on the adjacent FSSP and PSSF. In return, both the FSSP and PSSF contribute available manpower and machinery when needed to assist in fire management on the FPNWR and BCNP. The four sites also pursue joint research efforts that include monitoring water quality and quantity as well as wildlife inventories.

From these examples, it becomes clear that the cost of public ownership is influenced directly by the type and quantity of services that it provides. Costs of public ownership include purchase and annual management. Land purchase costs, operating budgets and staff numbers for each site are summarized in table 11.4. Land purchase costs ranged from $2,500 per acre in the western portion of PSSF (close in proximity to the city of Naples, Florida) to $315 per acre in the more remote FSSP (CARL, 1996). Land acquisition costs for the combined properties averaged $480 per acre. Assuming a nominal interest rate of 5 percent and the infinite life of public ownership, the annualized average cost of land purchase was $24 per acre.

Land management costs in south Florida are primarily related to fire, hydrology, removal of invasive exotics, and law enforcement to minimize poaching and other destructive human activity. An annual operating budget serves as one measure of management costs. For established properties, labor is the most significant cost item, ranging between 70 and 80 percent of the total operating budget (table 11.4). If management costs are calculated solely with respect to site-property boundaries, per-acre management cost varies widely, from $2.59 in FSSP to almost $28 in the FPNWR. However, for the sites considered in this study, this method would not accurately reflect true costs because many resources are shared across property-boundaries. For example, fire crews from the FPNWR conduct most of the prescribed burning program on the FSSP. Law enforcement and exotic plant control are other activities for which resources are shared among multiple sites. Therefore, combining the operating budgets for the four contiguous parcels would more accurately reflect per-acre management costs. During the 1997 fiscal year, the com-

TABLE 11.2 Projected and Acquired Public Land by County in Southwest Florida

County	Area	Total County Area acres	Acquired Area acres as of 1995	Projected Area acres	Projected Area % county area
Charlotte	All Inclusive	441,600	34,874	65,751	14.9
Collier	Big Cypress		385,157	575,042	
	Panther Refuge		23,400	26,400	
	Fakahatchee		60,334	74,374	
	Picayune		6,000	59,000	
	Other Public Land		101,587	146,770	
	Total Collier County	1,276,160	576,478	881,586	69.1
Glades	All Inclusive	488,320	4,585	51,316	10.5
Hendry	All Inclusive	744,320	20,235	38,373	5.2
Lee	All Inclusive	519,457	19,487	31,280	6.0
		3,469,857	655,659	1,068,306	30.8

Source: CARL (1996).

TABLE 11.3 Summary of Description, Mission Statements and Allowable Activities at Four Case Study Sites

	Big Cypress National Preserve	*Florida Panther National Wildlife Refuge*	*Fakahatchee Strand State Preserve*	*Picayune Strand State Forest*
Federal/State	Federal	Federal	State	State
Agency	National Park Service	U.S. Fish and Wildlife Service	Florida Dept. of Environmental Protection	Florida Division of Forestry
Established	1974	1989	1974	1985
Mission	Balanced Multiple Objectives	Preservation Applied Research	Preservation	Balanced Multiple Objectives
Habitat Management	Active Program	Active Program	Active Program	Active Program
Public Access	Controlled	Denied and Enforced	Limited Access	Uncontrolled
Commercial Activities	Permitted with Controls	Denied	Denied	Permitted
Environmental Education	Actively Promoted	Actively Promoted	Actively Promoted	No Programs in Place
Research	Active Support	Active Support	Passive Support	Passive Support

Source: Interviews with land managers.

TABLE 11.4 Operating Budgets and Recent Land Costs of Four Properties Managed by Public Authorities

Name	Agency[a]	Project Area (acre)	Recent Purchase Price	1997 Operating Budget	1997 Labor	1997 Expenses	1997 Fire Control	1997 Staff
Big Cypress [b]	USNPS	729,000	$530/acre	$4,250,890	$3,400,712	$850,178[c]	$495,748	74
Panther National Wildlife Refuge	USFWS	26,400	$420/acre	$737,300	$506,957[d]	$230,343	$292,200	13
Fakahatchee	FDEP	74,374	$356/acre	$192,800	$148,800	$44,000	$0	6
Picayune [e]	FDOF	59,000	$614/acre	$184,523	$68,748	$115,725	$0	1
Total		888,774	$480/acre	$5,365,513				94

[a]U.S. National Park Service (USNPS), U.S. Fish and Wildlife Service (USFWS), Florida Department of Environmental Protection (FDEP) and Florida Division of Forestry (FDOF).
[b]Information presented is for the entire FSSP, 575,042 acres in Collier County and the balance of the acreage in Monroe County.
[c]Estimated to be 20 percent of total operating budget (Hibbard, 1997).
[d]Includes non-fire personnel salaries ($256,957) and the estimated salaries of the fire crew ($250,000).
[e]PSSF is the combination of two acquisition projects—Southern Golden Gates Estates (41,000 acres) and Belle Meade (18,000 acres).

Source: CARL, 1996.

bined budgets of the four property managers totaled $5.35 million. The combined area under management totaled 890,000 acres, or a management cost of approximately $6 per acre. Adding the management and ownership costs, the annual public expenditure on land in the study area is roughly $30 per acre. It is important to emphasize that the $30 per-acre cost buys all the management objectives outlined in Table 11.3. In other words, the $30 per-acre cost is a weighted average of maintaining recreational amenities, applied research activities and law enforcement, as well as preserving wildlife habitats at each site.

Public land acquisition policy ensures that environmental objectives are prioritized and addresses the controversy associated with the ESA regarding compensations for restrictions on their private property rights. A larger question remains as to whether public ownership is the only, or the most cost-effective policy by which to conserve wildlife habitat. Public ownership represents a command-and-control approach toward preserving wildlife habitat. Once a property is under public ownership, land-use and management choices are limited within the context of environmental objectives and the economic benefits of alternative land uses may be ignored. The principle exception to agricultural exclusion on publicly owned lands is cattle grazing. Denying intensive agricultural activities implies that public land managers perceive a uniform per-acre marginal environmental value and a uniform per-acre damage function for conversion of native habitat into more intensive agricultural activities. However, if areas exist within a given property that could be commercially developed without the significant impairment of the property's environmental functions, then the overall cost of preserving environmental services within the landscape could be reduced. Deriving an efficient mix of commercial and preservation activities would require a more flexible approach that would depend upon the participation of private landowners.

PRIVATE OWNERSHIP WITH PRESERVATION INCENTIVES

Interest in developing more flexible policies toward habitat conservation is largely motivated out of a realization that public resources to buy and manage properties are limited. The goal of flexible incentives is to reduce the cost of achieving habitat protection (Batie and Ervin, this volume). The increase of flexibility would include private landowners in the conservation process and would capitalize on the specialized knowledge they have of their own properties. This would allocate resources more efficiently for the protection of habitat quality.

Privatization of habitat management is a concept being explored in southwest Florida. Landowners have proposed the idea of a conservation lease, in which the public would lease the landowner's conservation rights over wildlife habitat (Evans, 1995). Unlike conservation easements, the landowner would retain complete ownership rights. The proposed concept is similar to the U.S. Department of Agriculture's Conservation Reserve Program and the Wetland Reserve Program in that participating landowners forego further agricultural development during the term of the lease. In return, the landowner would receive an annual payment. Payment could be in the form of cash, tax credits, regulatory relief, or some combination of financial incentives. Interest in this proposal has increased to the point in which an active coalition of landowners, environmental activists and government agencies has been formed to work out the substantive details of a lease agreement and to promote its adoption.

Within the Batie and Ervin (this volume) typology of flexible incentives, the conservation lease described above could not be characterized as a flexible incentive because participating landowners would be precluded from expanding existing commercial enterprises during the lease period. Batie and Ervin (this volume) recognize, however, that subsidizing conservation efforts on private lands would add administrative flexibility. If $30 per acre was a reasonable estimate of annual public land ownership costs, then landowners who would accept annual lease payments of less than $30 per acre would represent a more cost-efficient solution to habitat conservation. In addition, a conservation lease may stimulate greater long-run economic flexibility by encouraging the development of alternative enterprises that are currently not being fully exploited. For example, a conservation lease may provide the seed money for eco-tourism. Already, several landowners have demonstrated that eco-tourism can be a successful business venture in south Florida. The Babcock Ranch in Charlotte County charges guests nearly $20 for a 90-minute motorized tour through its back country with an interpretive guide who explains its habitat, wildlife and local history. This type of commercial endeavor ensures the conservation of native habitats that might otherwise be converted into less environmentally friendly types of land use.

The enhancement of flexibility through incentive payments for habitat protection on privately owned property would depend on the linkage of payments to habitat quality and the adjustment of payment levels with changes in habitat quality. Basing incentive payments on habitat quality would ensure that landowners with higher quality habitat would receive prioritized consideration or be compensated with higher lease payments. Furthermore, if incentive payments could adjust to changes in habitat quality, landowners would be better able to weigh the economic values of commercial or agricultural development against the loss of income from conservation leases.

Biologists are developing assessment tools that would enable them to rank the relative ecological performance among various properties (Noss et al., 1997). Although prioritization of criteria may vary among different agencies or for different target objectives, ecological assessments would likely emphasize broad criteria by which to evaluate and prioritize candidate areas for inclusion into conservation lease agreements. For example, utilizing guidelines from Noss et al. (1997), criteria by which to evaluate the habitat for the Florida panther would include: (1) Location—Do the candidate lands occur in areas of known, suspected, or potential use by the Florida panther? (2) Size—What is the size of the contiguous area encompassed by natural areas within the candidate lands? Do these natural areas possess habitats known to provide for the needs of the panther? (3) Fragmentation—What is the patch density of natural areas and the ratio of natural areas to other land-use types in the area under consideration? (4) Connectivity—What is the inter-patch distance among natural areas within the candidate lands? What is the distance of candidate lands to other protected natural areas (public or private) that support panther habitat? Do barriers exist to panther movement among these

areas? (5) Quality—What is the condition of the candidate lands under consideration? Are these habitats in degraded ecological condition as a result of fire sup-
pression, invasion by exotic plants, clear-cutting, overgrazing, pollution, or other
factors? (6) Human Activity—What is the intensity of human activity within and
surrounding candidate lands, and will this negatively affects use or travel among
these areas by panthers? Can an effective habitat management program, such as
the use of prescribed fire, be implemented on the candidate lands? 7) Other ecological attributes (secondary criteria)—Do the candidate lands provide other ecological attributes, such as habitat for threatened and endangered plants and other
wildlife, important hydrological functions or other ecological attributes and services that are important components to the management of the ecosystem.

If the public were to finance habitat conservation, then they would expect public
access, to some degree, regardless if the land were publicly or privately owned.
Public access typically fosters environmental awareness and general support for
conservation efforts (Krakowski, 1997). However, public access may jeopardize
the very resource that is being protected. For example, the FSSP is home to a
number of endangered orchids. According to its public land manager and biologist, the harvest of these flowers, either by poachers or by uninformed tourists,
represents a real danger to preservation of the species (Owens, 1997). In another
example, panthers are known to be solitary and reclusive animals. Tourist traffic
through areas used as den sites could adversely affect their reproductive behavior
or the survival of cubs. Another potential problem with public access is the effect
of tourist traffic on applied research programs by disturbing wildlife patterns or
the physical elements of field experiments. Consequently, limiting public access
may be a desirable attribute under some circumstances.

CONCLUSION

The value of the conservation of native habitats and the maintenance of healthy
ecosystems is related to their long-term physical, chemical and biological functions (Costanza, 1994). While the environmental goods and services found in native habitats are considered public goods and are, therefore, outside direct market
allocation, the general public has expressed a willingness to pay for these services
by granting various government agencies the authority to purchase and manage
environmentally sensitive lands. Expenditures, either through government agencies or non-profit organizations for land acquisition and management, reflect values that the public has placed on these environmental resources. Public ownership
of environmentally sensitive land is a departure from direct land-use regulation. It
spreads the cost of environmental protection beyond private landowners who, until recently, have borne the entire cost of current land-use regulations. There remains a question of whether some of the environmental services, such as wildlife
habitat, could be delivered more efficiently under private land ownership. Limited
public resources are pushing policies toward more flexible approaches in which
private landowners assume more decision-making responsibility.

A first-best solution that would ensure flexibility and an optimal mix of commercial and environmental land-use activities suggests paying landowners the

marginal value of habitat services from their properties (Khanna et al., this volume). Incomplete information about the public's willingness to pay for habitat conservation and ecological measures of habitat quality hinder the adoption of a first-best solution. One objective of this chapter was to develop some insights, based upon current costs of managing wildlife habitat at sites now under public ownership, into the public's willingness to pay. Our findings reveal that, based upon data from an area encompassing more than 474,000 acres and managed by four separate public agencies, the average annual cost of the acquisition and management of land in southwest Florida is $30 per acre. This estimate reflects the combined value for all public services, which includes public accesses, that are being provided in the study region. Refinement of this value could serve as a basis for incentive payments to secure conservation leases with private landowners. An effective flexible incentive scheme for private landowners will depend upon the development of a measure of habitat quality (see Batie and Ervin, this volume) so that environmental managers and private landowners can more efficiently target public resources and can better approximate a first-best solution as suggested by Khanna et al. (this volume).

The Batie and Ervin (this volume) typology asserts that flexible incentives are means to an end and not ends in themselves. As such, the protection of an adequate share of wildlife habitat will require public as well as private land ownership. Habitat conservation under public ownership would provide an insurance policy. Alternatively, conservation leases could motivate landowners to utilize their knowledge of the landscape to protect and enhance habitat resources.

REFERENCES

Akerman, J.A. 1976. *The Florida Cowman, A History of Florida Cattle Raising*, p. 319. Kissimmee, FL: Florida Cattlemen's Association.

CARL (Conservation and Recreation Lands). 1996. *Annual Report*. Florida Department of Environmental Protection, Division of State Lands, Tallahassee, FL.

Costanza, R. 1994. "Valuation of Ecological Systems," in G. Meffe and C. Carroll, eds., *Principles of Conservation Biology*, pp. 449–450. Sunderland, MA: Sinauer Associates, Inc.

Evans, C. 1995. *A Landowners' Strategy for Protecting Florida Panther Habitat on Private Lands in South Florida: a Project Report*, pp. 64. Boca Raton, FL: Florida Stewardship Foundation, Inc.

Evans, C. and J. McGuire. 1996. *The Contribution of Agriculture to Collier County, Florida*, p. 115. Boca Raton, FL: Florida Stewardship Foundation, Inc.

Goldstein, J. 1996. "Whose Land is it Anyway? Private Property Rights and the Endangered Species Act." *Choices* Second quarter: 4–8.

Hibbard, W. 1997. Personal Communication. Superintendent, Big Cypress National Preserve, U.S. National Park Service, Ochopee, FL (21 March).

Krakowski, J. 1997. Personal Communication. Refuge Manager, Florida Panther National Wildlife Refuge, U.S. Fish and Wildlife Service, Naples, FL (22 March).

Logan, T., A.C. Eller, R. Morrell, D. Ruffner, and J. Sewell. 1994. *Florida Panther Habitat Preservation Plan, South Florida Population*, p 130. Report prepared by the Florida Panther Interagency Committee, Atlanta, GA.

Loomis, J.B. 1993. *Integrated Public Lands Management*, p. 474. New York, NY: Columbia University Press.

Maehr, D.S. and J.A. Cox. 1995. "Landscape Features and Panthers in Florida." *Conservation Biology* 9: 1008–1019.

Mazzotti, F.J., L.A. Brandt, L.G. Pearlstine, W.M. Kitchens, T.A. Obreza, F.C. Epkin, N.E. Morris, and C.E. Arnold. 1992. *An Evaluation of the Regional Effects of New Citrus Development on the*

Ecological Integrity of Wildlife Resources in Southwest Florida, p. 188. Final Report of the South Florida Water Management District, West Palm Beach, FL.

Mazzotti, F.J. 1991. *South Florida Wildlife in Danger of Extinction*. Cooperative Extension Service, Report No. SS-WIS-07, p. 9. Institute of Food and Agricultural Sciences, University of Florida, Gainesville, FL.

Meffe, G.K. and C.R. Carroll, eds. 1994. "Management to Meet Conservation Goals: General Principles," pp. 307–335, in *Principles of Conservation Biology* Sunderland, MA: Sinauer Associates, Inc.

Mulkey, W.D., R.L. Degner, S. Gran, and R.L. Clouser. 1997. *Agriculture in Southwest Florida: Overview and Economic Impact*. Food and Resource Economics, Staff Paper No. SP 97-3, p 42, Institute of Food and Agricultural Sciences, University of Florida, Gainesville, FL (April).

Noss, R.F., E.T. LaRoe III, and J.M. Scott. 1995. *Endangered Ecosystems of the United States: A Preliminary Assessment of Loss and Degradation*, p 58. Biological Report No. 28, U.S. Department of the Interior, National Biological Service, Washington, D.C.

Noss, R.F., O'Connell, M.A., and D.D. Murphy. 1997. The Science of Conservation Planning, p 58. Washington, DC: Island Press.

Owens, M. 1997. Personal Communication. Biologist, Fakahatchee Strand State Preserve, Florida Department of Environmental Protection, Copeland, FL (March).

Rohlf, D.J. 1989. *The Endangered Species Act, a Guide to its Protections and Implementation*, p 207. Stanford, CA: Stanford Environmental Law Society.

12 Environmental Policy and Technology Adoption in Animal Agriculture

Patricia E. Norris and Amy P. Thurow
Michigan State University, East Lansing, MI
Texas A&M University, College Station, TX

The increasing number and concentration of animals in beef, swine and poultry production units has led to heightened concerns over the environmental and nuisance impacts of such operations. Whether flexible incentives can be used effectively to reduce such environmental risks requires consideration of the economic and institutional factors driving the structural changes in animal agriculture. The design of environmental policy to address these concerns is complicated by disagreement over whether these animal operations are point or nonpoint sources of water pollution. The multidimensionality of environmental and nuisance concerns associated with animal agriculture suggest that two separate, but interrelated, policy issues exist—the location of these large animal operations and the management of the manure they generate. Policy responses incorporating flexible incentives are likely required at federal, state and local levels of government.

INTRODUCTION

The increasing size and concentration of animals in beef, swine and poultry production units has led to heightened concerns over the environmental and nuisance impacts of such facilities. Intense debates on how to address such concerns are being conducted in a policy setting characterized by a growing interest in more flexible environmental policy tools. An important question is whether flexible incentives can be used effectively to reduce environmental risks associated with animal production. This issue is complicated by ongoing confusion about whether animal operations are point source (PS) or nonpoint sources (NPS) of pollution. In addition, the development of any incentive program is complicated by increased vertical coordination in the animal industries and the associated implications for ownership of animals and responsibility for environmental protection.

First, we describe the U.S. animal agriculture industry, focusing on changes in the structure of the industry and the growing concentrations of large, intensive animal operations. Second, the concerns about the environmental impacts of concentrated animal production are presented. Third, the economic concepts that underlie the potential for environmental problems in animal agriculture are reviewed. Fourth, the current policy setting is described. Finally, opportunities for using flexible incentives to encourage adoption of environmentally sensitive technologies in animal agriculture are considered.

THE AGRICULTURAL SETTING

Animal agriculture in the United States is becoming industrialized. In production, industrialization means that specialized facilities are tended by specialized labor using routine methods (Rhodes, 1995). More generally, agricultural industrialization is characterized by larger farms, increased vertical coordination in production and processing, and regional shifts in location.

Larger farm size means a larger number of animals. Economies of size and scale in production have been a major factor driving the movement toward larger farms. New technologies, including improved disease control and feed programs, combined with a move toward confined operations and greater fixed investments, have led producers to increase output and lower per-unit costs of production. The increased size of U.S. hog and dairy operations over the past 10 years is illustrated in figures 12.1 and 12.2. An increase in animal density (more animals per acre of land) has been associated with the increase in farm size.

Vertical coordination in production and processing is also an important component of industrialization. The increasing use of contracting in production and marketing in the animal industries has been documented in several studies (USDA/ERS, 1996; Rhodes, 1995). Contracting offers opportunities for reduced transaction costs, increased responsiveness to consumer demand, improved quality control, risk shifting and risk reduction, and production efficiencies from specialization (Martin and Norris, 1998). Generally, production contracts mean that contractors control feed and animals, but contract growers own the production facilities and are responsible for manure management.

Shifts in location associated with the industrialization of animal agriculture have been rapid and region-specific. Percentage changes in December 1 hog and pig inventories by state, from 1987 through 1997, are shown in figure 12.3. Other animal industries have seen similar shifts. Before 1950, poultry production was distributed evenly across the United States as a backyard enterprise in the counties surrounding metropolitan areas, but by 1971 it was concentrated in the South (Martin and Zering, 1997; Reimund et al., 1981). In 1960, the Corn Belt and the Central Plains regions produced 56 percent of fed cattle, and the High Plains region produced 5 percent of fed cattle. By 1983, 20 percent of fed cattle production and processing was in the High Plains region. During this 23-year period, small cattle feeding operations in the Corn Belt and Central Plains regions exited

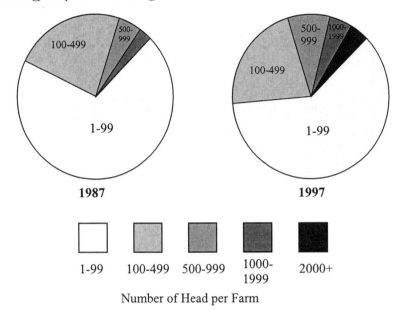

FIGURE 12.1 Change in Number of Operations by Size Category: Hogs and Pigs, 1987 and 1997

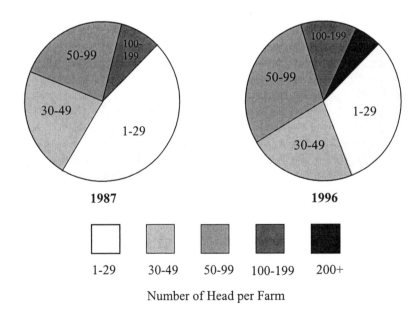

FIGURE 12.2 Change in Number of Operations by Size Category, Milk Cows, 1987 and 1996

FIGURE 12.3 Percent Change in December 1 Hog and Pig Inventory, 1987 through 1997

the industry and feedlots in the High Plains region increased in size (Thurow, 1998).

An increase in clustering is associated with these regional shifts in production and processing. In animal agriculture, there is a propensity for increasingly industrialized, vertically coordinated producers and processors to locate together and/or in close proximity to specialized infrastructure (Pagano and Abdalla, 1994). Many recent environmental conflicts have emerged because clusters of specialized facilities developed in areas traditionally characterized by smaller, diversified crop/livestock farms. These conflicts are expected to intensify when animal facilities are sited in locations with inadequate assimilative capacity for nutrients and/or when nuisance odor is a problem (Pagano and Abdalla, 1994).

ENVIRONMENTAL PROBLEMS IN ANIMAL AGRICULTURE

Threats of pollution and nuisance damages associated with clustering tend to compound over time as the density of animals in a region increases. Over the course of a year, the quantity of nitrogen in manure generated from a 200-cow dairy is the same as that found in sewage from a community of 5,000 to 10,000 people. The phosphorus generated from a 22,000-bird broiler house matches the quantity produced in sewage from a town of 6,000 people (Moffitt, 1995). Production units larger than 200 cows or 22,000 broilers are common to industrialized clusters. High levels of nutrients from swine and poultry production have been implicated in the recent outbreaks of the toxic microorganism *pfiesteria* in coastal North Carolina, Virginia and Maryland (EPA, 1998). There are also health risks from *e.coli* and from high nitrate levels found in ground water (Thu, 1995). In addition to water quality concerns from improper manure handling, odor and flies are often a problem on or near large animal production facilities (Van Horn, 1995; Thu, 1995).

According to the most recent statistics from the U.S. Environmental Protection Agency (EPA), animal feeding operations alone are responsible for 16 percent of the surface water impairment attributed to agricultural practices. This does not include the runoff from farms using manure as fertilizer (EPA, 1998). The EPA relies on reports and monitoring data from state environmental agencies to develop their estimates of pollution contributions by sector and by region (EPA 1993).

The 1972 Clean Water Act (CWA) defines livestock production facilities with 1,000 animal units or more as point source polluters. It defines livestock production facilities with fewer than 1,000 animal units as nonpoint-source polluters.[1] In some national- and state-level assessments of the contributions of animal agriculture to water quality problems, it is not clear whether the assessments describe the PS pollution, NPS pollution, or both. In addition, legal definitions are constantly evolving. For instance, in the *CARE v Southview Farms* case in New York (described later in this chapter), land application of manure was deemed a PS discharge.

Letson and Gollehon (1993) analyzed data from the 1992 Census of Agriculture to develop a national profile to delineate where animal agriculture is concentrated and to identify where nutrients from manure are, or could be, applied as an organic fertilizer to cropland. The central conclusion of their analysis is that large animal feeding operations are less likely to be located near areas of significant cropland acreage than are smaller animal feeding operations. With the trend toward specialized, concentrated animal production, traditional links between manure and cropland have been severed. This growing separation between crop and animal production, in turn, affects the economic pressures associated with adopting alternative manure-management technologies.

ECONOMICS OF LIVESTOCK AND THE ENVIRONMENT

Economics of Clustering

Clustering is a cumulative phenomenon. The establishment of a processing facility, for example, draws increasing numbers of producers. Once sited, producers tend to adopt production technologies that enhance profitability through economies of size. Additional animals are then purchased to generate the revenue to pay for such technological improvements (Outlaw et al., 1993). New entrants are attracted by the positive economic dynamics of a well-established cluster and its allied agribusinesses. Greater production volume from existing facilities and new entrants triggers expansion in processing capacity. The collective economic power of a cluster of livestock producers and processors is buttressed by this self-reinforcing pattern of growth.

For rural communities, this growth in animal agriculture is a two-edged sword (Pagano and Abdalla, 1994). In some areas the growth has been an important source of economic development (Brown, 1993; Jones et al., 1993). However, when the assimilative capacity of air and water is limited, the potential for environmental problems intensifies. Problems are particularly likely if manure management technologies are inappropriately designed or are poorly managed.

Incentives for Technology Adoption

Khanna et al. (this volume) review opportunities for precision technologies in crop production to reduce production costs while reducing sources of pollution. In animal agriculture, adoption of manure management technologies to meet compliance obligations is fundamentally different from adoption of production-enhancing technologies (Purvis and Outlaw, 1995). Most manure management technologies require significant up-front investments that, unlike production-enhancing technologies, do not generate revenue. It is unclear whether animal operations can be redesigned so that innovative technologies achieve improvements in both profits and environmental performance. Such innovation offsets would make compliance with environmental regulations less costly. Batie and Ervin (this volume) conclude that the evidence to support or refute the notion of innovation offsets is incomplete.

Thurow and Holt (1997) discuss the only purported case of innovation offsets in animal agriculture known to these authors. In complying with the Florida Dairy Rule (FAC 17-6.330 through 17-6.337), a subset of large-scale dairy producers in

Okeechobee, Florida, installed new technologies—for example, shades and sprinkler systems—that enhanced milk production and partially offset the costs of the mandatory phosphorus runoff abatement technologies. Thurow and Holt (1997) conclude that, because managers were already making major modifications to their dairies (which caused significant disruption to production activities), savvy managers decided to use this opportunity to make non-mandatory production-enhancing investments concurrent with the compliance-driven construction. Milk production in Okeechobee dairies was not enhanced, nor were costs reduced, by the mandated technologies. Rather, improved revenues were generated from production-enhancing technologies that were not associated with phosphorus runoff reduction. Thus, the Okeechobee investments do not fit a strict definition of innovation offsets (Porter and van der Linde, 1995).

In animal agriculture, pollution source reduction might also be achieved by modifying rations so the nutrient content of manure is reduced. For example, there are feed additives that can reduce the level of phosphorus in swine manure. Using such feeds could be expected to reduce phosphorus management costs for growers. However, in cases of production contracting where feeds are supplied by integrators who bear no responsibility for manure management, there is no incentive for the integrators to pay for the feed additive.

Economies and Externalities

Economies of size and scale in manure management technologies have received considerable attention for the hog sector (Roka et al., 1995; Martin and Zering, 1997) and for the dairy sector (Matulich et al., 1977; Boggess et al., 1991; OTA, 1991; Leatham et al., 1992; Lovell et al., 1992; Outlaw et al., 1993). Research has shown that pollution-averting technologies are not scale neutral and that, generally, large-scale facilities will have lower per-unit costs. This assertion holds except for land-constrained facilities (Martin and Zering, 1997), which may face higher costs of manure disposal if manure must be transported greater distances or if larger amounts of manure must be stored for longer periods of time.

Economies of size in manure management reinforce economies of size in production and further encourage a transition to larger facilities. However, if environmental externalities of improper manure management occur when animal numbers per facility exceed some threshold number (or, alternatively, when the assimilative capacity is reached), then theoretically social production cost curves are actually U-shaped. That is, economies of size mean per-unit production costs decline as animal numbers increase, but only to the point at which external costs are generated. When external costs are generated, the true per-unit costs of production increase accordingly.

Advocates of sustainable agriculture believe that land extensive farms with integrated crop and livestock operations would have significant cost advantages under a policy regime forcing large-scale operations to internalize their environmental costs. For state-of-the-art manure handling technologies in dairy and swine, current data and knowledge are insufficient (too anecdotal) to establish or disprove the U-shaped cost curve hypothesis. The only empirical work known to these authors to support this hypothesis was done by Matulich (1978) for California dairies.

A lack of reliable information about economies of size and scale leads to convoluted discussions about how large and small operations should be treated by environmental policies. The potential economies of size, combined with more modern technologies and potentially higher management skills associated with the newer large-scale operations, suggest that such operations are better equipped to adopt manure management technologies. In fact, investigations in North Carolina found that older, smaller hog operations tend to be greater sources of pollution (Zering, 1998). Currently, large-scale operations are regulated, but smaller operations are not. The potential magnitude of damages associated with discharges from large-scale facilities may explain this regulatory bias. In addition, economies of size in manure management have been cited as reasons why smaller, and often older, facilities cannot afford the costs of meeting the same regulatory requirements as larger facilities.

ENVIRONMENTAL POLICY FOR ANIMAL AGRICULTURE

Current policies that address environmental risks from animal agriculture are concerned primarily with water quality risks and are designed around the distinction between PSs and NPSs of pollution. In animal agriculture, PS operations are regulated, but NPS operations are not. Generally, environmental policies that address animal agriculture have originated at the federal level.

Point Sources

Federal regulation of large-scale animal agriculture started with the CWA of 1972. This act mandated that all concentrated animal feeding operations (CAFOs)—defined as PSs—maintain National Pollutant Discharge Elimination System (NPDES) permits. An NPDES permit requires that the CAFO build and maintain sufficient wastewater storage capacity to accommodate the amount of rainfall expected in a 24-hour period once every 25 years. The permit stipulates a performance standard: No discharges of wastewater (including runoff from rainfall) are allowed from a CAFO into U.S. waters. While specification of a performance standard can be characterized as a flexible incentive, such is not the case in this instance. Federal NPDES permit guidelines also include a technology standard. Design criteria are specified for anaerobic lagoons that hold wastewater and runoff—for example, the impermeability of clay liners and the recommended capacity of the lagoon. Criteria are also stipulated for best management practices (BMPs) to be used when applying manure from CAFOs to cropland.

Since EPA does not have sufficient personnel to issue and enforce individual CAFO permits, the authority to administer NPDES permits for CAFOs has been delegated to state environmental agencies in 42 states. Implementation procedures for NPDES permits and guidelines for applying manure to cropland vary considerably across states (EPA, 1993; Outlaw et al., 1993; GAO, 1995). The EPA critiqued existing NPDES permitting programs in 1993 (EPA, 1993), and the Congressional agriculture committees requested a follow-up study by the General Accounting Office (GAO) in 1995. This study estimated that 1,987 of an estimated 6,600 CAFOs in the United States hold federally administered NPDES permits (GAO, 1995). The remaining CAFOs received permits from state environmental regulatory authorities or do not hold permits at all.

Nonpoint Sources

Nonpoint sources of pollution traditionally have been addressed by encouraging voluntary adoption of BMPs, often with cost sharing assistance offered by federal and/or state programs for further incentive (Ribaudo and Caswell, this volume). The 1996 Federal Agriculture Improvement and Reform (FAIR) Act continues this tradition. The Environmental Quality Incentives Program (EQIP) makes cost sharing available to farmers who implement BMPs. One-half of EQIP funds are earmarked for manure management by livestock operations.

The provision of subsidies, or cost sharing, is potentially a flexible incentive. However, there are questions about the types of manure management practices eligible for cost sharing under the EQIP program. In the past, federal cost-sharing funds for manure management have been limited to a standard set of management practices developed by the U.S. Department of Agriculture, Natural Resource Conservation Service (USDA/NRCS) that satisfy that agency's design and management specifications. Livestock farmers interested in adopting manure management technologies not on the list of practices approved for cost sharing are generally forced, for financial reasons, to choose the USDA/NRCS-approved standard technologies. This is true even where the effectiveness of the alternative technologies has been proven in research trials or through experience in other countries. Cost-sharing programs for manure management practices have served as a *de facto* technology standard. Thus, these programs have created the same risks of inefficiency and high control costs experienced in other industries subject to technology standards for pollution control. The legislation specifies that EQIP should be used to encourage adoption of innovative technologies. However, in some cases, state EQIP Technical Advisory Committees appear to be simplifying the planning and approval process by making EQIP funds available to farmers who agree to adopt the same standard set of manure management practices (Batie et al., 1998). By doing so, they exclude the more innovative farmers from the program and sacrifice the economic opportunities afforded by a potentially more flexible approach.

A key design issue for EQIP is whether large scale operations will be eligible to receive federal cost-sharing funds to help them comply with regulations that they are already obligated to meet under the CWA. As authorized by FAIR, EQIP was developed to help farmers and ranchers address the environmental impacts of their activities. However, cost-share assistance to large livestock operations was forbidden and USDA was charged with defining the term large. As a baseline, the USDA has determined that operations with fewer than 1,000 animal units (the Clean Water Act point source designation) are eligible for EQIP funds. USDA's final EQIP program rules provide states with some flexibility—the NRCS State

Conservationist, in cooperation with the state's EQIP Technical Advisory Committee, can develop state-level criteria for defining the term large (USDA/NRCS, 1997). Critics have charged that such state discretion creates the possibility for 50 different interpretations of the term large, with livestock operations treated differently by different states with respect to their eligibility for EQIP funds. As a result, taxpayer dollars may be used to help corporate farms build pollution controls that the CWA already obliges them to install. Ferd Hoefner, of the Sustainable Agriculture Coalition (as quoted in a *Washington Post* article), expressed concern that, depending upon actions taken by individual states, EQIP could require "U.S. taxpayers to bribe large-scale polluters to obey the law." (WIAA, 1996).

The Coastal Zone Management Act, reauthorized as the Coastal Zone Act Reauthorization Amendments (CZARA), represents an attempt to move NPSs beyond the realm of voluntary BMP adoption. CZARA focuses on NPS discharges in the 29 coastal and Great Lakes states. Each of these states is required to define its coastal zone area. CZARA contains considerably tighter controls for NPSs than those stipulated in the NPS section of the CWA (section 308). Livestock operations with more than 50 animal units are targeted by CZARA. The proposed manure management requirements are similar to those for PSs. The law requires individual states to develop specific guidelines for how the CZARA requirements will be implemented by these NPSs (Morse, 1993). Because of state-level perceptions that CZARA requirements are inflexible and jeopardize existing, successful NPS programs, implementation of CZARA has been slow.

More Flexible State Approaches

In some states, right-to-farm laws are being used to address the environmental concerns associated with animal agriculture. Right-to-farm laws exist in all states and serve to protect farming operations—both crop and livestock, small and large—from nuisance complaints lodged by neighbors. Michigan provides an example of how some states have broadened the role of right-to-farm laws to encompass environmental protection objectives. Its law makes nuisance protection dependent upon farmers' use of generally accepted agricultural management practices. So long as farmers retain flexibility in their choice of management practices to meet environmental goals, then this extension of a right-to-farm law represents a more flexible approach to environmental protection policy. However, the enforcement of Michigan's right-to-farm guidelines becomes an issue only if a complaint is brought against an offending operation. After a complaint has been received, a determination is made as to whether an operation is, in fact, in compliance with the requirements for generally accepted practices.

FLEXIBLE INCENTIVES FOR TECHNOLOGY ADOPTION IN ANIMAL AGRICULTURE

As defined by Batie and Ervin (this volume), flexible incentives are "environmental management tools that specify objectives but allow choices as to response." The successful implementation of flexible incentives depends upon setting clear performance objectives (for example, standards) and may impose substantial transaction costs. Inflexibilities in policies to address environmental problems in animal agriculture arise when programs are based on size of operation, rely on technology standards, or fail to recognize the unique characteristics of an increasingly vertically coordinated industry. Opportunities exist to implement flexible incentives encouraging adoption of lower cost, effective manure management technologies. Before these opportunities are discussed, however, the constraints to flexibility should be identified.

Constraints to Flexibility

Clarification of how size affects the designation of an operation as a PS or a NPS of discharges to water is critical to the success of any environmental policy for animal agriculture. This is key to answering many questions raised by critics of policies that base the stringency of regulation on the size of an operation. Difficulties arise when policies are based on the assumption that a 49 animal unit operation is less risky environmentally than a 50 animal unit operation or that a 999 animal unit operation poses less risk to water quality than a 1,000 animal unit operation. Explicit consideration of assimilative capacity, perhaps through application of mass balance concepts to management of manure for land application, provides a more defendable approach to the size issue.

Policymakers grappling with the size issue are increasingly tempted to rely on ratios of number of animals per acre to reduce the risks of excess nutrients leaving the farm. However, the number of acres of cropland needed for storage or land application of manure (that is, the assimilative capacity) will differ depending on animal species and manure management technologies. Animal-per-acre ratio restrictions may inhibit more innovative producers from developing and adopting new management systems. Again, the application of mass balance concepts, with explicit recognition that different management technologies may impact nutrient content of manures differently, could provide significant flexibility in manure management.

Once a CAFO is designated as a PS based on number of animal units, it is subject to a no discharge performance standard. No other PS discharger regulated by the NPDES program is constrained by such a standard. Because of the inflexibility of this standard, the opportunities for CAFOs to adopt innovative technologies in manure management and manure treatment are limited. For example, CAFOs cannot take advantage of technologies that involve treatment of wastewater so that it satisfies water quality standards for discharge into surface water. Also, despite growing interest in flexible incentives, such as watershed-based effluent discharge credit trading (Batie and Ervin, this volume), CAFOs are prevented from participating in such a program by the no discharge standard. At the root of effluent allowance trading is the opportunity for regulated dischargers to trade control responsibilities to dischargers that have lower control costs. The no discharge standard for CAFOs eliminates them as potential generators of discharge credits (credits generated by reducing discharges below required levels) or as purchasers of discharge credits (credits that allow the purchaser to increase discharges since the increase is offset by the seller's decrease in discharges).

Flexible Incentives in Animal Agriculture

Establishing clear objectives is perhaps the most important requirement for improving flexibility in environmental policies that address animal agriculture. When performance standards differ depending upon the size of the operation, objectives become blurred. Similarly, when regulations require precise technologies, regardless of the effectiveness or the costs of other technologies, standards are sabotaged. The transition to a more flexible policy approach requires the setting of a clear, watershed-based performance standard—such as a standard for nitrogen and/or phosphorus levels in runoff—and enforcing it universally. As an alternative to setting standards for nutrient concentrations in runoff from agricultural operations, state water quality agencies, under EPA direction, are working to establish total maximum daily loads (TMDLs) for rivers, streams and their tributaries in a given watershed. This approach provides an opportunity for establishing a performance standard. Specifically, TMDLs provide a threshold for the ambient concentration of identified pollutants beyond which water quality is degraded.

Enforcement of performance standards is necessary to the integrity of a flexible program. Enforcement costs are included with information and administrative costs as types of transaction costs associated with implementing flexible incentives. These transaction costs can be quite high (Carpentier and Bosch, this volume). In the case of enforcing a watershed-level TMDL for nutrients or limits on nutrient concentrations in runoff from animal facilities, extensive (and expensive) monitoring would be required. However, an alternative to public monitoring is to place the burden of proof for compliance on the individual facility. Current implementation of federal CAFO requirements and most state permitting requirements essentially rely on this approach. Facilities are not inspected unless a pollution occurrence is observed. In order to avoid liability, the facility is required to demonstrate to the enforcement agency that it has satisfied the compliance requirements.

In an increasingly vertically coordinated industry, ownership of animals is often separated from ownership of production facilities. Current regulations place responsibility for manure management with the owners of the production facilities. This limits incentives for animal owners to address source reduction issues such as feed modifications or the assimilative capacity of the local environment. A recent proposal in federal legislation would make animal owners liable for environmental damages attributed to animal manure. This kind of change, combined with clear performance standards and flexibility in satisfying those standards, could mean significant changes in the patterns of adoption of innovative manure management technologies.

Multidimensionality

Khanna et al. (this volume) note that sources of environmental concern often have many dimensions, which complicates policy design. This complication is particularly marked in animal agriculture. In public policy, issues associated with environmental quality and the industrialization of animal agriculture are inextricably linked. During the past 20 years, a disturbingly predictable pattern of negative environmental externalities and community outrage has repeated itself in several states where the clustering of poultry, beef, dairy and swine producers and processors has occurred. (Pagano and Abdalla, 1994; Smith and Kuch, 1995; Thurow, 1998).

The policy objective of both state and federal environmental regulations governing the operation and management of livestock facilities has been proper nutrient manage to protect water quality. Not all complaints against the industrialization of animal agriculture, however, are quelled with assurances of reliable water pollution prevention. Nuisance damages (in particular, odor and flies) provoke neighbors to stage "not in my back yard" (NIMBY) protests. Neighbors and other community members resist clustering of animal agriculture on grounds that it threatens their way of life. They argue that industrialization undermines established rural customs and culture. They also argue that factory farms with corporate profit motives replace multigenerational family farms and, thus, destroy heritage, lifestyle and livelihood. Disputes do not occur just between industrialists and environmentalists, or between industrialists and nonagricultural neighbors. Often, the most vehement opposition to clustering comes from established farmers who fear that they will be put out of business by competition for local resources, competition for market access, and increased input prices. In some cases, bitterness is most extreme against local producers who abandon their traditional operations and become contract producers.

It is common for livestock producers and their neighbors to reach an impasse about whether and how to site a production facility. It is even more common for them to disagree about the requirements for satisfactory coexistence. When disputes cannot be settled informally or through locally staged mediation, the plaintiffs go to court (Vukina et al., 1996). Lawsuits are often characterized by a legal dilemma: the only policy instruments that give neighbors legal standing against CAFOs are the federal NPDES permit (written to assure surface water quality protection) or state regulations on nutrient management (designed to prevent water pollution). Accordingly, the lawsuit is about the adequacy of the CAFO's management to avert water pollution, even if the actual problem is odor, flies, or, more abstractly, the disruption of a way of life. The problem for CAFO managers and for the design of flexible incentives, however, is that such arguments do not send signals that motivate changes in behavior or technology to address the root problems causing resistance to large-scale animal agriculture.

A recent lawsuit in rural New York State illustrates this instrumentation problem. Dick Popps owns and manages Southview Farms, a 2,200-cow dairy farm. Popps' neighbors and local environmental organizations sued him on the grounds that he was guilty of noncompliance with his NPDES permit. Allegations included groundwater contamination affecting drinking water and surface water contamination from dairy effluent. According to the New York Commissioner of Agriculture and Markets, however, "this case was really about odor" (Merrill, 1995). Originally, a local jury decided the case in favor of the dairy farmer—it was not convinced that dairy effluent was responsible for the alleged groundwater and surface water pollution—but in an appeal that ruling was overturned (Merrill, 1993). The U.S. District Court of Appeals (New York District) ruled that a 2,200-cow dairy farm was a PS of pollution. It also ruled that the adjacent fields—in which forage crops were raised using manure from Southview Farms as an organic fertilizer—were also designated as PSs (Merrill, 1995). Popps appealed to the U.S. Supreme Court, but the appeal was not heard (Martin, 1996).

A Local Policy Role

The multidimensionality of environmental concerns associated with animal agriculture suggests that two separate, but interrelated, policy issues exist—the siting of large animal operations and the management of the manure they generate. While federal and state laws have addressed the management issue, siting of livestock operations is clearly a local land use issue. Traditionally, local planning and zoning authorities have resolved land use issues. Given their experience at guiding specific land uses within their jurisdictions to minimize conflicts between incompatible uses, local governments clearly have a comparative advantage in addressing the local impacts of CAFOs.

In those areas where rural zoning is used, local governments have begun to modify zoning ordinances to guide, or in some cases restrict, the siting of livestock operations. Zoning approaches have included the permission of such operations by right, permitting by right but subject to site suitability or management conditions, and requiring operations to apply for special use permits that stipulate specific restrictions or conditions of operation. Rural zoning has not been adopted by local governments in all states where it is authorized. Local communities in Texas, Oklahoma and Florida have resisted adoption of rural zoning. As a result, such communities are often ill equipped to deal with siting issues and, in some states, local governments have pressed for a state policy response to address siting.

Acceptance of local controls on the siting of livestock operations is not universal. Several states (for example, Iowa and Missouri) preclude local governments from using zoning powers to restrict agricultural production practices. There is evidence that state-level actions to restrict the local role can be circumvented when localities are intent upon managing the issue locally. In Pennsylvania, local governments developed an effective patchwork of fairly sophisticated approaches to site livestock operations and to address local nutrient management issues. Under pressure from its agricultural industry, Pennsylvania passed a nutrient management law that, among other things, prohibits local governments from enacting nutrient management requirements more stringent than those included in the state law. In response, local governments abandoned attempts to address nutrient management and turned, instead, to ordinances that address odors, flies, dust and other concerns associated with livestock operations.

The role of local governments in enforcing manure management requirements is problematic. Where local governments have chosen to use zoning authority to regulate management of new facilities, serious questions are raised about their ability to enforce such regulations when they have limited expertise and resources. In Michigan, some local governments require special use permit applicants to comply with the generally accepted agricultural management practices developed under the state's right-to-farm law. Similar requirements could also be included in local ordinances in the event of federally or state enforced performance standards. In any case, there is a role for local involvement in ongoing state and federal efforts to police the management of CAFOs. Successful enforcement requires trust and consistency, which means that there must be a local presence of authority. Inherent in the choice of where to place policy responsibilities for manure management in animal agriculture is the trade-off between the desire to create a level playing field for producers across regions and the need to provide enough flexibility so that local preferences can be articulated in the policy design (Thurow and Holt, 1997).

Targeting Flexible Incentives

The effectiveness of state and federal flexible incentive programs that include local participation can be enhanced by targeting such programs to regions that would produce the greatest benefit per-dollar invested. This type of targeting differs significantly from previously implemented soil conservation and water quality protection programs—for example, the targeted erosion control areas identified by the Soil Conservation Service in the early 1980s or the hydrologic unit area water quality programs of the late 1980s and early 1990s. Recently, the USDA/NRCS State Conservationist in Michigan expressed the belief that the EQIP program would improve upon previous resource conservation programs because it would target funds to those areas identified as critical by county and district USDA/NRCS staff. Whether such targeting will benefit the livestock issue depends upon the criteria used for identifying critical areas. Rather than merely focusing resources on areas that face particular environmental problems or need significant investments in manure management to prevent environmental problems, targeting may be more effective if a broader vision of where funds should be allocated is developed. For example, a critical area could be identified as an area in which the long term viability of the livestock industry can be assured with some degree of confidence.

A litmus test for industry viability that can be used early in establishing targeting criteria is the determination of whether the animal industry is perceived to be valuable by a particular state. A financial commitment on the part of state policymakers to support the livestock industry through research and education is one example of evidence that the industry is valued. In 1994, the Michigan Legislature funded an initiative to revitalize animal agriculture in the state, providing more than $70 million dollars to Michigan State University to modernize and improve research facilities, strengthen research resources and personnel, and expand outreach capacity. One of the catalysts for this grassroots initiative was the recognition that, in a grain surplus state, Michigan farmers could support an animal industry while adding value to locally produced crops.

Without local support for the siting of animal operations or clusters, however, state-level objectives may be stymied. Building upon the experience of local governments in addressing issues of incompatible land uses, while enhancing local expertise and resources, may be another way in which states can target policy efforts. Given that natural features and social preferences can vary widely across a state, general acceptance of animal agriculture, especially clusters of large operations, cannot be expected. Instead, program resources may best benefit the industry and the environment if clusters can be established in areas where rural economies welcome the financial boost and the physical and climatic features make environmental and nuisance concerns less problematic.

The implication for subsidy programs, such as EQIP, is that funds may well be wasted if they are used to install manure management technologies on farms that, because of inadequate local commitment or intense conflict with other land uses, face an increasingly uncertain future. Programs like EQIP could be used both to help offset farm-level costs of compliance with environmental standards and to steer animal production to those areas that are more environmentally and economically suitable. This kind of targeting could ensure that increasingly scarce resources are not squandered on paying for environmental protection from animal agriculture in areas where it may be a waning industry.

CONCLUDING REMARKS

Effective policy design will require more than operation-by-operation decisions on expenditure of limited cost share funds. Flexible policies will be most effective for protecting environmental quality in animal agriculture if they include broader, multi-level institutional collaboration. Important objectives include informing local siting decisions, minimizing restrictions inherent in requiring specific management practices (either by dictating technologies, by funding specific technologies or by enforcing questionable performance standards), and providing support to industry growth in geographic areas in which animal agriculture makes economic and environmental sense.

ENDNOTES

1. In some cases, depending upon how discharges emanate from smaller operations (a function of manure and storm water management), operations with less than 1,000-animal units may be designated as point source.

REFERENCES

Batie, S.S., M.A. Schulz, and D. Schweikhardt. 1998. *The Environmental Quality Incentives Program: Locally Managing Natural Resources.* Department of Agricultural Economics, Staff Paper

No. 98-03, Elton R. Smith Endowment Policy Series, Michigan State University, East Lansing, MI.

Boggess, W.G., J.Holt, and R.P. Smithwick. 1991. *The Economic Impact of the Dairy Rule on Dairies in the Lake Okeechobee Drainage Basin*. Food and Resource Economics Staff Paper No. 91-39, University of Florida, Gainesville, FL.

Brown, D. 1993. *Changes in the Red Meat and Poultry Industries: Their Effect on Non-metro Employment*. Agricultural Economics Report No. 665, USDA/ERS, Washington, DC.

EPA (U.S. Environmental Protection Agency). 1992. *Managing Nonpoint-source Pollution: Final Report to Congress on Section 319 of the Clean Water Act (1989)*. Office of Water, Report No. EPA-506/9-90, Washington, DC.

_____. 1993. *The Report of the EPA/State Feedlot Working Group*. Office of Wastewater Enforcement and Compliance, Washington, DC (September).

_____. 1998. *Draft Strategy for Addressing Environmental and Public Health Impacts from Animal Feeding Operations*. http://www.epa.gov/owm (4 March).

GAO (General Accounting Office). 1995. *Animal Agriculture: Information on Waste Management and Water Quality Issues*. Nutrition and Forestry Briefing Report to the Committee on Agriculture, U.S. Senate, Report No. GAO/RCED-95-200BR, Washington, DC.

Jones, L.L., A.J. Wyse, R.B. Schwart, A.P. Pagano, and R.D. Lacewell. 1993. *Economic Analysis of the Dairy Industry in the Cross Timbers Region of Texas*. Department of Agricultural Economics, Information Report No. 93-3, Texas A&M University, College Station, TX.

Leatham, D.J., J.F. Schmucker, R.D. Lacewell, R.D. Schwart, A.C. Lovell, and G. Allen. 1992. "Impact of Texas Water Quality Laws on Dairy Income and Viability." *Journal of Dairy Science* 75: 2846–2856.

Letson, D. and N. Gollehon. 1996. "Confined Animal Production and the Manure Problem." *Choices*. Third Quarter: 18–24.

Lovell, A., B. Schwart, R. Lacewell, J. Schmucker, D. Leatham, J. Richardson, and G. Allen. 1992. "Is There Financial Life After Mandatory Dairy Waste Systems?" *The Dairyman* 73(12): 22–25.

Martin, J.H. 1996. "C.A.R.E. v Southview Farms: A Review," in *Animal Agriculture and the Environment: Nutrients, Pathogens, and Community Relations*. Proceedings of "Animal Agriculture and the Environment," pp. XXX. Northeast Regional Agricultural Engineering Service Research Bulletin, NRAES-96, Ithaca, NY.

Martin, L.L. and P.E. Norris. 1998. "Environmental Quality, Environmental Regulation and the Structure of Animal Agriculture." *USDA Agricultural Outlook Forum '98*, Washington, DC (February).

Martin, L.L. and K.D. Zering. 1997. "Relationships Between Industrialized Agriculture and Environmental Consequences: The Case of Vertical Coordination in Broilers and Hogs." *Journal of Agricultural and Applied Economics* 29(1): 45–56.

Matulich, S.C. 1978. "Efficiencies in Large-scale Dairying: Incentives for Future Structural Change." *American Journal of Agricultural Economics* 60(4): 642–647.

Matulich, S.C., H.F. Carman, and H.O. Carter. 1977. *Cost-size Relationships for Large-scale Dairies with Emphasis on Waste Management*. Giannini Foundation Research Report No. 324, California Agricultural Experiment Station, University of California, Berkeley, CA (October).

Merrill, L.S. 1993. "No Clear Winners in Southview Farm Manure Suit." *Hoard's Dairyman* (25 August).

_____. 1995. "New York Manure Case Reversal Raises Regulatory Concerns." *Hoard's Dairyman* (10 January).

Moffitt, D.C. 1995. *Animal Manure Management*. Resource Conservation Act, USDA/NRCS Issue Brief No. 7 (p. 6), Washington, DC (December).

Morse, D. 1993. "Here's how the Coastal Zone Management Act will Affect your Dairy." *Hoard's Dairyman* (25 September).

OTA (Office of Technology Assessment). 1991. *U.S. Dairy at a Crossroad: Biotechnology and Policy Choices—Special Report*. U.S. Government Printing Office, Report No. OTA-F-470, Congress of the United States, Washington, DC (May).

Outlaw, J.L., R.B. Schwart, Jr., R.D. Knutson, A.P. Pagano, A. Gray, and J. Miller. 1993. *Impacts of Dairy Waste Management Regulations*. Agricultural Food Policy Center, Working Paper No. 93-4, Texas A&M University, College Station, TX (May).

Pagano, A.P. and C.W. Abdalla. 1994. "Clustering in Animal Agriculture: Economic Trends and Policy," in Daniel E. Storm and Kareta G. Casey, eds. *Balancing Animal Production and the Envi-*

ronment, pp. 192–199. Proceedings of "The Great Plains Animal Agriculture Task Force," Great Plains Publication No. 151, National Cattlemen's Association, Denver, CO.

Porter, M.E. and C. van der Linde. 1995. "Toward a New Conception of the Environment—Competitiveness Relationship." *Journal of Economic Perspectives* 9(4): 97–118.

Purvis, A. and J. Outlaw. 1995. "What We Know About Technological Innovation to Achieve Environmental Compliance: Policy Issues for an Industrializing Animal Agriculture Sector." *American Journal of Agricultural Economics* 77(5): 1237–1243.

Reimund, D.A., J.R. Martin, and C.V. Moore. 1981. *Structural Change in Agriculture: The Experience for Broilers, Fed Cattle, and Processing Vegetables*. Economics and Statistics Service, USDA/ESS Technical Bulletin No. 1648, Washington, DC (April).

Rhodes, V.J. 1995. "The Industrialization of Hog Production." *Review of Agricultural Economics* 17(2): 107–118.

Roenfedt, S. 1995. "Save Money Now." *Dairy Herd Management* 32(8): 26–28.

Roka, F.M., D.L. Hoag, and K.D. Zering. 1995. "Making Economic Sense of Why Swine Effluent is Sprayed in North Carolina and Hauled in Iowa," in K. Steele, ed., *Animal Waste and the Landwater Interface*. Boca Raton, FL: Lewis Publishers.

Smith, K.R. and P.J. Kuch. 1995. "What We Know About Opportunities for Intergovernmental Institutional Innovation: Policy Issues for an Industrializing Animal Agriculture Sector." *American Journal of Agricultural Economics* 77(5): 1244–1249.

Thu, Kendall, ed. 1995. *Understanding the Impacts of Large-scale Swine Production*. Proceedings from "Interdisciplinary Scientific Workshop," Des Moines, IA (June 29–30).

Thurow, A.P. 1998. "An Industrializing Animal Agriculture: Challenges and Opportunities Associated with Clustering," in Stephen Wolf, ed., *Privatization of Information and Technology and Agricultural Industrialization*. Ankeny, IA: Soil and Water Conservation Society.

Thurow, A.P. and J. Holt. 1997. "Induced Policy Innovation: Environmental Compliance Requirements for Dairies in Texas and Florida." *Journal of Agricultural and Applied Economics* 29(1): 17–36.

USDA/ERS (U.S. Department of Agriculture, Economic Research Service). 1996. "Farmers Use of Marketing and Production Contracts." Farm Business Economics Branch, Agricultural Economic Report No. 747, Rural Economy Division, Washington DC.

USDA/NRCS (U.S. Department of Agriculture/Natural Resource Conservation Service). 1997. "Environmental Quality Incentives Program—Final Rule." *Federal Register.* http://www.nhq.nrcs.usda.gov/OPA/FB96OPA/EQIPfinal.html (22 May).

Van Horn, H.H. ed. 1995. *Nuisance Concerns in Animal Manure Management: Odor and Flies: Proceedings of a Conference*. Florida Cooperative Extension Service, University of Florida, Gainesville, FL.

Vukina, T., F. Roka, and R.B. Palmquist. 1996. "Swine Odor Nuisance: Voluntary Negotiation, Litigation and Regulation, North Carolina's Experience." *Choices*. Fourth Quarter: 26–29.

WIAA (Wallace Institute for Alternative Agriculture). 1996. "Campaign Charges Proposed Rule on EQIP Favors Large-scale Polluters." *Alternative Agriculture News*. Greenbelt, MD (November).

Zering, K.D. 1998. Personal Communication. Professor, Department of Agricultural and Resource Economics, North Carolina State University, Raleigh, NC.

13 FLEXIBLE INCENTIVES AND WATER QUALITY CONTROL TECHNOLOGIES FOR THE EVERGLADES AGRICULTURAL AREA

Donna J. Lee and J. Walter Milon
University of Florida, Gainesville, FL

The degradation of the unique wetland ecosystem of the Everglades can be characterized as a nonpoint-source pollution (NSP) problem that has a large number of emitters. A significant proportion of these emitters is from the farming enterprises located in the Everglades' Agricultural Area (EAA) of South Florida. Under plans to restore the ecosystem of the Everglades, phosphorus concentrations in water discharged from the EAA must be reduced below 50 parts per billion (ppb), and perhaps as low as 10 ppb. The cost of this approach is estimated to be at least $700 million. Two questions remain at the forefront of this issue: (1) Who should be financially responsible for this clean up? and (2) How clean is clean? The resolution of these issues is complicated by historically undefined property rights; the NSP nature of the water pollution; and uncertainty about the regenerative capacity of the Everglades at lower nutrient levels. Providing farmers with flexible incentives to adopt on-site water conserving technologies and to switch to low input production practices may be a more efficient and less expensive means of achieving long-term water quality improvement goals for restoring the ecosystem of the Everglades. This chapter examines the adoption of decentralized water quality control mechanisms to reduce the total costs of lowering phosphorus concentrations in EAA discharges.

INTRODUCTION

The dominant water quality issue in the Everglades is a nonpoint-source pollution (NSP) problem, which has a large number of polluters who contribute to the degradation of a unique wetland ecosystem. There are two questions on the forefront:

(1) Who should be financially responsible for the clean up? and (2) How clean is clean? The answers to these questions are muddled by historically undefined property rights, by the nonpoint-source nature of the pollutant and by the uncertainty about the regenerative capacity of the Everglades at lower nutrient levels.

The Everglades Agricultural Area (EAA) comprises 718,400 acres of some of the most productive agricultural land in the nation. In 1990, the EAA generated $1.5 billion in agricultural sales (HSEES, 1992). The dominant crops are sugarcane (453,000 acres); multi-cropped vegetables (50,000 acres); sod (33,000 acres); and rice and other crops (6,000 acres). More than one-half of all U.S. sugar production is in Florida's EAA. The EAA is reclaimed wetland that requires intensive water management. Average rainfall is 52 inches per year; supplemental irrigation is 0.385 million acre-feet per year; and, during the rainy season, storm-water runoff and agricultural drainage average 0.931 million acre-feet per year.[1] Storm-water runoff and agricultural drainage carry excess phosphorus fertilizer and mineralized phosphorus from the exposed muck soils to the Water Conservation Areas (WCAs) and the Everglades National Park (ENP)[2] (figure 13.1). Years of high nutrient drainage from the EAA have contributed to the eutrophication of Lake Okeechobee and to the alteration of plant communities in the WCAs.

This chapter examines water quality management in the EAA. The present management approach is reviewed with respect to its flexibility, inflexibility and uncertainty to achieve near and long-term water quality goals. We proffer a novel alternative—a two-tiered tax with emissions trading as a more flexible and efficient means of attaining water quality goals.

WATER QUALITY AND THE EVERGLADES FOREVER ACT

On private land in the EAA, landowners regulate water on 220 drainage basins that vary in size from 27 acres to 22,900 acres. These drainage basins utilize more than 300 structures that include gated culverts and large pump stations (SFWMD, 1997b). On public lands that surround the EAA, the South Florida Water Management District (SFWMD) manages a system of canals, pump stations and water control complexes. The SFWMD controls drainage flow from the EAA to WCAs at seven stations along the EAA border.

Under Florida's water law, there are no appropriative rights for either water quantity or water quality. Throughout the 1970s, drainage water from the EAA was back-pumped into Lake Okeechobee. As sugarcane acreage increased, water quality in Lake Okeechobee deteriorated. Efforts to restore Lake Okeechobee's water quality to its former level included reducing the volume of back-pumping and redirecting drainage to the south. Soon thereafter, algal blooms appeared in the WCAs, and cattails became the dominant vegetation in some areas (Davis, 1994). By 1986, water quality problems in the Everglades were a matter of national attention. In 1988, the U.S. Department of Justice sued the State of Florida for violating state and federal water quality laws. After years of litigation and negotiations, a settlement was reached and the State of Florida enacted the Everglades Forever Act (EFA) of 1994 (Stone and Legg, 1992; John, 1994).[3]

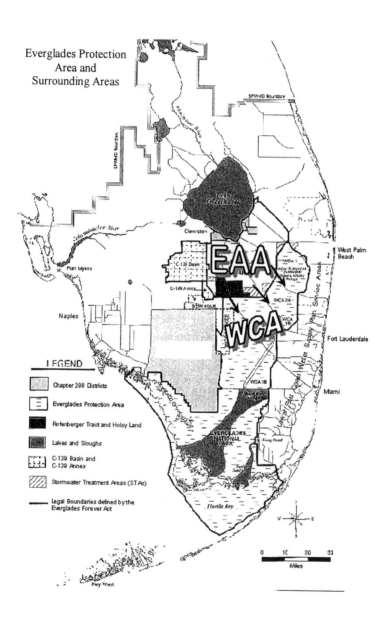

FIGURE 13.1 Map of the Everglades Agricultural Area and the Water Conservation Area

The Everglades Forever Act Approach

The EFA of 1994 established a comprehensive plan for improving the quality and distribution of fresh water, for removing exotic plant species in the Everglades National Park, and for restoring natural habitats. In the initial component of the plan, Phase I, the Florida legislature adopted a water, fertilizer and sediment management strategy. The initial incentive mechanisms of these strategies included on-farm best management practices (BMP); the removal of 100,000 acres of land from agricultural production; off-farm constructed filter marshes (storm-water treatment areas (STAs) to treat 1 million acre-feet of agricultural drainage; a land tax to generate funds for off-farm treatment; and an incentive credit to encourage additional on-farm control. Using the Batie and Ervin typology (this volume, table 5.1), these flexible incentive mechanisms are categorized as *other, charges* and *subsidies*. The combined strategy is intended to reduce the phosphorus content in drainage flows to 50 parts per billion (ppb) by the year 2002. The second component of the plan, Phase II, begins in 2001 when the Florida Department of Environmental Protection (FDEP) must determine the final water quality standards for water that flows into the WCAs. Phase II comprises the strategies that will be necessary to achieve the final water quality standards. The default standard for phosphorus under Phase II is 10 ppb.

Phase I

On-farm phosphorus reduction is the first line of abatement in the EFA. All landowners are required to reduce load rates by at least 25 percent of their 1979–1988 base. To encourage additional on-farm reduction, the EFA included an incentive tax credit that could be used to offset the per-acre land tax called the agricultural privilege tax. The agricultural privilege tax is added to the existing ad valorem tax and is used to pay for part of the construction cost of the STAs.

Between 1998 and 2001, farmers will receive an annual credit of $0.54 per acre for each percentage point reduction in farm phosphorus loads that exceed the mandated 25 percent reduction of their base load. The credit will be applied toward the $27 per acre annual privilege tax, but may not exceed $2.11[4] per acre by the year 2001. Unused credit may be carried forward to reduce taxes in subsequent years. Under this schedule, privilege tax and incentive credit values appear in table 13.1. Between 2002 and 2005, the tax credit rises to $0.61 per percentage point up to a maximum of $6.11 per acre per year. By 2006 and 2013, the land tax incentive credit rises to $0.65 per percentage point reduction and a credit up to $10.11 per acre per year. In 2014, the agricultural privilege tax drops to $10 per acre with no incentive credit (SFWMD/FDEP, 1996).

Segerson (this volume) addresses the difficulty of observing emissions from nonpoint-source pollution (NSP). In the EAA, targeted phosphorus reduction is assessed through demonstrated implementation of approved BMPs. Water quality monitoring devices at outflow points throughout the basin allow for the near field-level measurement of phosphorus loads.

TABLE 13.1 The Everglades per-acre Base Tax, Agricultural Land Tax and Incentive Credits, 1998–2014 and Later

Tax Period (Nov 1 – Oct 31)	EAA Base Tax[a]	Incentive Credit	Incentive Credit Cap	Target Load Reductions	
				EAA Basin	Individual Farm
	dollar	$ per acre	dollar	percent	
1994–1997	24.89	n.a.		25	30
1998–2001	27.00	0.54	2.11	25	35
2002–2005	31.00	0.61	6.11	25	40
2006–2013	35.00	0.65	10.11	25	45
2014–later	10.00	0	0	—	—

[a] Referred to as an "agricultural privilege tax" on land farmed in the EAA.

Source: SFWMD/FDEP (1996); Aumen (1997).

The Implementation of Best Management Practices

To achieve on-farm reduction targets, SFWMD provided parcel owners with a list of BMPs from which they could choose. (The owners could, however, propose an alternative BMP that would be subject to the approval of the SFWMD.) The list specified fertilizer application rates; fertilizer application methods; sediment retention methods; water detention methods; pasture management; physical control of drainage; and xeriscaping (SFWMD, 1997b). Parcel owners began adopting these BMPs in 1994. The total phosphorus load (table 13.2) dropped from 268 tons in 1995 to 119 tons in 1997, which was a 50 percent drop from the base period loads. As of 1997, BMPs were implemented on all of the acreage in the EAA, and the average phosphorus concentration of agriculture drainage was 97 ppb (SFWMD, 1997a).

While on-farm control may be the most efficient means of reducing phosphorus loads, the combined strategy comprising BMPs, taxes and credits is insufficient to achieve the Phase I water quality standard of 50 ppb. To remove additional phosphorus from drainage flows, the EFA authorized the SFWMD to design and construct STAs.

Storm-water Treatment Areas

Storm-water Treatment Areas (STAs) are large constructed marshes designed to slow water movement by directing drainage through a series of shallow holding ponds to allow time for marsh plants to absorb phosphorus. Six STAs with a combined area of more than 46,600 acres will be built in the EAA, to treat up to 1 million acre-feet of drainage each year (SFWMD, 1996). In a demonstration project on 3,813 acres, outflow phosphorus concentrations were consistently lower

**TABLE 13.2 Implementation of Best Management Practices and Phosphorus
Load Reduction from 1994 to 1997**

		Average Annual Phosphorus		
			Load	
Year	*Implementation of BMPs*	*Total Load*	*Reduction*	*Concentration*
	% total acreage	*metric tons*	*percent*	*ppb*
1994	15	132	17	112
1995	63	268	31	116
1996	100	162	68	98
1997	100	119	50	97

Source: SFWMD (1997b).

than 50 ppb but higher than 10 ppb (SFWMD, 1997a). At the end of 1996, 26,000 acres of land were acquired for STA construction (SFWMD, 1996). Construction of STAs is scheduled to begin in 1999 and is to be completed by 2002 (SFWMD, 1996). Total costs for land acquisition, construction and operation are estimated to be $685 million over the next 20 years (Davis and Sprague, 1997).

Phase I Funding

Subsequent to the EFA of 1994, a statewide referendum approved an amendment to the Florida Constitution (1996)[5] that required:

> . . . those in the EAA who cause water pollution within the Everglades Protected Area (EPA) shall be primarily responsible for paying the costs of the abatement of that pollution. (section 7 (b))

The Florida Supreme Court's interpretation of this amendment was that polluters in the EAA "would bear their share of the costs of abating the pollution found to be attributable to them" (SCF, p. 4, 1997). Funding for Phase I was provided by the Everglades Forever Act. State sources include two land taxes (the agricultural privilege tax and ad valorem taxes); toll revenues from Alligator Alley (I-75); Preservation 2000 funds; and Surface Water Improvement and Management (SWIM) funds. Partial federal funds for STA design and construction will also be available. While small-scale demonstration trials of STA technology appear promising, the technology has never been tested on the planned scale, and the STAs alone will be inadequate for the reductions that are expected in Phase II.

Phase II

Under 1997 land uses and BMPs, reduction of phosphorus levels *below* 50 ppb may require superior technologies. Research in superior technologies includes assimilation of phosphorus by algae, bacteria and plants to form peat; chemical coagulation in conjunction with mechanical or plant filtration to physically remove inorganic phosphorus precipitate (Parker and Caswell, this volume); and the implementation of STAs with periphyton and microscopic algae in place of cattails (Aumen, 1996). If 10 ppb were to become the standard of water quality that would enter the WCAs under Phase II, it would not be clear if superior technologies would suffice. The reduction of phosphorus concentrations to 10 ppb for the volume of drainage that comes from the EAA (up to 1 million acre-feet per year) is unprecedented anywhere in the world (Aumen, 1997).

FLEXIBILITY AND INFLEXIBILITY OF THE EFA APPROACH

The Phase I system for managing water quality in the EAA is flexible in that it provides parcel owners with incentives for reducing phosphorus loads and achieving lower cost solutions. It is inflexible, however, in that it limits opportunities for lower cost outcomes, and it may be inadequate to achieve further emission reductions. Listed below are some of the inherent flexibilities and inflexibilities of the existing EFA approach.

Flexibility of the EFA Approach

- The system allows for nearly point-source water quality monitoring.
- The choice of BMP is flexible.
- Parcel owners may propose an alternative BMP for approval by the SFWMD.
- Positive incentives to reduce loads beyond the minimum requirements are offered.
- Credits can be carried over to reduce taxes in future periods.

Inflexibilities

- Base tax is levied on all farmers without regard to phosphorus loading.
- Exogenous factors, such as rainfall, substantially influence the loading rates that create uncertainty.
- There are uniform phosphorus load targets across multiple discharge locations that disregard differences in receiving area tolerance to nutrient rich drainage flows.
- Credit that can be received in any one year is capped, which limits opportunities to achieve lowest cost solutions.
- Load reductions in one year cannot be banked for use in a future year.

Since much remains unknown regarding the effectiveness and environmental consequences of the proposed off-farm superior technologies, achieving water quality goals may require the pursuit of more aggressive on-farm phosphorus controls. Except for the incentive credit, the current scheme provides relatively little flexibility or incentive for individual parcel owners to achieve additional phosphorus reduction or to innovate alternative phosphorus control strategies

(Khanna et al., this volume). We propose an alternative tax strategy to add flexibility, which makes it possible to achieve higher water quality standards.

A TWO-TIER TAX APPROACH

Our proposal builds on an alternative that was initially proposed to deal with the water quality problems in the EAA. Hahn (1992) outlines an emissions allowance system with trading that is similar to the sulfur dioxide (SO_2) allowance trading program that was created under Title IV of the 1990 Clean Air Act (Burtraw, 1996). Under Hahn's system for the EAA, a total phosphorus load would be established and allowances or shares (one share equals one pound of phosphorus) would be issued to emission sources.[6] The shares could then be used to emit phosphorus loads equal to the number of shares held; to sell some or all of the shares and reduce loading to equal the remaining share balance; or to bank shares for use against future emissions. Parcel owners who would not have sufficient shares could be assessed a fixed fee for their excess emissions that would pay for control technologies, such as STAs. Boggess (1992) was critical of Hahn's proposal because he believes it understated the complexity of the pollution load problem and the political difficulties of establishing an initial allocation of emission shares. Boggess (1992) also believed that the proposal did not consider the full costs of establishing, monitoring and enforcing an emission share trading system for the EAA.

A Two-tier Tax

A strategy that could increase the flexibility of the existing management scheme *and* help achieve Phase II phosphorus emission targets is described in this chapter. Segerson (this volume) discusses the shortcomings of an input tax and an ambient tax that is used to control NSP when emissions are difficult to measure. The tax approach described in this chapter, however, differs from the usual instruments recommended for mitigating NSP (Xepapadeas, 1995) because of the extensive basin-level monitoring within the EAA. The first step of the strategy is to define an emission quota (E^*) for the entire basin and to allocate shares of the quota, q_i, to individual parcel owners, such that $\Sigma q_i = E^*$. Second, the strategy builds on the existing tax to form a two-tier tax with a base tax rate, b, and a penalty tax rate, t, (Roberts and Spence, 1976) such that the total emission tax paid by each parcel owner is given by,

$$Total\ Tax_i = be_i + t(e_i - q_i) \quad b > 0,\ t \geq 0\ for\ e_i > q_I,\ t = 0\ for\ e_i \leq q_i.$$

Here e_i is the actual emission from a parcel i, and q_i is the number of shares allocated to the parcel. Parcel emission levels above q_i trigger the penalty tax. Thus, parcel owners pay the base tax, b, for positive levels of phosphorus emissions and incur the penalty tax, t, for emissions that exceed their share allocation. With a two-tier tax, parcel owners will allocate resources between reducing emissions and paying taxes, depending on their marginal control cost and the tax structure. As illustrated in figure 13.2, a parcel owner with high control costs (shown by the marginal control cost function, MC_h) will choose to reduce emissions to e_h, (where $e_h > q$) and will pay a total tax equal to area A+B+C, which equals $be_h + t(e_h - q)$. A parcel owner with medium control costs (MC_m) will find it more cost-effective to reduce emissions to meet the quota than to pay the penalty tax. The parcel with medium control costs will reduce emissions to $e_m = q$ and will pay a total tax equal to area B+C = be_m. A parcel owner with low control costs, MC_L, will emit e_L, a load that is less than the allowance q, and will pay a total tax equal to area C, which equals be_L. In application, emissions can be measured either in terms of total load (for example, pounds per acre) or concentration (for example, parts per billion). Tax rates can be based on estimated damages from emissions, to internalize the cost of emission or, alternatively, the tax rate can be established to pay the marginal cost of off-site treatment technology.[7]

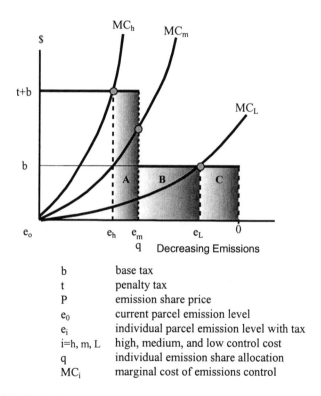

b	base tax
t	penalty tax
P	emission share price
e_0	current parcel emission level
e_i	individual parcel emission level with tax
i=h, m, L	high, medium, and low control cost
q	individual emission share allocation
MC_i	marginal cost of emissions control

FIGURE 13.2 Emission Level with Two-tier Tax

Two-tier Taxes with Trading

The two-tier tax would provide parcel owners with the proper incentive for complying with emission quotas. If emission control costs differed by parcel, emission trading between parcels would increase the flexibility of the system and improve overall outcome efficiency. Parcel owners with low costs may choose to reduce emissions below their quota (e < q), accumulate emissions shares (if e < q), and sell the excess shares (equal to q - e) in the market. Parcel owners with high control costs may opt to emit at a rate higher than their initial share allocation (e > q) and either pay the penalty tax for excess emissions or purchase emission shares in the market to reduce their tax liability.

Emission Share Market

In an efficient market, emission shares will be traded and the price of an emission share will equal the basin-wide marginal cost of reducing phosphorus, as illustrated in figure 13.3. The basin-wide marginal cost function will comprise the lowest cost phosphorus load-reduction technologies throughout the basin. Under a tax structure of base tax (b), penalty tax (t) and emission quota (E*), if the basin-

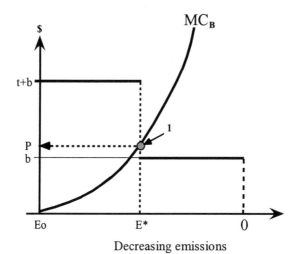

b	base tax
t	penalty tax
P	emission share price
Eo	current parcel emission level $0 = \Sigma e^o$
E*	emission quota, $E* = \Sigma qi$
MC_B	basin wide marginal cost of emission

FIGURE 13.3 Market for Emission Shares

wide marginal control costs were MC_B with emission share trading, the price of an emission share would equilibrate at P. As depicted in figure 13.4, if shares were traded at price P, parcel owners with high control costs will choose to emit e_m', purchase $e_h' - q$ emission shares to offset their penalty tax obligation, and pay bq in total taxes. Parcel owners with medium control costs will reduce emissions to e_m', purchase $e_m' - q$ shares in the market, and pay bq in total taxes. Parcel owners with low control costs will emit less than their share allocation ($e_L < q$), sell the q - e_L excess shares in the market for P per share, and pay be_L in total taxes.

Two-tier Tax Revenue

In addition to providing parcel owners with incentives to reduce emission loads on-site, the two-tier tax structure will generate revenue for funding off-site phosphorus reduction facilities. For example, if the combined penalty tax and base tax (t+b) were greater than the equilibrium share price P, then parcel owners would meet their share allocations by either reducing on-site or by purchasing shares at point 2 in figure 13.5, and tax revenue will equal bE*. Alternatively, if the tax structure were such that the share price P was greater than the penalty tax plus the base tax, as illustrated in figure 13.6, then parcel owners would reduce emissions on-site to point 3 and tax revenues would equal $bE* + (b + t)(E - E*)$.

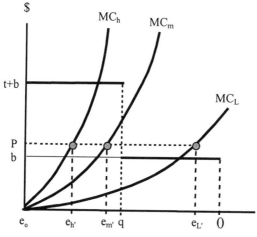

b	base tax
t	penalty tax
P	emission share price
e_0	current parcel emission level
e_i'	individual parcel emission level with tax and trading
i=h, m, L	high, medium, and low control cost
q	individual emission share allocation
MC_i	marginal cost of emissions control

FIGURE 13.4 Emission Level with Two-tier Tax and Trading

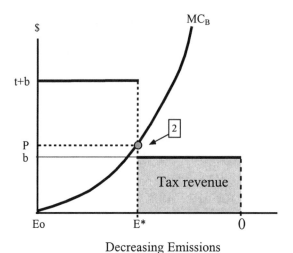

Decreasing Emissions

b	base tax
t	penalty tax
P	emission share price
E0	current parcel emission level
E*	emission quota
MC_B	basin wide marginal cost of emission control

FIGURE 13.5 Two-tier Tax Revenue

Two-tier Taxes with an Auction

An alternate method to the distribution of emission quotas among parcel owners is the auction of emission shares. Parcel owners could bid for shares based on the penalty tax and of their control costs. Other groups, such as conservation organizations, may also bid for shares to reduce total loads by purchasing emission shares and choosing not to use them. Proceeds from the auction would provide another revenue source for off-site control technologies.

Two-tier Tax in the Everglades Agricultural Area

The multi-tiered tax approach has had limited application in the United States but has been adopted by several of the new market-oriented countries in Eastern Europe (Farrow, 1995). The two-tier tax with trading is a reasonable approach for managing water quality in the EAA for several reasons. First, the Everglades Best Management Practices Program (FAC, chapter 40E-63) requires water quality monitoring at each landowner's drainage basin (with costs borne by the owner). Thus, a system for monitoring emissions is already in place. It has been observed

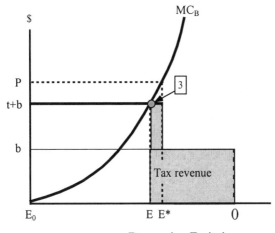

Decreasing Emissions

b	base tax
t	penalty tax
P	emission share price
E_0	current parcel emission level
E^*	emission quota
MC_B	basin wide marginal cost of emission control

FIGURE 13.6 Two-Tier Tax Revenue

that parcel emissions vary considerably: Annual loads have ranged from higher than 500 ppb to lower than 50 ppb (SFWMD, 1997b). Second, with tradable emission shares, individual parcel owners with lower emission control costs have additional incentive and flexibility to reduce emissions. Rather than comply with mandated management practices, parcel owners may select the mix of control strategies based on their marginal costs of control and on the price of emission shares. In this way, pollution costs are internalized, and parcel owners pay the full cost of their own emissions—they cannot free ride on off-site controls with costs that are borne by all parcel owners. This is consistent with the intent of the 1996 Amendment 5 to the Florida Constitution, since emitters pay the costs that are attributable to them. Third, the tax strategy can reduce emissions to levels below those of the existing program while generating revenues to fund the STAs (mandated under the EFA). Fourth, the tax strategy and emission quota can be designed to be fully consistent with either Phase I or Phase II water quality targets. For example, if the target is 10 ppb—the default standard under the EFA—the base tax rate could then be set to yield an amount equal to the agricultural privilege base tax rate for a given year.[8] Fifth, the two-tier tax with trading allows for additional flexibility to achieve water quality goals through state purchases (or sales) of emission shares in the open market. For example, a rough calculation indicates that if the price of an emission share for one pound of phosphorus were less than $374, the state of Florida could remove phosphorus from the system more cheaply by purchasing shares than by constructing STAs.[9]

The most difficult and contentious aspect of establishing a tiered tax with trading strategy is to select an allowable total load for the EAA and then to distribute emission shares. Presently, drainage flows from the EAA into the WCA contain 97 ppb of phosphorus (table 13.2), which is nearly double the Phase I goal of 50 ppb. Concentrations that are required under Phase II are likely to be lower. The total allowable load under Phase II cannot be determined until the Phase II standard is established and the effectiveness of STAs and other off-site control technologies is known. Once the total allowable load is determined, an initial allocation of shares must be made. We described an auction as a means of allocating the emission quota. An alternative option to an auction is to allocate the quota in proportion to existing phosphorus loads. While this approach may seem simple to implement, it fails to provide the necessary incentive for landowners to reduce loads. Discussions of these and other quota allocation mechanisms are left for future work.

CONCLUSION

In 1994, the EFA mandated a multi-pronged multi-year strategy for systematically reducing phosphorus loads from the EAA. Three lines of abatement were used: (1) farm source reduction through the use of mandated controls and incentive tax shares; (2) construction of wetland STAs to filter the phosphorus from agricultural drainage; and (3) superior technologies to chemically and mechanically filter phosphorus from drainage water to meet, as yet undetermined, Phase II standards.

The combination of the EFA and of the 1996 Amendment 5 to the Florida Constitution have created a unique place in the history of NSP controls in the United States: This is the first occasion in which a tax mechanism was used as part of a regulatory strategy. The current tax approach with incentive credits, however, does not provide parcel owners with sufficient incentive or flexibility to seek innovative control strategies, and it does not produce the least-cost solution. An emission quota with the two-tier tax offers policymakers an opportunity to refine the existing tax scheme to comply with phosphorus load objectives and to meet funding needs. The two-tier tax with permit trading is presented as a mechanism to endow parcel holders with an interest in the quality of the water in their system, so they perceive their opportunity cost of emitting. As the parcel owners' opportunity cost of emitting would increase, their incentive to abate would also increase. Tradable shares increase the opportunity cost of emitting, thereby enhancing the incentive to abate or to transfer the burden of abatement to parcel owners with lower cost control technology. In this way, technology adoption and innovation are encouraged, and phosphorus reduction is least-cost per parcel and across parcels. Nonetheless, many details must be addressed prior to adopting this proposed system of taxes and of emission trading within the EAA. Difficulties with previous water quality emission trading programs should not be overlooked (Crutchfield et al., 1994).

Despite the many advantages of a two-tier tax with trading system, it is important to note that this approach, even if fully implemented, could only provide second-best outcomes. Land-use activities in the EAA are distorted by water allocations made under Florida's reasonable beneficial-use standard.[10] Since water users in the EAA do not pay the true marginal cost of water supply, there is no assurance of the efficient use of water (Saarinen and Lynne, 1993). Moreover, sugarcane, the dominant crop in the EAA, is protected from foreign competition by import quotas and by market stabilization prices (Alvarez et al., 1994; GAO, 1993; Schmitz and Polopolus, this volume). A more thorough evaluation of all policies that influence land uses in the EAA may be necessary to determine a socially effizcient solution to NSP problems in the region.

ENDNOTES

1. Runoff and drainage volumes range from 0.276 to 1.56 million acre-feet per year (Windemuller et al., 1997).
2. Of the phosphorus in discharged water, 56.4 percent comprises organic and inorganic particulate form—a form not readily accessible to plants.
3. Florida Statutes, Chapter 373
4. $2.11 = $27 − $24.89. Farmers must pay a minimum of $24.89 per acre privilege tax each year.
5. Also referred to as the 1996 Amendment 5 and the polluter pays amendment.
6. A similar trading approach that is based on regional waste load allocation has also been proposed for the Lake Okeechobee watershed (Armstrong et al., 1995). This area is north of the EAA and is mainly comprised of dairy operations and pastureland. The EAA privilege tax does not apply in this area.
7. For example, if the emissions quota for the EAA were set at 50 ppb in 1999 and the base tax rate was set at $0.53 for each part per billion of phosphorus, a parcel owner who would met the parcel quota would pay $26.50 per acre in taxes. This amount would equal the agricultural privilege base tax (with incentive credits) under the existing tax structure (table 13.1). If penalty rates were set at $0.75 and the emission rates from the parcel were 100 ppb, the parcel owner would pay a total tax of $64. Unlike the existing tax approach, the two-tier tax would provide a strong incentive to reduce emissions equal to or below the quota, and those complying with the quota would pay no more than the current tax.
8. Using 1999 rates (table 13.1) as an example, the base tax rate would be $2.65 for each ppb of emissions per acre. Since the tax proceeds from the existing approach are being used for off-site controls, it is questionable whether an emissions rate of 10 ppb should be taxed. An alternative approach would be to set the base tax rate equal to zero and to assess only the penalty fee on emissions above the quota. Following this approach, with 1999 privilege tax rates, a parcel owner who emitted 50 ppb would pay a penalty tax rate of $0.66 for each ppb of emissions over the 10 ppb quota. This would leave the parcel owner no worse off than he would be under the existing program. An additional tier could be added to the tax structure for emissions over 50 ppb.
9. In 1997, approximately 1 million acre feet of water were drained from the EAA into the WCAs with a total phosphorus load of 119 tons (table 13.2). Treatment of this volume of water to meet the Phase I target of 50 ppb would require removal of about 55 tons, or 128,000 pounds, of phosphorus. The cost of meeting the 50 ppb target, using STA technology, is estimated as follows. The STAs will cost $575 million for the land and construction, and $110 million for the annual operation for a total of $685 million to be paid over the next 20 years (Davis and Sprague, 1997). If we were to assume that land and construction costs were paid in equal installments over a 20-year period, operating costs of $5.5 million would be incurred annually, and 128,000 pounds of phosphorus would be removed each year. The annual cost of removing phosphorus using STA technology would be $374 per pound. The present value-compounding factor of 13.5903 represents the present cost of paying $1 per year for 20 years with a prevailing discount rate of 4 per-

cent. The average cost of phosphorus removal using STAs was computed as follows: [($575 mil ÷ 13.5903) + $5.5 mil] ÷ 128,000 pounds = $374 per pound.

10. Florida Statutes 373.019(4).

REFERENCES

Alvarez, J., G. Lynne, T. Spreen, and R. Solove. 1994. "The Economic Importance of the EAA and Water Quality Management," in A. Bottcher and F. Izuno, eds., *Everglades Agricultural Area: Water, Soil, Crop, and Environmental Management.* Gainesville, FL: University Press of Florida.

Armstrong, N., G. Cooke, L. Huggins, F. Humenick, L. Shabman, R. Ward, and R. Wedepohl. 1995. *Final Report to the South Florida Water Management District.* Paper prepared for the SFWMD Review Panel on Phosphorus Control in the Lake Okeechobee Basin, West Palm Beach, FL (11 October).

Aumen, N. 1996. Meeting summary notes prepared by Nick Aumen, Research Program Director, Ecosystem Restoration Department, South Florida Water Management District. West Palm Beach, FL (December).

_____. 1997. Research Program Director, Ecosystem Restoration Department, South Florida Water Management District. Written correspondence, West Palm Beach, FL (July).

Boggess, W.G. 1992. *On the Use of Marketable Emission Credits to Help Preserve the Everglades: Observations and Suggestions.* Food and Resource Economics, Staff Paper SP92-13, Food and Resource Economics Department, University of Florida, Gainesville, FL (June).

Burtraw, D. 1996. "Trading Emissions to Clean the Air: Exchanges Few but Savings Many." *Resources* 122 (Winter): 3–6.

Crutchfield, S., D. Letson, and A. Malik. 1994. "Feasibility of Point-nonpoint Source Tradings for Managing Agricultural Pollutant Loadings to Coastal Waters." *Water Resources Research* 30: 2825–2836.

Davis, S.M. 1994. "Phosphorus Inputs and Vegetation Sensitivity in the Everglades," in S.M. Davis and J.C. Ogden, eds., *Everglades: The Ecosystem and Its Restoration*, Chapter 15. Delray Beach, FL: St.Lucie Press.

Davis, D. and D.E. Sprague. 1997. *Oversight Combined with Performance and Fiscal Controls Should Help Implement the Everglades Construction Projects on Time and Within Budget.* Florida Tax Watch Research Report. Tallahassee, FL (August).

FAC (Florida Administrative Code) Chapter 40E-63

Farrow, S. 1995."The Dual Political Economy of Taxes and Tradable Permits." *Economic Letters* 49: 217–220.

Florida Constitution. 1996. Article II, Section 7(b)

GAO (U.S. General Accounting Office). 1993. *Sugar Program: Changing Domestic and International Conditions Require Program Changes.* U.S. General Accounting Office, RCED-93-84. GAO/RCED Washington, DC (April).

Hahn, R.W. 1992. "Saving the Environment and Jobs: A Market-based Approach for Preserving the Everglades." Unpublished Manuscript. Washington, DC: Economists Inc. (8 April).

HSEES (Hazen and Sawyer Environmental Engineers and Scientists). 1992. *Evaluation of the Economic Impact from Implementing the Marjory Stoneman Douglas Everglades Restoration Act and U.S. Versus SFWMD Settlement Agreement.* Contract Completion Report, South Florida Water Management District, Contract No. C-3172, Hollywood, FL (October).

John, DeWitt. 1994. *Civic Environmentalism Alternatives to Regulation in States and Communities.* Washington DC: Congressional Quarterly Press.

Roberts, M. and M. Spence. 1976. "Effluent Charges and Licenses under Uncertainty." *Journal of Public Economics* 5: 193–208.

Saarinen, P. and G. Lynne. 1993. "Getting the Most Valuable Water Supply Pie: Economic Efficiency in Florida's Reasonable Beneficial Use Standard." *Journal of Land Use and Environmental Law* 8: 491–520.

SCF (Supreme Court of Florida). 1997. *Advisory Opinion to the Governor, 1996 Amendment 5, p 4. (Everglades).* p. 4 90,042, (26 November).

SFWMD (South Florida Water Management District). 1997a. *Everglades Nutrient Removal Project—Year 2 Synopsis.* South Florida Water Management District, West Palm Beach, FL (March).

_____. 1997b. *Everglades Best Management Practice Program—Water Years 1996 and 1997.* South Florida Water Management District, West Palm Beach, FL (11 September).

SFWMD/FDEP (South Florida Water Management District and the Florida Department of Environmental Protection). 1996. Everglades Program Implementation, Program Management Plan, Revision 2. West Palm Beach, FL (14 November).

Stone, J.A. and J. Legg. 1992. "Agriculture and the Everglades." *Soil and Water Conservation*. May–June: 207–215.

Windemuller, P., D.I. Anderson, R.H. Aalderink, W. Abetew, and J. Obeysekera. 1997. "Modeling Flow in the Everglades Agricultural Area Irrigation/Drainage Canal Network." *Journal of the American Water Resources Association* 33(1): 21–34.

Xepapadeas, A.P. 1995. "Observability and Choice of Instrument Mix in the Control of Externalities." *Journal of Public Economics* 56:485–498.

14 ADOPTION OF WATER CONSERVING TECHNOLOGIES IN AGRICULTURE: THE ROLE OF EXPECTED PROFITS AND THE PUBLIC INTEREST

Frank Casey and Gary D. Lynne
Northwest Economic Associates, Vancouver, WA
University of Nebraska, Lincoln, NE

Incentive policies that promote the adoption of drip irrigation technology on to-mato fields in southwest Florida are evaluated. A conceptual framework built upon the theory of derived demand was expanded to incorporate farmers' percep-tions of profitability and of public interest norms. The interaction between ex-pected profitability and public interest norms was found to positively affect the initial decision to adopt drip irrigation technology. The intensity of adoption was determined by field-level agronomic and physical conditions. Two primary con-clusions are offered: (1) incentive policies to promote the adoption of environ-mental technologies must incorporate the values of the target group and (2) any typology of flexible incentives should be expanded to encompass public interest norms.

INTRODUCTION

The use of incentives to promote the adoption of environmental technologies is discussed in this chapter. Flexible incentives are compared to current policies that compel agricultural producers to employ drip irrigation technology (DIT) as a means of water conservation. First, a brief background and problem focus is pro-vided. Second, a conceptual framework for considering the adoption decision that is associated with an environmental technology is presented. Third, economic and physical factors that influence the adoption decision are identified and modeled using the theory of derived demand. Fourth, the analytical framework is expanded to incorporate the role that a grower's subjective expectations of DIT performance and profitability play in the adoption decision process. This expanded framework also incorporates the influence of a grower's sense of public interest that reflects a social norm for the adoption of DIT and the conservation of water. The influence of social norms gives rise to the concept of multiple utility, which represents the irreducible self-interest and public interest. Fifth, this analytical framework is empirically tested when tomato growers in southwest Florida apply it to their adoption of DIT. Sixth, the results from this case study are discussed to reveal important implications for the types of incentives that are needed to promote the increased use of environmental technologies in general.

BACKGROUND AND PROBLEM FOCUS

The demand for fresh water in Florida for use by humans and for sustaining local ecosystems is escalating, and in some areas there is increasing evidence that groundwater supplies are being severely depleted. There has been an increasing reliance on groundwater sources, particularly by agriculture. High population growth rates that were estimated at 33 percent between 1980 and 1990 (Marella, 1992), and the attendant requirements of the domestic, agricultural, industrial and service sectors are the driving forces behind increased competition and conflict over Florida's water.

Agriculture is the industry that uses the greatest quantity of Florida's freshwater. A number of high value crops in the state, which includes fruits, vegetables, florals and sod, are highly dependent on supplemental irrigation. This level of dependence is due to the types and drainage properties of Florida soils, the high evapo-transpiration rates during the prolonged summer season, and the inadequate rainfall during some stages of the growing season. Relative to other water uses (for example, public supply, thermo-electric power generation, domestic use and self-supplied commercial industrial use), agriculture accounts for roughly 62 percent of freshwater use from surface sources and 43 percent from ground resources (Marella, 1992).

The Florida Water Resources Act (the Act) provides an institutional framework for the allocation of the state's water resources. This framework is based on reasonable beneficial use of water (Maloney et al., 1972; Florida Statutes, 1972). The Act mandates a central role for the state in water allocation and gives water resources the status of a state-owned good. Thus, private property rights for water do not exist in Florida, only water use rights for a specified time period are granted by the state.

The Act, as amended in the mid-1980s, requires that each of the state's Water Management Districts (districts) establish its own water use permitting process. Districts assign individuals, firms and public water suppliers qualified use rights in the form of consumptive use permits. Districts have the authority to define and enforce permit regulations that are designed to fulfil the Act's conservation objectives (Lynne et al., 1991). For Florida's agricultural producers, these use-permitting regulations entail technical efficiency and requirements for the reduction of water use. In order for prospective water users to obtain a permit, the proposed use must also be shown to be reasonable and beneficial to the district and to the state.

To promote the state's water conservation goals in the agricultural sector, districts set crop-specific water quotas and, in some cases, have mandated irrigation performance standards. Renewal of a consumptive use permit is sometimes contingent on meeting this standard.

Although performance standards often allow those being regulated to meet a stated environmental objective in a voluntary and flexible manner (Batie and Ervin, this volume), some districts have proposed irrigation efficiency standards of 80 percent. Currently, the only available technology capable of achieving such a high level of efficiency is drip irrigation. Consequently, the 80 percent performance standard is, in essence, a *de facto* design or a technology standard.

Most of the economic literature that addresses the adoption of modern irrigation technologies rests on the behavioral assumption of profit maximization. Adoption is induced whenever the grower's choice leads to increased profits—either through reduced costs, increased revenue or both. For drip irrigation, revenue can be enhanced either directly through improved yields or by switching to higher valued crops that respond better to drip irrigation. Decreased water costs can result from pumping less water (that is, increased water use efficiency) or from a more efficient use of complementary inputs, such as fertilizer. Modern low-volume irrigation technologies can extend the range of soil types and water quality can be effectively used in agricultural production.

Production functions that include detailed field-level information on physical environmental characteristics are used in most current economic models that investigate DIT adoption decision behavior (Caswell, 1991; Dinar et al., 1992; Green et al., 1996). Human capital variables (for example, age, farming experience and education) or farm structure variables (such as firm size, firm organization, land tenure and irrigation management) are sometimes included when modeling adoption decisions.

While profit is undoubtedly an important motivating factor in technology adoption, this chapter considers the possibility that social behavior, as suggested by Sen (1977), Hirschman (1985) and Etzioni (1986), could also influence the decision. The motive behind this social behavior is the grower's desire to fulfill a community norm or to satisfy the public interest.

Empirical results from Hodges et al. (1993), Lynne et al. (1995) and Lynne and Casey (Forthcoming) support the contention that social norms do affect DIT adoption behavior. These studies indicate that the affect of a grower's community interest on the adoption of DIT by both strawberry and tomato growers is as significant as the profit motive. Earlier research in Florida also suggests that the presence of a conservation ethic and/or a sense of social responsibility may influence the adoption of soil resource conserving technologies (Lynne, et al., 1988). Weaver (1996) employs a similar modeling approach and reinforces the importance of a conservation ethic.

The major questions posed in this chapter are the following: What factors motivate growers to adopt DIT? What role does a grower's subjective expectations

about the profitability of DIT play in the adoption decision? Is the adoption decision motivated solely by profit maximization or by the desire to fulfill a perceived
community norm, or both? What are the implications for the design and selection
of policy instruments to promote the adoption of environmental technologies in
agriculture?

MULTIPLE UTILITY AS A CONCEPTUAL FRAMEWORK

The multiple utility conceptual framework posits that a decision-maker pursues
both a self-interest utility and an irreducible public interest utility. A multiple utility framework offers two notions that build and elaborate on the standard theory of
derived demand. First, the adoption of environmental technologies involves the
pursuit of utility as well as of profit. Secondly, the decision-maker has a dualistic
nature in that he or she pursues a separate utility that is related to fulfilling the
public interest.

The general works of such well-known economists as Adam Smith, Hume,
Edgeworth, Walras, Bentham and J.S. Mill consider the dualistic character of human nature. For example, it is well known that Smith established self-interest as
the foundation for economic behavior in The Wealth of Nations, but he also addressed the existence and role of socially benevolent behavior in the Theory of
Moral Sentiments (Khalil, 1990; Evensky, 1992). More recently, several works
suggest the need for restructuring single utility models with the broader concepts
embodied in choice behavior.[1] The common thread, found throughout this literature, is an attempt to incorporate commitment, values and morality into standard
utility theory. The development of this framework begins with the recognition of
the importance of these concepts in choice behavior and ends with a formal theory
of multiple utility.[2]

Water conservation is a social issue that involves different groups that compete
for a common resource. The application of multiple utility theory to the adoption
of DIT, in particular, and to environmental technologies, in general, implies that
adoption is influenced by the perception of the public interest as well as by profitability.

Social norms are the means by which social psychologists assess the influence
of an individual's public interest norms on choice behavior. Consistent with Ajzen
(1988) and Fishbein and Ajzen (1975) there is a need to incorporate grower's
community norms into the behavioral equations. That is, if growers consider what
others in their community think about their water conservation practices, then it
will influence their decisions to adopt DIT.

Case studies suggest that there are community-related phenomena at work in
farmers' soil conservation decisions (Lynne and Rola, 1988; Lynne et al., 1988)
and water saving technology decisions (Lynne et al., 1995; Casey and Lynne,
1997) that cannot be explained by self-interest alone. This finding is corroborated
by Lynne et al. (1988) and further tested by Kalaitzandonakes and Monson (1994)
and Weaver (1996). Kalaitzandonakes and Monson (1994) determine that Missouri farmers only pursue self-interest. Weaver (1996) concludes that the larger
community does influence the adoption decisions of Pennsylvania farmers. These

empirical findings suggest that a multiple utility model (Lynne, 1995) is more appropriate than standard utility theory for evaluating technology adoption decisions.

Consider figures 14.1 and 14.2. In figure 14.1, two sets of indifference curves occupy the All Other Technology (AOT) and Conservation Technology (T) spaces. The utilities are jointed in that one utility cannot be achieved without having some of the other. The self-interest or I-Utility is represented by the set U^I, while the public interest or We utility is represented by the set U^{We}. The public interest utility is the grower's utility that is received from fulfilling his or her perceptions of the public interest. There are two preferred paths for the grower to examine when deciding on the mix of technologies in which to invest. Along OAZ, the farmer is maximizing U^I, which is primarily associated with maximizing profits, while along OCZ the grower is paying more attention to public interest utility, U^{We}. On either path, there is some fulfillment of the other utility. At the optimum level of the I-Utility, represented at point A, the farmer experiences U_1^{We} of the We Utility. Similarly at the optimum level of the We Utility at point C, the farmer experiences U_1^I of the I-Utility.

There is also a degree to which the two sets of indifference curves overlap. For example, if a grower were to choose U^{We} as representing the only legitimate utility function on which to base production decisions, then U^I would be superimposed onto U^{We}, and pursuing U^I would be accomplished along OCZ. In this case, the classic Adam Smith public interest would be achieved through the pursuit of unconstrained self-interest.

Because there are two paths the farmer could take, it is possible that he or she would select some intermediate point B, rather than an extreme. To understand why this might be the case, consider figure 14.2, which is the joint utility space derived from moving along the budget line, A'C', in figure 14.1. In moving from point A' to A in figure 14.2, both U^{We} and U^I increase and represent an area of complementary utilities. In this area of joint utility space the producer would choose point A. Similarly, both types of utility increase in segments C'C in figures 14.1 and 14.2 and the grower would move at least to point C. Therefore, the two paths OAZ and OCZ in figure 14.1 and the competitive segment, AC, of figure 14.2 represent the bounds for rational choice. Notice that at point A in figure 14.2, the farmer does not pursue the public interest at all, yet he or she still achieves the level U_1^{We}. Similarly, at point C there is still some satisfaction of the self-interest shown by U_1^I. What this model demonstrates is the degree to which the public interest U^{We} can be achieved by the pursuit of the private interest, U^I.

It is reasonable to expect that most growers would rationally take an intermediate position, such as point B, in the two figures. The conceptual model suggests that growers would indeed look at the relative costs of various technologies, but demonstrates the possibility that a grower's sense of public interest could also be important to the adoption decision. This model also suggests that incorporating appropriate measures of UI and UWe into the standard derived demand model can enhance the authors' understanding of the decision problem for the adoption of environmental technologies.

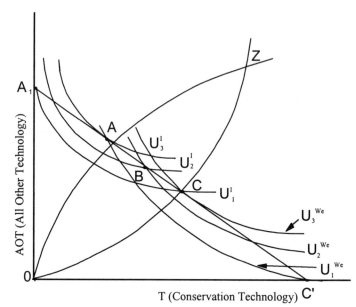

**FIGURE 14.1   Joint Indifference Curves Representing the Pursuit of Multiple
Utility in the Self and Public Interest**

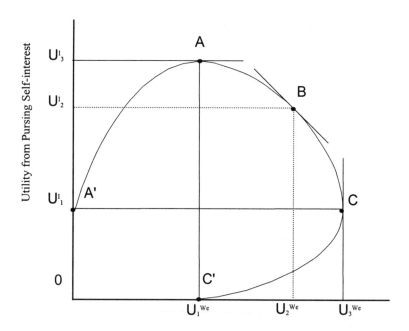

**FIGURE 14.2   Joint Utility Outcomes from the Pursuit of Multiple Utility
in Self and Public Interest (Lynne, 1995)**

A form of expectancy value, or valence model, is employed to represent how U^I and U^{We} influence growers' DIT adoption decisions. Valence models evaluate the consequence of some action and serve as the link between the concepts of attitude or social norm and of utility in the field of social psychology.[3] Valence models are equivalent to subjective expected utility (SEU) models in economics and are a method of empirically representing SEU (Feather, 1982).[4] The valence model is composed of two elements. The first element is the value or utility that an individual attaches to a particular attribute of the choice subject. The second element is an individual's subjective expectation that such a value or utility will actually be forthcoming. Feather (1982) employed a theory of behavior that was based on the concept of positive and negative valences that are held by an individual.

Fishbein and Ajzen (1975) place valence, or SEU, within the context of attitude. Attitude describes a grower's belief as positive or negative. For example, a grower could hold a positive valence (or SEU) regarding whether DIT adoption would result in the decreased use of fuel or electricity for pumping water. This positive belief, however, could be offset by a negative valence that is associated with an expected need to increase labor and management. Before making the decision to adopt DIT (or any technology) a grower would conduct an internal analysis that would weigh the positive and negative valences associated with the technology's various attributes. An overall positive valence across the selected attributes of a particular technology would result in the decision to adopt DIT.

The field of social psychology offers us a means to develop an empirical model to test the theoretical SEU concept. An empirical, multiple utility economic decision model is constructed by building onto the theory of reasoned action. The theory of reasoned action assumes that individual choice behavior is influenced by two distinct factors: (1) self-interest and (2) social norms (Fishbein and Ajzen, 1975; Ajzen, 1988; Ajzen, 1991). The empirical model employed here encompasses self-interest, which is represented by pre-adoption expectations of profitability, U^I, and individuals' social norms, which are represented by public interest (U^{We}) norms, as motivational factors in DIT adoption decisions. An additive version of the model is shown in equation 1 and is based on the work of Lemon (1973), thus

(1) $B \sim \text{Intentions} = \alpha\Sigma\, b_i e_i + \beta\Sigma\, n_j m_j$

$$= \alpha\Sigma\, A + \beta\Sigma\, SN$$

in which:

B	=	Economic behavior or action to invest in DIT,
Intentions	=	Intention to act (adopt DIT),
b_i	=	Belief (probability) about achieving the ith attribute of DIT,
e_i	=	Evaluation utility associated with the ith attribute of DIT,
n_j	=	Belief (probability) about achieving the social norms of referent groups (Sen's claims of others) for the jth ref-

erent group,

m_j = Motivation (utility) from complying with norms of the jth referent group,

A = Overall attitude toward the profitability of DIT,

SN = Overall social norm (public interest) associated with the adoption of DIT, and

α and β = The parameters representing self-interested and public interest utility, respectively.

Equation 1 is the equivalent of an economic SEU model. The belief proxies, bi and nj, represent expectations, and the evaluation proxies, ei and mj, represent utility (unobservable and latent).

EMPIRICAL MEASUREMENT OF EXPECTED PROFITABILITY

In the context of the valence model, if a grower has a strong attitude (A) toward DIT, it indicates that he or she would gain utility from its adoption. In other words, attitude is revealed (and therefore can be used as a proxy for) latent utility. As noted, the attitude and social norm measures are constructed by multiplying the subjective probability that an event will occur (belief) by the utility (evaluation) of the event.[5]

Specifically, b_i is a probability that measures the degree of belief about whether a consequence will occur for attribute i, while e_i is the utility received if the consequence occurs. Thus, b_i*e_i measures the underlying (latent) SEU for a specific attribute. The individual attributes are then summed to establish an overall Attitude index (A).[6]

Measures of belief and evaluative components for thirteen selected attributes that are relevant to the efficacy of drip irrigation are used in the empirical analysis. Various production input attributes (such as fuel and electricity for pumping water, fertilizer, pesticides, labor and management, water withdrawn and pump engine or motor size for irrigation) impact the efficacy of DIT through changes in costs. Output attributes that are associated with increasing yield and product quality reflect the technology's revenue enhancing effects. All together, these attributes represent the grower's pre-adoption subjective expectation toward the profitability of DIT.

Empirical Measurement of the Social Norm

Resource use and resource conserving behavior are social endeavors because individual actions affect others. This is especially true for natural resources that often have many competing and conflicting uses. The decision to adopt water conserving technologies and management practices could be influenced by a grower's perception of community expectations. Weaver (1996), who uses the general approach of Lynne et al. (1988), suggests that growers would respond to the influence of various community members with whom they identify.

The second term in equation 1, $(n_j m_j)$, represents the expected public interest or social norm. Thus, njmj is a proxy for U^{We} of figure 14.1. This social norm index (SN) represents public interest and is composed of n_j grower perceptions of community beliefs or norms, which are multiplied by the utility arising from acting on these beliefs. The community groups in the index could include university and experiment station personnel; soil conservation services; water management districts; county government personnel; local growers' associations; nearby homeowners; irrigation equipment dealers; environmental groups; family; other growers in the area; and business associates. A large SN would suggest heavier commitment to these groups. It would also reflect sharing of the same norms.

AN EMPIRICAL DERIVED DEMAND MODEL OF DRIP IRRIGATION TECHNOLOGY ADOPTION[7]

To test the theory of multiple utility using expectancy value models, DIT adoption is empirically modeled for tomato production in south Florida. Economic reasoning and prior research (Caswell, 1991; Dinar et al., 1992; Green et al., 1996) identify economic and physical variables that affect the decision to adopt DIT. The unconstrained derived demand model for an input to production is

(2)   B = f (product prices, technology costs, other input prices),

in which behavior, B, represents the amount of the input used.

Financial and economic variables are important supplements to subjective expectations in modeling technology adoption (Lynne and Rola, 1988; Lynne et al., 1988; Lynne et al., 1995). Tomato growers in south Florida face approximately the same prices for all inputs except for irrigation water pumping costs. Growers receive different product prices that depend on market timing and skill. Given these assumptions, the model becomes

(3)   B = f (technology costs, pumping costs).

Previous research on DIT adoption also suggests the need to incorporate physical and environmental variables because these influence technology costs (Caswell, 1991; Caswell and Zilberman, 1985 and 1986). The actual physical conditions on growers' fields may not be completely accounted for by subjective expectations about the profitability of DIT. When these environmental variables are included, the equation becomes

(4)   B = f (technology costs, pumping costs, physical environmental conditions).

Field level physical environmental conditions for this case study are represented by soil and water quality. Soil quality reflects field drainage characteristics (poorly drained, well drained and very well drained) and are assigned discrete values of 1, 2, and 3, respectively. Water quality is measured with a 0,1 dummy variable that corresponds to the absence, or presence, of problems that are related to the quality of the irrigation water. These problems included high pH, iron, sulfur, sediments, calcium or algae, which are in the water source. It is hypothesized that soil drainage capacity and the presence of water quality problems will positively relate to the probability and intensity of decisions to adopt DIT.

A water quota constraint, often used in other studies of DIT adoption (Dinar et al., 1992), is not applicable to this analysis. Although agricultural water quotas exist in the form of consumptive use permits in Florida, these quotas are not determined independent of the physical environment or the irrigation system used.

A differentiation is made between whether a field is used to produce tomatoes exclusively or to produce tomatoes and another crop over two consecutive seasons. It is hypothesized that fields, in which growers produce tomatoes exclusively, are more likely to have DIT because there would be no costs associated with having to reconfigure the irrigation system for a different crop. Crop choice is represented by a 1, 2 dummy variable with 2 equal to the decision to grow two different crops. Thus, crop choice is assumed to have a negative influence on DIT adoption.

There are two types of cost that affect the adoption decision. First, there is the variable cost of pumping water. Because detailed data on pumping costs are not obtained, depth-to-water table (DEPTH) measured in feet is used as a proxy. It is hypothesized that the greater the depth from which growers must pump irrigation water, the greater the likelihood that they would adopt DIT. The per-acre capital investment cost of DIT is the second cost element. For non-adopters of DIT, an average investment cost for a drip irrigation system is used to represent what the grower would be required to spend if he or she were to adopt DIT. It is hypothesized that capital investment costs would be negatively related to the probability of DIT adoption for tomatoes in South Florida. Investment costs are measured in dollars per acre at the field level. Capital costs are measured in dollars.

Thus, equation 5 is

$$(5) \quad Y = B_0 + B_1 \, WQP + B_2 \, SOIL\_TYP + B_3 \, DEPTH + B_4 \, COSTAC + B_5 \, CROP + e,$$

in which Y simultaneously represents the probability and intensity (percentage of field area) of adopting DIT on a specific field on the farm. WPQ and SOIL_TYP represent the presence of water quality problems and soil type, respectively. DEPTH represents pumping costs as proxied by the depth-to-water table, CROP indicates cropping pattern, and COSTAC is the per-acre, capital investment cost.

In addition to the classical derived demand variables, the empirical model includes the grower's subjective expectations about the profitability of DIT (represented by A), the social norm index (represented by SN) and an interaction term, A*SN, which allows for the testing of the jointness in the two utilities. It is assumed that the influence of the SN index, and its interaction with A, are only operative in the first stage of the decision process. Whereas a grower is sensitive toward the influence of the community in the initial adoption decision, the intensity of adoption is primarily dictated by physical conditions on the farm and on potential profitability.

Equation 6 presents the complete empirical model

(6) $Y = B_0 + B_1 WQP + B_2 SOIL\_TYP + B_3 DEPTH. + B_4 COSTAC$
$+ B_5 CROP + B_6 A + B_7 SN + B_8 A*SN + e.$

To solicit growers' subjective expectations about profitability and social norms, a questionnaire was developed following a design developed by Ajzen (1988) and Ajzen and Fishbein (1980). Bagozzi (1984) outlines important considerations in designing the scaling and testing for salient properties of the (A) and (SN) indexes.[8] To construct the A and the SN indices, it is necessary to determine if all the elements of these respective indices are salient. Bagozzi (1984) suggests that not all the component parts of the attitude and/or community norm indexes may be salient, and proposes a method for testing for this saliency. This method is referred to as the "Bagozzi test" and a detailed description of the method is provided in Casey and Lynne (1997). The statistically salient pre-adoption attitudes and community groups that resulted from the Bagozzi test are shown in tables 14.1 and 14.2, respectively.

From table 14.1, most of the significant attributes of the A index for this field-level analysis are associated with expected cost reductions (for example, fuel, electricity, fertilizer, labor, management, pump size and overall production costs). With regard to revenue, increasing yield is an important characteristic. Improvement of tomato quality or overall farm profit is only marginally significant. Growers also expect that the use of DIT would significantly improve the quality of irrigation water runoff and leaching. Thus, favorable expectations toward DIT are associated mostly with the impact that DIT is expected to have on reducing the costs of complementary inputs and on decreasing externalities related to water pollution.

The salient community groups of the SN index are identified in table 14.2. The most significant community groups include the University Extension Service, the University Experiment Station, the Natural Resource Conservation Service, business partners, the County Government and homeowners. Family influence is of only marginal importance. Extension Service/Experiment Stations and business partners support growers' farming operations. The importance of the Resource Conservation Service, County Government and surrounding homeowners suggest that growers are sensitive to the water conservation desires of these groups in field-level decisions.

Adoption behavior is modeled as a two-stage decision process. The first stage is a discrete dichotomous choice to adopt or not to adopt DIT for a particular field. The farm-level decision process is analyzed in Lynne and Casey (Forthcoming). Importantly, attitude and social norms play substantially different roles at the field level in contrast to the farm level. Social norms are significant forces at the farm level. This highlights the importance of context and target (Ajzen and Fishbein, 1977). If the adoption decision were affirmative, the second-stage decision would be to choose the level of effort or intensity of the adoption decision on that same field. Intensity is measured by the percentage of the field that is put into DIT. The two separate decision processes are essentially a 0,1 discrete behavior followed by

TABLE 14.1 Salient Attributes of the "A" Index

Attribute	Parameter Estimate	Standard Error
Reduces Fuel & Electricity	.25[a]	(.08)
Reduces Fertilizer	.17[a]	(.06)
Reduces Labor & Management	.21[a]	(.08)
Reduces Overall Production Costs	.14[b]	(.06)
Reduces Motor Size for Pumping	.58[b]	(.25)
Improves Water Quality	.70[b]	(.34)
Improves Tomato Quality	1.51[c]	(.81)
Increases Yield	.16[a]	(.07)
Increases Farm Profit	.75[c]	(.41)

[a] Significant at the .01 probability level.
[b] Significant at the .05 probability level.
[c] Significant at the .10 level.
Source: Authors' compilation.

TABLE 14.2 Salient Community Groups Derived of the SN Index

Community Groups	Parameter Estimate	Standard Error
Family	.52[c]	(.29)
Extension Service/Experiment Stations	1.27[a]	(.43)
County Government	.68[b]	(.31)
Non-farmers: e.g. Homeowners	1.06[a]	(.32)
Business Partners	1.41[a]	(.37)
Natural Resource Conversation Service	.81[a]	(.31)

[a] Significant at the .01 probability level.
[b] Significant at the .05 probability level.
[c] Significant at the .10 level.
Source: Authors' compilation.

the level of effort (Blundell and Meghir, 1987). The Time Series Processor Sample Selection Program is used to derive the parameter estimates for the two-step decision process.

ANALYTICAL RESULTS

The Adopt/Not to Adopt Decision

The results of the two-stage adoption model are shown in table 14.3. The Mills Ratio is a measurement of whether the first decision stage is significantly different from the second decision stage (Davidson and MacKinnon, 1993). A t-test of the Mills Ratio indicates that the adopt/not to adopt decision is significantly different from the intensity decision, which implies that modeling the decision as a two-stage process is justified.

In the first-stage adopt/not to adopt decision, both environmental variables (water quality and soil type) are significant. As the quality of water for irrigation deteriorates, the probability of adopting DIT on a specific field increases, which confirms our expectations. The sign on soil type, however, is inconsistent with previous findings that soils with high drainage capacity are positively related to the probability of adopting DIT. The sign indicates that, as soils increase in drainage capacity, there is a lower probability that growers will adopt DIT. This may be because all soils in southwest Florida, in contrast to soils in other parts of the United States, are all well-drained and that the most porous are not suitable for DIT.

TABLE 14.3 Parameter Estimates for the Multiple Utility Model of Drip Adoption

| Parameter | First Stage: Adopt/Not to Adopt | | Second Stage Proportion of Field | |
	value	standard error	value	standard error
Constant (B0)	-0.03	(2.53)	2.21	(5.61)
WQP (B1)	1.52^a	(0.58)	0.54	(1.24)
SOIL_TYP (B2)	-0.41^b	(0.27)	1.71^b	(0.90)
DEPTH (B3)	$0.61*10^{-3b}$	$(0.59*10^{-3})$	$0.20*10^{-2b}$	$(0.13*10^{-2})$
COSTAC (B4)	$0.21*10^{-3}$	$(0.59*10^{-3})$	$0.25*10^{-2b}$	$(0.14*10^{-2})$
CROP (B5)	$0.18*10^{-2}$	(0.32)	-2.51^a	(0.92)
(A)(B6)	$-0.71*10^{-2}$	$(0.99*10^{-2})$	0.01	(0.01)
(SN)(B7)	-0.02	(0.02)		
(A*SN)(B8)	$0.12*10^{-3b}$	$(0.79*10^{-4})$		
MILLS RATIO			2.84^a	
Degrees of Freedom	78		78	

[a] Significant at the .01 probability level.

[b] Significant at the .05 probability level.

Source: Authors' compilation.

Cropping pattern and the capital costs of DIT investment are not significant factors in the initial adopt/not to adopt decision. The fact that the capital investment cost is not significant suggests that it is not perceived as a severe constraint to the probability of adoption. DEPTH (pumping cost) is important in the first-stage adoption decision. This lends some support to the contention that increasing well depth (and, thus, pumping costs) has a positive effect on the probability of DIT adoption.

In the linear version of the model, pre-adoption subjective expectations and social norm indexes alone are insignificant. The interaction term, A*SN, however, is significant. The parameter sign and the level of significance for the interaction term show that the expected profitability of drip irrigation and of social norms reinforced one another at the field level. This indicates that the utilities that are associated with the expected profitability of DIT and the fulfillment of a public interest social norm are joint in the decision process. That is, a grower who simultaneously has a stronger positive pre-adoption expectation toward the profitability of DIT and gains utility from paying attention to what he or she perceives that community groups prefer, is more likely to install more drip irrigation. Also, for this particular case study, the social norm, or public interest, appears important only within the context of those factors that affect expected profitability. This result does not seem surprising given that two (business partners and university extension/research personnel) of the five community groups that significantly influence a grower's adoption decision support the profitability objectives.

Intensity of Adoption Decision

The second-stage adoption decision reflects the intensity of adoption, which is represented by the percentage of the total area of a field that is put into DIT. The significant factors in the second-stage decision include soil type, cropping pattern, per-acre capital investment cost and DEPTH. The sign on the soil type indicates that the percentage of area installed in DIT increases with drainage capacity.

The sign on the cropping pattern indicates that the intensity of DIT adoption increases when tomatoes are mono-cropped. The sign on COSTAC suggests that once a grower has installed DIT on a particular field, an increase in the per-acre capital investment cost results in a greater proportion of the field that is put into DIT. Although this appears counter intuitive, it may be interpreted to mean that growers who adopt DIT are attempting to capture some economies of scale because of the system's large initial capital cost. Similar to the initial adoption decision, DEPTH also affects the percentage of the field area that is allocated to DIT.

POLICY IMPLICATIONS

Policy implications for flexible incentives are discussed from two perspectives: (1) implications for the promotion of the adoption of DIT, and more generally, (2) for the identification of incentives for the adoption of environmental technologies.

Incentives for the Adoption of Drip Irrigation Technology

The results of the adoption model confirm the importance of first considering environmental and crop management variables. This supports the findings of previous DIT adoption studies. In the first stage, non-adopters of DIT who have irrigation water quality problems and high DEPTH could be targeted for educational and/or extension activities that would promote DIT. For growers who have already adopted DIT, targeting could result in more intensive use. Targeting would also allow for a more efficient use of funds to promote increased adoption of DIT.

The significance of DEPTH as a proxy for pumping costs suggests that imposing a fuel or utility tax to raise the cost of pumping would encourage the adoption of DIT. The implementation of an energy tax aimed only at non-adopters of DIT, however, could entail high transaction costs. It is not likely, however, to be economically (or politically) feasible (Segerson, this volume).

Water management districts in south Florida acknowledge that physical environmental factors are important in growers' technology adoption decisions. Some districts have instituted methods to promote increased irrigation efficiency that are more voluntary and educational in nature. For example, mobile irrigation labs have been deployed to some districts to help growers identify ways to improve irrigation efficiency on a field-level basis. Targeting these extension-type efforts toward promoting more intensive use of DIT in locations in which soils have a high drainage capacity and/or DEPTH could result in higher proportions of fields that are allocated to DIT.

The positive Attitude/Social Norm interaction term suggests three policy orientations: (1) research and development should be aimed at decreasing the costs of complementary inputs (for example, fertilizer, fuel and pesticides) that would enhance the expected profitability of DIT and would increase adoption rates; (2) extension efforts to demonstrate the profitability of DIT are necessary complements to research and development; (3) because social public interest norms are important, appeals to support water conservation efforts through public involvement programs and moral suasion would positively affect adoption rates. The interaction between expected profitability and public interest norms suggests that growers' subjective expectations of the profitability of DIT are changing in the same direction as the social norm. Thus, an appeal to growers to support community water conservation goals through public involvement programs, in addition to reinforcing the profit motive, would help increase adoption rates. Likewise, there must also be an appeal to regulators to become more aware of the consequences of design standard type environmental policies on the farming community.

There are potential benefits to farmers and regulators through working together to build environmental social norms. In some districts, the establishment of Agricultural Advisory Groups is a good indicator of the importance of cooperation, as districts begin to realize that public interest norms need to evolve with farmers' input and cooperation.

Soil type, cropping pattern and DEPTH are the important determining factors for decisions that are related to the intensity of adoption. Targeting the extension efforts toward promoting more intensive use of DIT in locations where soils have a higher drainage capacity and/or DEPTH could result in higher proportions of fields being adapted to DIT.

INCENTIVES FOR ADOPTION OF ENVIRONMENTAL TECHNOLOGIES

What are the general implications for flexible incentives from the findings in this chapter? First, blanket performance standards across heterogeneous physical environments are not found to result in an efficient allocation of on-farm resources (Batie and Ervin, this volume; Carpentier and Bosch, this volume). The results of the analysis reveal that it would be more efficient to target areas that have the requisite environmental conditions that are conducive to the benefits that a specific technology can offer.

Second, setting performance standards that can only be met with one technology (in essence, a design standard), reduces growers' flexibility when developing new technologies or management practices that conserve water resources. This makes growers and society worse off.

Third, there is a need for continued research on the decreasing costs and the uses of complementary inputs, such as fertilizer and pesticides, so that the efficiency of capital intensive technologies can be enhanced. This would also help meet goals of sustainability (Khanna et al., this volume).

Fourth, increased educational efforts could positively influence the subjective expectations of potential adopters about the profitability of environmental technologies. These efforts could be combined with public involvement programs to promote resource protection and conservation.

Fifth, and perhaps most important, is that the Batie/Ervin typology (this volume) of flexible incentives should be expanded to include the consideration of public interest norms as an additional mechanism to encourage the adoption of environmental technologies. The results herein show that growers pay attention to the claims of others. The importance of social norms is also supported by Randall's (this volume) suggestion that greater cooperative efforts and information sharing occur between regulators and the regulated in order to address the isolation paradox. There is also a need to develop mechanisms whereby regulators, farmers, and perhaps, citizen interest groups can discuss and debate together what environmental norms should be. These groups can then jointly decide on the technologies to be used to achieve these social norms. Understanding the best mix of flexible incentives to encourage the adoption of environmentally sustainable technologies requires the knowledge of physical conditions and expectations of a technology's profitability, as well as how farmers relate to the broader public interest.

ENDNOTES

1.   See Sen's (1977) Rational Fools: A Critique of the Behavioral Foundations of Economic Theory; Hirschman's (1985) Against Parsimony: Three Easy Ways of Complicating Economic Discourse; Etzioni's (1986, 1987) The Case for a Multiple Utility Conception and his Moral Dimension, Toward A New Economics; and tests of multiple utility by Lynne et al. (1995), Casey (1996) and Casey and Lynne (1997).

2.   For a detailed description of the evolution and the foundation of a multiple utility concept see Casey (1996) and Casey and Lynne (1997). For a discussion of the theoretical merits of the multiple utility concept see Brennan (1989, 1993), Lutz (1993), Hausman and McPherson (1993), and Stewart (1995).

3.   Lynne, et al., (1988, p.2) cite Mitchell (1982, p.294) who notes that "the valence of an outcome for a person is defined conceptually as the strength of his positive or negative affective orientation toward it."

4.   Feather (1982) stated that,

> The terms expectation and expectancy are used interchangeably and they are indexed in terms of the perceived likelihood that an action will be followed by a particular consequence—that is, by a subjective probability that the consequence will occur given the response. Similarly, there is a high degree of overlap in the concepts used to refer to the subjective value of the expected consequences. Among the concepts that have been employed are incentive values, utilities, valences and reinforcement values. (p. 1)

5.   For a detailed description of how the belief and evaluative statements were formulated and scaled for the Attitude and Social Norm Variables, see Casey and Lynne (1997).

6.   Support for including attitudes in a behavioral decision model has come from Nowak and Korsching (1983) who tested attitudinal variables among Iowa farmers and found that risk attitudes have a significant influence on soil conservation decisions. Adesina and Zinnah (1993) found that "perceptions of technology specific characteristics significantly condition technology adoption decisions" (p. 297) among rice producers in West Africa. Lynne et al., (1988) and Lynne and Rola, (1988), showed that soil conservation behavior is influenced by grower's attitudes toward conservation and by context variables (for example, income and farm terrain). Hodges et al., (1993) concluded that strawberry growers' pre-adoption attitudes significantly affected their investment decisions in drip irrigation. East (1993) has shown that attitudes are important to general investment decisions by individuals. However, there is mixed evidence in the literature. Ervin and Ervin (1982) included attitudinal variables in a model of soil conservation decisions, but their empirical results did not indicate that these variables were very important among Missouri farmers. Likewise, Kalaitzandonakes and Monson (1994) did not find attitudes or norms toward soil conservation important in explaining Missouri farmer's technology or management decisions.

7.   The theoretical production function and derived demand model are specified in Casey (1996).

8.   For detailed description of how the belief and evaluative statements were formulated and scaled for the Attitude and Social Norm variables, see Casey and Lynne (1997).

REFERENCES

Adesina, A.A. and M.M. Zinnah. 1993. "Technology Characteristics, Farmer's Perceptions, and Adoption Decisions: A Tobit Model Application in Sierra Leone." *Agricultural Economics* 9: 297–311.

Ajzen, I. 1988. *Attitudes, Personality, and Behavior*. Chicago IL: Dorsey Press.

_____. 1991. "The Theory of Planned Behavior." *Organizational Behavior and Human Decision Processes* 50: 179–211.

Ajzen, I. and M. Fishbein. 1977. "Attitude Behavior Relations: A Theoretical Analysis and Review of Empirical Research." *Psychological Bulletin* 84: 888–918.

_____. 1980. *Understanding Attitudes and Predicting Social Behavior*. Englewood Cliffs, NJ: Prentice-Hall Press.

Bagozzi, R.P. 1984. "Expectancy Value Attitude Models: An Analysis of Critical Measurement Issues." *International Journal of Research in Marketing* 1: 295–310.

Blundell, R. and C. Meghir. 1987. "Bivariate Alternatives to the Tobit Model." *Journal of Econometrics* 34: 179–200.

Brennan, T. 1989. "A Methodological Assessment of Multiple Utility Frameworks." *Economics and Philosophy* 5: 189–208.

_____. 1993. "The Futility of Multiple Utility Frameworks." *Economics and Philosophy* 9: 155–164.

Casey, C.F. 1996. "A Multiple Utility Approach to Understanding Conservation Technology Adoption: Application to the Florida Tomato Industry." Unpublished Ph.D. Dissertation. Department of Food and Resource Economics, University of Florida, Gainesville, FL.

Casey, C.F. and G.D. Lynne. 1997. *Factors Affecting Investment in Drip Irrigation Technology Adoption: A Multiple Utility Approach.* Economics Report, No. ER 97-1, Food and Resource Economics Department, University of Florida, Gainesville, FL.

Caswell, M.F. 1991. "Irrigation Technology Adoption Decisions: Empirical Evidence," in A. Dinar and D. Zilberman, eds., *Economics and Management of Water and Drainage in Agriculture.* Boston, MA: Kluwer Academic Publishers.

Caswell, M. and D. Zilberman. 1985. "The Choice of Irrigation Technologies in California." *American Journal of Agricultural Economics* 67: 224–234.

_____. 1986. "The Effects of Well Depth and Land Quality on the Choice of Irrigation Technology." *American Journal of Agricultural Economics* 68: 798–811.

Davidson, R. and J.G. MacKinnon. 1993. *Estimation and Inference in Econometrics.* New York, NY: Oxford University Press.

Dinar, A., M. Campbell, and D. Zilberman. 1992. "Adoption of Improved Irrigation and Drainage Reduction Technologies under Limiting Environmental Conditions." *Environmental and Resource Economics* 2: 373–398.

East, R. 1993. "Investment Decisions and the Theory of Planned Behavior." *Journal of Economics and Psychology* 14: 337–375.

Ervin, C.A. and D.E. Ervin. 1982. "Factors Meeting the Use of Soil Conservation Practices: Hypotheses, Evidence and Policy Implications." *Land Economics* 58: 277–292.

Etzioni, A. 1986. "The Case for a Multiple Utility Conception." *Economics and Philosophy* 2: 159–183.

_____. 1988. *The Moral Dimension, Toward A New Economics.* New York, NY: The Free Press.

Evensky, J. 1992. "The Role of Community Values in Modern Classical Liberal Economic Thought." *Scottish Journal of Political Economy* 39: 21–38.

Feather, N.T., ed. 1982. *Expectations and Actions: Expectancy Valence Models in Psychology.* Hillsdale, NJ: Lawrence Erlbaum Associates.

Fishbein, M. and I. Ajzen. 1975. *Belief, Attitude and Behavior.* Reading, MA: Addison-Welsey Publishing Co.

Florida Statutes. 1972. *Florida Water Resources Act,* Chapter 373.

Green, G., D.Sunding, D. Zilberman, and D. Parker. 1996. "Explaining Irrigation Technology Approaches: A Microparameter Approach." *American Journal of Agricultural Economics* 78: 1064–1072.

Hausman, D.M. and M.S. McPherson. 1993. "Taking Ethics Seriously: Economics and Contemporary Moral Philosophy." *Journal of Economic Literature* 3(1): 671–731.

Hirschman, A.O. 1985. "Against Parsimony: Three Easy Ways of Complicating Some Categories of Economic Discourse." *Economics and Philosophy* l: 7–21.

Hodges, A., G.D. Lynne, M. Rahmani, F. Casey, and L. LaForest. 1993. *Factors Influencing Adoption of Energy and Water Conserving Technologies in Florida: Revised Final Report to the Florida Energy Extension Service.* Food and Resource Economics Department, University of Florida, Gainesville, FL.

Kalaitzandonakes, N.G. and M. Monson. 1994. "An Analysis of Potential Conservation Effort of CRP Participants in the State of Missouri: A Latent Variable Approach." *Journal of Agricultural and Applied Economics* 26: 200–208.

Khalil, E.L. 1990. "Beyond Self-interest and Altruism: A Reconstruction of Adam Smith's Theory of Human Conduct." *Economics and Philosophy* 6: 255–273.

Lemon, N. 1973. *Attitudes and Their Measurement.* New York, NY: John Wiley and Sons.

Lutz, M. 1993. "The Utility of Multiple Utility: A Comment on Brennan." *Economics and Philosophy* 9: 145–154.

Lynne, G.D. 1995. "Modifying the Neo-classical Approach to Technology Adoption with Behavioral Science Models." *Journal of Agriculture and Applied Economics* 27: 67–80.

Lynne, G.D. and R. Rola. 1988. "Improving Attitude Behavior Prediction Models with Economic Variables: Farmer Actions toward Soil Conservation." *Journal of Social Psychology* 128(1): 19–28.

Lynne, G.D., J.S. Shonkwiler, and L.R. Rola. 1988. "Attitudes and Farmer Conservation Behavior." *American Journal of Agricultural Economics* 70(1): 12–19

Lynne, G.D., J.S. Shonkwiler, and M.E. Wilson. 1991. "Water Permitting Behavior Under the 1972 Florida Water Resources Act." *Land Economics* 67(3): 340–351.

Lynne, G.D., F.C. Casey, A. Hodges, and M. Rahmani. 1995. "Conservation Technology Adoption Decisions and the Theory of Planned Behavior." *Journal of Economic Psychology* 16: 581–598.

Lynne, G.D., and F.C. Casey. Forthcoming. "Regulatory Control of Technology Adoption Decisions by Individuals Pursuing Multiple Utility." *Journal of Socio-economics.*

Maloney, F.E., R.C. Ausness, and J.S. Morris. 1972. *A Model Water Code (with Commentary).* Gainesville, FL: University of Florida Press.

Marella, R.L. 1992. *Water Withdrawals, Use and Trends in Florida, 1990.* Water Resources Investigation, Report No. 92-4 140, U.S. Geological Survey, Tallahassee, FL.

Mitchell, T.R. 1982. "Expectancy Value Models in Organizational Psychology," in N.T. Feather, ed., *Expectations and Actions: Expectancy Value Models in Psychology*, pp. 293–312. Hillsdale, NJ: Lawrence Erlbaum Associates.

Norris, P.E. and S. Batie. 1987. "Virginia Farmers' Soil Conservation Decisions: An Application of Tobit Analysis." *Southern Journal of Agricultural Economics* 19(1): 79–90.

Nowak, P.J. and P.J. Korsching. 1983. "Social and Institutional Factors Affecting the Adoption of Agricultural BMPs," in F. Schaller and H. Bauley, eds., *Agricultural Management and Water Quality*, pp. 349–373. Ames, IA: Iowa State University Press.

Sen, A.K. 1977. "Rational Fools: A Critique of the Behavioral Foundations of Economic Theory." *Philosophy and Public Affairs* 6: 317–344.

Stewart, H. 1995. "A Critique of Instrumental Reason in Economics." *Economics and Philosophy* 11: 57–83.

Weaver, R.D. 1996. "Pro-social Behavior: Private Contributions to Agriculture's Impact on the Environment." *Land Economics* 2: 231–247.

15 TECHNOLOGICAL INNOVATION TO REMOVE WATER POLLUTANTS

Douglas D. Parker
Margriet F. Caswell
University of Maryland, College Park, MD
Economic Research Service
U.S. Department of Agriculture, Washington, DC

Irrigation and precipitation runoff from agricultural fields contributes to topsoil loss and to the sedimentation of river basins. The movement of suspended soil particles from fields to streams can also serve as a transport mechanism for agricultural chemicals. These agricultural nonpoint sources of pollution (NSP) are receiving increased attention from water quality regulators. Potential actions to curb such pollutants include regulating runoff and return flows, restricting chemical applications, and imposing agricultural best management practices (BMP), such as catch-basins and tailwater recovery systems. An improved technology, polyacrylamide polymer (PAM), may offer a cost-effective alternative for reducing the transport of chemicals on soil particles. PAM can be added to irrigation water or spread directly on fields to flocculate the suspended soil particles, causing them to settle out of the water column. This chapter explored the conditions under which the on-farm benefits (such as reduced topsoil loss, increased infiltration, reduced water use and reduced chemical use) will be sufficient to promote the adoption of PAM, and the conditions under which additional incentives may be needed to encourage innovative technological adoption.

INTRODUCTION

The quality of water in the nation's surface waterways has improved significantly since the 1960s. Through technological innovation, strict regulations, improved management and the concentrations and outflows of pollutants have been reduced.

While many sources of pollutants have been identified, the majority of their reductions have been achieved through the treatment of point-source pollution (PSP), which are known to have originated from specific sites or from individual firms. Because these outflows can be pinpointed, they are more easily identified and regulated. Nonpoint-source pollution (NSP) consists of pollutants that enter the environment through diffuse and nonspecific means. Their nature makes them difficult to quantify and to control. NSP are organic or inorganic materials that enter a system in such quantities as to disrupt the natural ecosystem. There are many sources of NSP that affect water quality (such as surface runoff, precipitation, atmospheric deposition, drainage and seepage). NSP may affect surface water and groundwater, and may come from agricultural or urban sources. This chapter is concerned with the control of agricultural NSP that affects surface water quality.

NSP is recognized as a significant contributor to surface water deterioration. Difficulties with the identification or quantification of these sources of pollution have slowed efforts to control and to regulate them. Some forms of NSP have been controlled through better information and management that has enabled regulators to reclassify them as PSP. Other forms of NSP remain unregulated and efforts to control them often require the cooperation of large groups of polluters in order to reduce emissions. Enforcement is difficult because of the common pool nature of the problem.

In order to control NSP, regional cooperation among those who generate the pollutants is often necessary (Randall, this volume). For example, storm drainage management agencies are sometimes created in urban areas to control the runoff from city streets. Urban runoff often contains motor oil and other automobile-related pollutants; therefore, education of the urban dwellers that pertains to the proper disposal of these potential pollutants can reduce NSP pollutants. More extensive management could require settling basins or the partial treatment of storm water. In agricultural regions, the education of growers (with regard to the management of agricultural inputs) has reduced the nutrient and pesticide runoff that ends up in the sensitive waterways. In the irrigated west, drainage management agencies could take the lead in the education of growers and in the promotion of BMP (such as catch basins, cover crops and tailwater return systems) to reduce the level of runoff and the concentration of pollutants in the runoff.

Agricultural NSP is generally controlled through field-level BMP or through reductions in inputs that are related to the NSP problem. For irrigated agriculture, modifications in the use of chemical inputs and of irrigation water can reduce the NSP of groundwater (Helfand and House, 1995). The relationship between NSP and agricultural inputs has been modeled in the drainage literature (Dinar et al., 1991). While the majority of the literature has focused on the NSP of groundwater from agricultural pesticides, most literature on the NSP of surface waters has focused on nutrients and salinity from agricultural runoff (Wu and Segerson, 1995; Jacobs and Casler, 1979; Gardner and Young, 1988). Most of the theoretical literature on agricultural pesticides considers the potential of various forms of regulation and of taxation to control NSP (Shortle and Dunn, 1986; Griffin and Bromley, 1982).

This chapter considers forces that affect technological adoption as a means to control NSP. Many potential pollutants become problems only after they are transported from an environment in which they cause no harm (such as a field) to an environment that cannot assimilate them (such as a stream). Technological innovations that control the movement of pollutants can reduce NSP loadings. By keeping agricultural chemicals (such as fertilizers and pesticides) in the field, technological innovations contribute to greater production levels and reduce the need for additional inputs to replace those lost through percolation and runoff. BMP that control sediment runoff include sediment basins and more advanced irrigation methods. Most of these BMP have been available for many years. This chapter looks at the introduction of an improved technology, polyacrylamide polymers (PAM), to control runoff.

METHODOLOGY AND THEORETICAL BACKGROUND

The quantity of irrigation runoff or drain water from farm fields depends upon the quantity of water applied, soil types and slopes, and types of irrigation technology that are used. The quantity of NSP from agricultural drain water is determined primarily by the choice of irrigation technology and by other management practices. Rules, regulations and other incentives that influence the growers' choice of irrigation technology and of management practices can be used to change runoff levels and, thus, NSP. There is extensive literature on irrigation technology adoption that relates water-use efficiency to irrigation technology choice (Caswell and Zilberman, 1985; Dinar and Zilberman, 1991). The conclusions from these literary works will be followed when we model both irrigation efficiency (water available to crops) and other input efficiencies (nutrients that are available to the crop) in this chapter. The proportion of the input that is ultimately made available for crop uptake defines effective water and chemical use. For simplicity, deep percolation is ignored and the focus of this chapter is on surface runoff. Factors that determine effective water and chemical use include irrigation technology choice (for example, sprinklers, furrows and gated pipes) and BMP (for example, laser leveling, low or no tillage and cutback irrigation). Thus, for a given irrigation technology or BMP, i, effective water, e_i^w, will equal applied water, a_i^w, minus runoff, R_i^w. Similarly, fertilizer availability is dependent upon the water parameters so that effective fertilizer, e_i^f, is equal to applied fertilizer, a_i^f, minus fertilizer runoff, R_i^f, where R_i^f is a function of a_i^f, and R_i^w, or

(1) $\quad R_i^f = g\,(a_i^f, R_i^w)$.

Thus, increasing water (fertilizer) application and/or reducing water (fertilizer) runoff will increase effective water (fertilizer) availability.

Crop production is determined by water and fertilizer availability. Holding all other factors constant, the relationship can be expressed as

(2) $\quad Q = h(e_i^w, e_i^f)$.

The grower's profit function for using a given technology can be written as

$$(3) \qquad \pi_i = PQ - P_i^w a_i^w - P_i^f a_i^f - C_i,$$

in which P is the price of the crop, P_i^w is the cost of water when using technology i, P_i^f is the cost of the fertilizer when using technology i and C_i is the fixed cost that is associated with technology i. For simplicity in notation, variable input application costs are included in the price of the input. Lower operating and maintenance costs for technology will lower the effective price of the input, which will result in the input price being technology specific. Therefore, more efficient irrigation technology can impact grower profits through the yield function (by having more water or nutrients available), the input costs (by reducing the level of applied water and nutrients) and the fixed costs of technology.

Environmental damage from agricultural drainage into a waterway will depend upon the quantities of drainage water and the concentrations of various pollutants in that drainage. Agricultural runoff of water, R_i^w, fertilizer, R_i^f, and concentrations of pollutants will depend upon the water and fertilizer applied and the technology, or BMP, used. Additional methods to control pollutants in agricultural runoff, which do not affect crop production but do add costs to the grower's profit function, are available. For instance, sediment basins remove sediments from runoff water before it enters a stream. The sediments can then be removed from the basin and returned to the field. While sediment basins may not change the amount of water leaving the field, they can reduce the damage from agricultural runoff to the extent that the sediment and the chemicals attached to the sediment cause environmental damage in streams. The impact of sediment basins can be represented by including the variable, O_i^f, in the function for fertilizer runoff as

$$(4) \qquad R_i^f = g(a_i^f, R_i^w, O_i^f).$$

Therefore, fertilizer runoff can be reduced through reductions in fertilizer applications, reductions in water runoff and the implementation of alternative controls—such as sediment basins. Sediment basins do not change effective water use, but they do add a cost of operation (for example, land taken from production and cleanout costs) to the grower's production function.

The damage to environmental resources or to human health from degraded water quality will depend upon the amounts of water and nutrients that leave the field, and upon the concentration of the nutrients

$$(5) \qquad D = m(R_i^w, R_i^f / R_i^w).$$

Thus, the effectiveness of practices that reduce pollutant loads may depend upon their impact on water runoff. This was demonstrated in a study of the San Joaquin River Basin (SJRB) (USDA et al., 1995).

WEST STANISLAUS HYDROLOGIC UNIT AREA

The San Joaquin River NSP problem is an example of a situation in which environmental policy can allow flexibility to achieve environmental objectives. According to Batie and Ervin (this volume), there are four requirements that policymakers need to achieve agro-environmental goals within flexible public policies: (1) measurable objectives; (2) knowledge of the pollutant sources; (3) knowledge of the pollution transport mechanisms; and (4) knowledge of the impact of pollutants on the environment. For the West Stanislaus Hydrologic Unit Area (HUA), these requirements are met. Therefore, if sufficient forces were to exist within the HUA to control and to monitor pollution, flexible policies could be used to effectively achieve its agro-environmental goals.

The SJRB drains the northern portion of California's San Joaquin Valley. A portion of the river is located in Stanislaus County—about 100 miles south of the city of Sacramento. The California State Water Resources Control Board (SWRCB) has declared 100 miles of the San Joaquin River as an impaired waterway. Water and sediments in the river were found to contain excessive levels of pesticides and fertilizers, allegedly deposited there by local agricultural runoff and by drain water. The SWRCB determined that the levels of organochlorine pesticide residues (DDT, DDD, DDE and Dieldron) in the river's sediment exceeded the acceptable levels. Sediment in the river exceeded the aquatic life criterion for DDT concentrations of 0.001 micrograms per liter. The SWRCB determined that the source of the DDT was the runoff from the agricultural fields of Stanislaus County. DDT was used legally on these fields more than 25 years ago. Today, DDT remains in the fields, adsorbed to the soil particles. Excess runoff or drainage from the fields allegedly carried these soil particles, with the DDT attached, into the river. From there, the DDT entered the lower end of the food chain through sediment-ingesting invertebrates. The chemical then bio-accumulated through the food chain and reached toxic levels for fish and bird species.

In response to the SWRCB demand that this waterway be cleaned up, the Stanislaus County's University of California Cooperative Extension Office (CEO), the U.S. Department of Agriculture's Soil Conservation Service (USDA/SCS), the Agricultural Stabilization Conservation Service (USDA/ASCS), and the West Stanislaus Resource Conservation District (RCD) collaborated to form the West Stanislaus HUA. The HUA has taken the lead to promote the voluntary adoption of BMP to reduce NSP in the area. The SWRCB has allowed the HUA (and the growers in its region) to attempt voluntary compliance. As long as progress toward a well-defined goal is being made, the SWRCB has indicated that it will allow the local landowners and land users the leeway to solve the problem themselves. If consistent progress is not made, the SWRCB will intercede and will require land users to adopt specific management practices to address the problem. The threat of the SWRCB to impose mandatory management practices forces farmers to choose between self-compliance or mandatory controls. This voluntary approach by the state allows growers to choose to meet the goals through a flexible, self-governing system. As noted by Segerson (this volume), to the extent that a voluntary system allows greater flexibility to meet the objectives, it has the potential to reduce costs through innovation.

The HUA represents 134,000 acres of land in Stanislaus County, California, of which 129,000 acres are irrigated cropland. More than 40 major crops are grown in the area, which generate more than 100 million dollars in annual revenues. Sediment loss on this land has been estimated at more than 1.5 million tons per year, with 95 percent of this sediment settling into the SJRB. Water is supplied to the land from the Federal Central Valley Project, the State Water Project and directly from the SJRB. A small portion of water also comes from the local groundwater aquifer. Eight creeks and 18 agricultural drains return agricultural runoff to the San Joaquin River. For several months per year, this agricultural runoff makes up the entire water flow of the river. Thus, the quality of water that leaves the fields is extremely important for the preservation of the river's ecosystem. In addition to damage from agricultural chemicals, sediment runoff (from the irrigated fields) damages the ecosystem directly through the reduction of light infiltration and the clogging of the fish spawning beds. Irrigation induced erosion in the HUA has varied from 1.9 tons per acre per year to 14.7 tons per acre per year, with an average erosion rate of 11.6 tons per acre per year (USDA et al., 1991).

The goal of the HUA is to satisfy the SWRCB mandate by reducing sediment loads. Specifically, the HUA is striving to reduce the suspended solids in agricultural drainage from every field to a maximum level of 300 milligrams per liter. On average, this represents an area-wide reduction of 80 percent, from a baseline of 1,500 milligrams per liter. The management of the HUA seeks to achieve this goal through the voluntary adoption of BMP. The plan is consistent with the 1991 California Inland Surface Water Plan that utilizes three levels of implementation: (1) voluntary implementation of BMP; (2) regulatory-based encouragement of BMP use; and (3) regulatory implementation (such as wastewater discharge permits).

An excellent summary of sediment control techniques can be found in Carter et al. (1993). The effectiveness and costs of the HUA's recommended BMP are shown in table 15.1. The BMP are divided into three categories: (1) moderate sediment reduction (less than 50 percent); (2) significant sediment reduction (50–89 percent); and (3) nearly complete control of off-farm sediment (90–100 percent). The cost of implementing the BMP has averaged $15 per acre for moderate control, $20 per acre for significant control and $50 per acre for nearly complete control. The benefits from implementing BMP depend upon the specific BMP that are used. All of the BMP provide public benefits in terms of reduced erosion, sediment, chemical runoff and increased environmental quality. Some of the BMP could provide private benefits, such as improved irrigation efficiency. This would allow growers to decrease water use and save on water costs. Similarly, BMP could increase the efficiency of other chemicals, such as fertilizers. They could also reduce the costs of these inputs and increase the yields and revenues to the growers. In the moderate control category, tarps in ditches and filter strips are expected to provide no additional on-site benefits. Cutback streams would decrease applied water costs. Gated pipes, conventional tillage, and gated pipes with tarps could affect water use and yield. In the significant control category, all three BMP could affect water use and yield. In the nearly complete

TABLE 15.1 Cost and Effectiveness of Various Management Practices to Reduce Sedimentation for Row and Field Crops.

Practices	Costs Per Acre		Sediment Rate
	Range	*Net average* [a,b]	
	dollars per year		*tons per acre*
Baseline	—	—	5.0–19.0
Moderate Reduction (≤50%)			
Gated Pipe–No IWM	57.00–89.00	71.00	11.0
Conservation Tillage	0.00–42.00	12.00	10.5
Tarps in Ditches	7.00–21.00	14.00	6.0
Filter Strips	2.00–9.00	6.00	9.0
Gate Pipe with Tarps	72.00–96.00	83.00	6.0
Cutback Streams	4.00–9.00	6.00	4.5
Gated Pipe with Cutback Streams	—	61.00	4.5
Significant Reduction (51–89%)			
Surge Irrigation	—	99.00	2.0
Cutback Stream with Tarps	8.00–26.00	16.00	2.0
Gated Pipe with Tarps and Cutback Streams	55.00–87.00	70.00	2.0
Nearly Complete Elimination (90–100%)			
Sediment Basins	18.00–95.00	53.00	0.5
Tailwater Return	26.00–92.00	55.00	0.1
Sediment Basin with Tarps	16.00–74.00	42.00	0.5
Sediment Basin with Surge	93.00–125.00	108.00	0.5
Sediment Basin with Tarps and Cutback Streams	12.00–44.00	26.00	0.5
PAM	8.00–40.00	18.00	< 0.5

[a] The net average annual cost is determined based on how the practice system changes the typical farm operation. Some changes in the costs must be annualized while other costs are annual expenses. All changes in costs and returns have been discounted and amortized over a 20-year evaluation period using a 12 percent interest rate.

[b] There are no yield improvements factored into these values. Some practices may improve yields depending on the individual farming operation.

Source: USDA Soil Conservation Service et al., 1992.

control category, the sediment basins and the sediment basins with tarps would provide no additional on-farm benefits. The sediment basins with tarps and with cutback and tailwater return could reduce water cost. Sediment basins with surge irrigation could reduce water cost and improve yield.

Any gains in yield from the adoption of BMP are expected to be modest. Furthermore, it is difficult to determine whether these gains in yield could result from the BMP themselves or from more careful management that would result from the process of implementing them. Thus, gains in yield could be possible through a more intensive management of the existing systems.

Some BMP have the potential to reduce water applications from one-quarter acre-foot per acre per season to nearly one and one-quarter acre-feet per acre per season. Local water rates vary from almost $5 per acre-foot to nearly $25 per acre-foot. In most cases, the growers' benefits, in terms of water savings, would not cover the costs of the BMP.

The BMP most likely to cover its cost is the method of cutback irrigation. This practice involves returning irrigation water back to the field once it has reached the end of the furrow and reducing the flow rates to reduce runoff. Unfortunately, it also requires water providers to reduce the flow of water into the source canal or ditch. In most cases, the institutions (water districts) that would do this have neither the personnel nor the equipment to provide growers with this extra flexibility. Thus, in most cases, the implementation of cutback irrigation would reduce on-farm profits.

In order to address the problems that are associated with the use of cutback irrigation, the HUA has obtained funds through its member agencies to provide educational programs, demonstration research and cost sharing to growers. In this way, they are better able to induce growers to adopt some set of BMP and, thus, to improve off-farm environmental quality in the SJRB.

Structural (managerial) BMP (USDA et al., 1995) has been adopted by 24 percent (31,000 acres) of the area in the HUA. The use of BMP is estimated to have reduced DDT runoff by 996 pounds, sediment runoff by 526,000 pounds and total applied water by 31,000 acre-feet. BMP previously had been implemented in 42 percent of the region. The remaining 34 percent of the area have minimal BMP use and have land in need of significant improvements.

There is concern within the HUA that they will be unable to entice individuals, who control the remaining 34 percent of the acreage, to voluntarily adopt the BMP that is necessary to further reduce sediment loads. Recent regulations by several water supply districts should lead to a small increase in the adoption of BMP practices. These districts face water quality concerns within their supply systems because runoff from some fields becomes the water supply to others. Thus, these local water supply agencies are going to require their members to meet quality standards.

Some growers in the HUA, who have voluntarily adopted BMP and cleaned up their runoff, are considering the formation of a local drainage authority. This agency would require growers to adopt BMP to improve water quality. The growers who have voluntarily complied are concerned that their efforts to solve the problem locally may not count if the quality of water in the river does not meet

California standards. If the growers fail to clean up the river, California authorities have indicated that they would issue waste discharge requirements (WDR) to individuals or to an agency who represents the area (UCCE, 1996b). Local growers believe that this would be expensive for them and that it could seriously affect their ability to operate.

The main factor to discourage growers from voluntarily adopting BMP is cost. Some BMP can seriously reduce profits while others can increase costs only marginally. With local cost-sharing assistance, BMP can usually be implemented with only small short-run reductions in growers' profits. Many of the BMP are profitable in the long run because they improve soil conditions.

A NEW TECHNOLOGY

A new technology is being tested in the region that could allow growers to reach the sediment reduction goals. Properly used, this product could reduce sediment runoff to acceptable levels while it could increase grower profits. Increased profits could result from water savings and from yield increases. Many fields in the HUA have very low infiltration rates once the furrow is wetted, which results in runoff rates of up to 80 percent of the water applied. This innovation can significantly increase infiltration rates throughout the irrigation system and is being used on several fields in the HUA. Researchers are closely monitoring its use to determine the impact it has on growers' profits for several different crops that are under a variety of conditions. If successful, this innovation could help growers in the HUA meet California goals and could eliminate the threat of state regulation and control. The new technology is polyacrylamide polymers (PAM), which can settle sediments out of the water column.

Polyacrylamide Polymer

PAM is a polymer that was first used during World War II to hold together the soils of new roads and landing strips. From the early 1950s through the 1970s, there have been several waves of enthusiasm for the use of PAM as a soil stabilizer in agriculture. Recent improvements in the formulation and manufacture of PAM have made it more effective and less expensive, which makes it economical for some agricultural uses. Sample formulations of the improved PAM have been approved by the Food and Drug Administration (FDA) and by the Environmental Protection Agency (EPA) for sensitive uses—such as food preparations and potable water treatment. These formulations also have been declared safe for aquatic life. PAM decomposes into water and carbon dioxide in the soil at a rate of about 10 percent per year. Improved formulations of PAM for use in reducing agricultural erosion have allowed its application rates to be reduced by at least 200-fold. The combination of lower application rates, lower costs and increased environmental concerns over the impact of agricultural drainage has led to new research into the application and benefits of PAM (Lentz, 1996; Sojka and Lentz, 1994).

The majority of research to date has been on furrow irrigated crops in the west. PAM is also effective at reducing erosion on sprinkler irrigated fields, although the gains are smaller. Erosion, induced from furrow irrigation, is estimated to be between 2 and 22 tons per acre per year (Sojka and Lentz, 1996a). This occurs because of the velocity of the water in the furrow, which determines the level of shear or drag force that pulls on the soil particles. The velocity of the water also determines its sediment transport capacity. Initial furrow irrigation loosens particles from the rough soil. These particles drift downstream and fill in the cracks and pores of the soil. This increases water velocity and reduces the infiltration rates, thus increasing the velocity even further. PAM holds the rough soil in place, slows the water velocity and keeps the soil pores open to better absorb the water. The slower water velocity causes less erosion and transports less sediment.

PAM is also more effective at reducing the transport of finer soil particles. Practices (for example, sediment ponds and filter strips) that reduce erosion losses as much as 60 percent to 70 percent still leave most of the finer clay particles suspended in the drain water. These are the very particles that are most likely to be contaminated with DDT and agricultural chemicals. These fine particles cause additional environmental harm when they disrupt stream ecosystems by diffusing natural light.

In 1995, PAM was used on an estimated 50,000 U.S. acres (Sojka and Lentz, 1996b) that reduced erosion by an estimated one million tons. On Idaho soils, PAM has been shown to increase water infiltration from 15 percent to 50 percent (Sojka and Lentz, 1996a). With the proper management, PAM use could translate into direct savings for applied water. The increased capacity of lateral wetting, which would result from the reduced erosion of the furrow, would imply even greater water savings early in the season when plants have a small root structure (Sojka and Lentz, 1996b). In Arizona, PAM reduced per-acre water use on cotton by one-half per acre-foot, from 4.2 acre-feet per acre to 3.7 acre-feet per acre. Growers also experienced higher yields from better soil wetting and reduced wind erosion. One grower estimated that at $5 per acre, PAM reduced his water consumption by three-fourths of an acre-foot. This can be calculated as an implicit value of water of $6.67 per acre-foot, which is well below the cost of water in Arizona.

Polyacrylamide Polymer and Nonpoint-source Pollution

There has been a limited amount of research to determine the effectiveness of PAM for the control of agricultural chemical NSP. In repeated trials on three Idaho fields, PAM reduced nitrate and total nitrogen runoff by up to 86 percent (Bahr and Stieber, 1996). Nitrate losses were high at the beginning of the irrigation in both cases, but dropped off significantly in the PAM-treated furrows. Overall, nitrogen losses decreased from 20 pounds per acre to 2 pounds per acre in a single irrigation: Approximately 90 percent of the nitrogen that was lost in the control furrow were adsorbed to soil particles. PAM also was found to reduce phosphorus losses by 79 percent. This discovery translates to an increased level of effective nutrients left in the soil that, in turn, can reduce grower costs (because growers need to use fewer inputs) and increase yields (because more of the inputs remain in the soil).

PAM was also shown to be effective in the reduction of pesticide transport and runoff. Significant reductions in runoff of EPTC, Trifluralin and Bromoxynil also have been documented. Bromoxynil in the tailwater was reduced from 3.1 parts per billion to 0.4 parts per billion (Bahr and Stieber, 1996). This study also found significant reductions of chlorpyrifos, oxyflourfen and pendimethilin in the sediments that were transported off the fields.

PAM in the HUA

PAM field trials in the West Stanislaus HUA demonstrated increased infiltration efficiency from 15 percent to 47 percent (Lilleboe, 1995). PAM reduced suspended solids by 99.7 percent and sediment concentration by 60–75 percent (McCutchan et al., 1993). Thus, for the HUA, some additional sediment reduction controls may be necessary. The use of PAM could significantly reduce the extent and cost of other alternatives.

The decision rule regarding the use of PAM, using only the water savings benefit, depends upon the price of applied water, the total cost to apply the PAM (in cost per acre-foot of applied water), and the percent increase in infiltration is shown in equation (6). A single formula shows that grower profits will increase when

$$(6) \qquad P^w > P^{PAM} \left(1/IR - 1 \right),$$

in which P^w is the price of water, P^{PAM} is the price of PAM and IR is the increase in percentage of irrigation efficiency.

In a recent field trial of PAM in the HUA, one grower used the new power block formulation at a rate of only 0.5 parts per million in applied water (UCCE, 1997). This treatment reduced runoff by 50 percent and reduced the sediment load by 75 percent. Using the water savings formula and a cost for PAM of $6 per acre-foot, the growers' decision rule would be to use PAM if the cost of his water were greater than $12 per acre-foot. In another local field trial, PAM was used on a tomato field to test infiltration and yield during an entire season. In this trial, the cost of PAM was approximately $3 per acre-foot. The study found an increase in the infiltration rate of 30 percent (UCCE, 1996a). Using the formula previously described, the water savings could cover the cost of PAM if the cost of water were greater than $7 per acre-foot. Yields increased a surprising 12 percent, or 6 tons per acre, in this trial. At a farm price of $53 per ton, grower revenue increased by $305 per acre. The total cost of PAM was only $10 per acre!

None of these studies, however, quantified the savings that PAM could create for other inputs. Using information from the Idaho case (Bahr and Stieber, 1996), PAM could reduce nitrogen runoff by as much as 86 percent. Growers could adjust for this reduction in nutrient runoff by simply reducing their nutrient applications. The savings to the grower would depend upon the original level of runoff he or she would have and on the quantity of the nutrient that would remain in his soil, which would be readily available to the crop. A decision rule similar to that for water savings could be devised to determine the profitability of using PAM to reduce nutrient applications.

CONCLUSION

Irrigation and precipitation runoffs from agricultural fields contribute to topsoil loss and to the sedimentation of river basins. The movement of soil particles from fields also transports agricultural chemicals, such as pesticides and fertilizers, to streams and to other natural bodies of water. This chapter assesses the potential of polyachrylamide polymers (PAM) to reduce NSP. PAM could be added to the alternative management practices and technological innovations for the control of NSP. Long-term research on the use of this product in other applications has improved its efficacy and reduced its costs. Environmental concerns have led to new research on the application of the improved PAM to control NSP.

The use of this innovation is progressing from the small research stage to a stage of use and testing by private growers. Extension and education continue to play important roles to promote this product and its proper use. At this stage in its promotion, each application of PAM must be adjusted to account for variations in field conditions and of grower management practices. The product is simple to use, and most growers have learned how to use it effectively within just a few irrigation trials. The risks from improper use are small. Under-use leads to no new benefits while overuse wastes the product: Overuse does not, however, add to the NSP problem.

Currently, the incentives to use PAM include cost savings for water, cost savings for crop nutrient inputs, and the mitigation of the threat of more strict state controls if erosion and NSP are not otherwise reduced. To the extent that water institutions in the region distort the opportunity cost for water, the incentive to use this product will be less than optimal.

Many people expect that, in the long run, Pest Control Advisors (PCAs) or other farm consultants will provide producer education concerning the benefits and uses of PAM (Khanna et al., this volume). Some PCAs have already begun advising customers on the use of PAM. Eventually, growers may find that the routine method for the supply and application of PAM will be similar to that now used for nutrient applications: A consultant would provide the correct mixture in a tank and the irrigator would just turn on the switch and adjust the flow.

Additional savings from the use of PAM need to be further explored. Use of PAM keeps soil on the field that, in turn, reduces on-farm maintenance of drainage ditches and sediment basins. For some fields, savings are also gained from the reduced transport of soil from the head of the furrow to the tail of it. As soil is moved down the furrow and deposited at the lower end of the field, the overall slope of the field is reduced, because the sediment builds up. Decreasing the sedimentation effect can create long-run savings by postponing field grading maintenance.

Finally, this chapter demonstrates the benefits from a flexible institutional control structure. The status of the HUA as an organization that promotes voluntary compliance toward water quality goals allows for flexibility in the testing and adopting of new NSP control alternatives. Because PAM is approved for use in agriculture, field experimentation and the use of PAM has not required agency approval as a means for meeting the goals of the HUA for NSP reduction. The HUA provides information and education and does not force growers to adopt any given set of BMP. This allows growers to choose methods that are best for them. This case study demonstrates the benefits of maintaining a flexible institutional structure that can be used to meet local environmental goals.

REFERENCES

Bahr, G.L. and T.D. Stieber. 1996. "Reduction of Nutrient and Pesticide Losses Through the Application of Polyacrylamide in Surface Irrigated Crops," in Sojka, R.E. and R.D. Lentz eds., *Managing Irrigation Induced Erosion and Infiltration with Polyacrylamide*, pp. 41–48. Miscellaneous Publication No. 101-96, University of Idaho, Twin Falls, ID.

Carter, D.L., C.E. Brockway, and K.K. Tanji. 1993. "Controlling Erosion and Sediment Loss from Furrow Irrigated Cropland." *Journal of Irrigation and Drainage Engineering* 119 (6): 975–988.

Caswell, M. and D. Zilberman. 1985. "The Choices of Irrigation Technologies in California." *American Journal of Agricultural Economics* 67: 224–233.

Dinar, A., S.A. Hatchett, and E.T. Loehman. 1991. "Modeling Regional Irrigation Decisions and Drainage Pollution Control." *Natural Resource Modeling* 5 (2): 191–212.

Dinar, A. and D. Zilberman. 1991. *Effects of Input Quality and Environmental Conditions on the Selection of Irrigation Technologies*, pp. 229–250. Norwell, MA: Kluwer Academic Publishers.

Gardner, R.L. and R.A. Young. 1988. "Assessing Strategies for the Control of Irrigation Induced Salinity in the Upper Colorado River Basin." *American Journal of Agricultural Economics* 70 (1): 37–49.

Griffin, R.C. and D.W. Bromley. 1982. "Agricultural offoff as a Nonpoint Externality: A Theoretical Development." *American Journal of Agricultural Economics* 64: 547–552.

Helfand, G.E. and B.W. House. 1995. "Regulating Nonpoint Source Pollution under Heterogeneous Conditions." *American Journal of Agricultural Economics* 77 (4): 1024–1032.

Jacobs, J.J. and G.L. Casler. 1979. "Internalizing Externalities of Phosphorus Discharges from Crop Production to Surface Water: Effluent Taxes versus Uniform Reductions." *American Journal of Agricultural Economics* 61 (2): 309–312.

Lentz, R. D. 1996. "Irrigation," pp. 162–165. *McGraw-Hill Yearbook of Science and Technology, USA*. New York, NY: McGraw-Hill, Inc.

Lilleboe, D. 1995. "PAM: An Ally for Furrow Irrigators." *The Sugarbeet Grower*, pp. 8–10. (February)

McCutchan, H., P. Osterli, and J. Letey. 1993. "Polymers Check Furrow Erosion, Help River Life." *California Agriculture* 47 (5): 10–11.

Shortle, J.S. and J.W. Dunn. 1986. "The Relative Efficiency of Agricultural Source Water Pollution Control Policies." *American Journal of Agricultural Economics* 68: 668–677.

Sojka, R.E. and R.D. Lentz. 1994. "Time for Yet Another Look at Soil Conditioners." *Soil Science* 158(4): 233–234.

_____. 1996a. "A PAM Primer: A Brief History of PAM and PAM-related Issues," in R.E. Sojka, and R.D. Lentz, eds., *Managing Irrigation Induced Erosion and Infiltration with Polyacrylamide*, pp. 11–20. Miscellaneous Publication No. 101–96, University of Idaho, Twin Falls, ID.

_____. 1996b. "Polyacrylamide for Furrow Irrigation Erosion Control." *Irrigation Journal* 64: 8–11.

USDA et al. (U.S. Department of Agriculture; Soil Conservation Service; U.S. Department of Agriculture; Agricultural Stabilization and Conservation Service; University of California; Cooperative Extension; and West Stanislaus Resource Conservation District.) 1991. *West Stanislaus Hydrologic Unit Area Project: 1991 Progress Report*. CA: Stanislaus County.

_____. 1992. *West Stanislaus Hydrologic Unit Area Project: 1992 Progress Report*. CA.

_____. 1995. *West Stanislaus Hydrologic Unit Area Project: 1995 Progress Report*. CA.

USDA/SCS (U.S. Department of Agriculture, Soil Conservation Service). 1992. *West Stanislaus Sediment Reduction Plan*. Stanislaus County, California.

UCCE (University of California, Cooperative Extension). 1996a. *Formation of Drainage Authority Becomes a Possibility*, p. 1. Modesto, CA: Westside Water.

_____. 1996b. *Preliminary Polymer Trial Results Promising but Inconclusive*, p. 5. Modesto, CA: Westside Water.

_____. 1997. *Powerblocks Revisited*, pp.1–2. Westside Water, Modesto, CA.

Wu, J. and K. Segerson. 1995. "The Impact of Policies and Land Characteristics on Potential Groundwater Pollution in Wisconsin." *American Journal of Agricultural Economics* 77 (4): 1033–1047.

16 HEALTH RISK INFORMATION TO REDUCE WATER POLLUTION

Scott M. Swinton, Nicole N. Owens and Eileen O. van Ravenswaay

Michigan State University, East Lansing, MI
U.S. Environmental Protection Agency
Washington, DC
Michigan State University, East Lansing, MI

A growing body of empirical studies indicate that farmers are concerned about how agricultural practices may affect health risks and environmental quality. These studies suggest that farmers are not simply profit maximizers. Instead, they have multiple objectives that include health and environmental concerns. As a result, their privately optimal behavior can result in less use of polluting inputs than would result from straight profit maximization. A recent survey of Michigan corn growers found that many do care about herbicide risks, but that growers often lack adequate information about associated health and environmental risks. Results on willingness to pay (WTP) for reduced risk from herbicide leaching, carcinogenicity and fish toxicity suggest that better information could induce crop farmers to reduce nonpoint-source pollution.

INTRODUCTION

Since the 1970s, a variety of regulatory actions have targeted nonpoint-source pollution (NSP) of water from crop agriculture. Most of those regulatory actions were motivated by the assumption that farmers and agribusinesses would not regulate themselves, and so needed to be coerced or bribed to meet public water quality standards (Ribaudo and Caswell, this volume; Ogg, this volume). We contend that it is a mistake to assume that self-regulation cannot occur. We develop a behavioral model and offer empirical evidence of the potential for voluntary NSP abatement by farmers based on their concerns about health and environmental quality.[1]

Regulatory Attempts to Reduce Crop-related Water Pollution

NSP of water associated with crop agriculture in the United States can result from the use of crop nutrients and pesticides or from soil erosion. Agricultural water pollution typically occurs via leaching or erosion. Water-soluble pesticides and nitrates can leach into ground and surface waters. Insoluble pesticides and fertilizer nutrients can adsorb to eroding soil particles and be carried with them into surface waters (Parker and Caswell, this volume). Recent studies at the U.S. Department of Agriculture (USDA) and the U.S. Environmental Protection Agency (EPA) documented that substantial areas of the United States have experienced, or are at risk of experiencing, groundwater or surface water contamination from agricultural nutrients and pesticides (Kellogg et al., 1992; GAO, 1993; NRC, 1993; EPA, 1990; Hallberg, 1989). These areas correspond to those parts of the country where intensive crop production takes place, where soils are permeable and/or where aquifers are relatively shallow.

The crops most closely associated with NSP are those that cover the greatest acreage—corn, soybeans, cotton and wheat. Due to their widespread cultivation, these crops receive the majority of the agrochemical inputs applied. Results from the 1995 USDA Cropping Practices Survey showed that herbicides were applied to 97 percent of U.S. corn, soybean and cotton acres and to 56 percent of winter wheat acres (Padgitt, 1997), while insecticides were applied to 28 percent of corn, 76 percent of cotton, and under 6 percent of winter wheat and soybean acres. In the same year, commercial fertilizers were applied to 98 percent of corn, 87 percent of cotton, 28 percent of soybean, and 87 percent of wheat acres (Taylor, 1996).

Evidence of direct agricultural water pollution and vulnerability to further contamination has triggered a number of regulatory abatement efforts in the United States. Most of them either encourage or coerce farmers to curtail practices that cause NSP. Since 1985, the U.S. farm bills have followed a *carrot* and *stick* approach. As examples of *carrot*-style positive incentives, minimal crop residue on fields over the winter has been required in order to qualify for deficiency payment benefits (conservation compliance). Likewise, the Conservation Reserve Program has leased highly erodible cropland to keep it out of production. More coercive approaches undergird pesticide policy in the Federal Insecticide, Fungicide and Rodenticide Act, as amended, and in the Food Quality Protection Act of 1996 (Benbrook et al., 1996). These acts specify a rigorous pesticide registration process. Pesticides that fail to meet established standards (or whose manufacturers decline to perform the required tests) are withdrawn from the market. From a farm level perspective, the impact is equivalent to a ban on specified uses. Certain states ban selected pesticides locally. Wisconsin, for example, forbids the use of atrazine herbicide over aquifers vulnerable to leaching (Nowak et al., 1993). A nonregulatory, coercive approach to the abatement of environmental externalities from agricultural production comes from nuisance lawsuits against farmers, in which "nuisance" is defined as injury to the health or enjoyment of property rights (Hamilton, 1990).

How Valid Are Assumptions Behind NSP Abatement Theory?

The conceptual framework behind government programs that bribe or coerce farmers to abate agricultural environmental contamination is one that assumes farmers are well-informed profit maximizers. The implication is that they are oblivious to health and environmental risks from their input choices, because these risks are economic "externalities" that do not affect their bottom line. An alternative view is embodied in conceptual models that view farmers as pursuing several goals (in a multi-attribute utility function), which includes health or environmental quality. In these models, abatement or risk-averting behavior activities are shown to be socially optimal. In some of them, abatement can be privately optimal if the producer cares about health or environmental quality (Zilberman and Marra, 1993; Swinton, 1998; Cropper and Freeman III, 1991; Owens et al., 1995). These two conceptual approaches lead to rather different prescriptions for policy, so it is important to answer the empirical question of whether health and environmental quality matter to farmers' objectives. Specifically:

Do farmers care about health and environmental quality? If so, how much?

If the answer to this question is yes, then at least some environmental and health issues are internal to the farm, so there is a potential for selfish abatement or environmental risk-averting activities. The amount of this depends in part on the accuracy of information that farmers have at hand. The multi-attribute utility models mentioned above all presuppose perfect information about health and environmental risks. Alternatively, consider the case where the farmer decision-maker perceives less than the "true" level of environmental risk. It is easy to show that an optimizing farmer does less to reduce pollution than if he or she were fully informed. This raises a second empirical question:

How do farmers perceive the health and environmental risks that are related to agricultural water pollution?

If risk perception is reasonably accurate, then selfish behavior should function to ameliorate some of the water quality risks that are not due to externalities. However, if farmers under-perceive risk, then they may do too little to abate or avert environmental risk, even for their own private welfare.

A simple theoretical model can illustrate how attitudes toward health and the environment affect farmers' choices of production practices. Standard neoclassical microeconomic studies have begun from the assumption that the producer seeks to maximize profit subject to a technology constraint. If, instead, we assume that the producer cares about health or environmental quality as well as profit (or the consumption attainable from profit), then the model structure will affect the optimal level of input demanded when inputs exist that affect health status or environmental quality. In particular, fertilizers or pesticides that can enter ground or surface water and thereby harm human health should be used at lower levels than if either (a) health does not matter, or (b) the inputs do not affect health (Swinton, 1998; Zilberman and Marra, 1993). Whether this logic is realistic hinges on the validity of these last two assumptions, raising the questions:

Are certain agricultural inputs perceived to threaten health?

Do health effects from agricultural inputs matter to farm households?

How much do farmers value reductions in risk to health or environmental quality?

These questions motivated a recent study of Michigan corn grower attitudes toward corn herbicides and associated safety issues.

PRIOR EVIDENCE ON VOLUNTARY NSP ABATEMENT

Previous attempts to measure farmer concern about the health and environmental effects of agricultural inputs have utilized a variety of empirical methods. Abdalla and his collaborators (Abdalla, 1990; Abdalla, 1994; Abdalla et al., 1992) interviewed users of nitrate contaminated groundwater supplies to determine how much they spend in order to avert exposure to nitrates. This approach is attractive in that it is grounded in market transactions, but it has two flaws. First, it may miss benefits that do not accrue through drinking water purchases. For example, if someone felt added satisfaction from knowing the water was safe and clean for others to consume, this satisfaction would not be measured. Second, it is quite difficult to ascribe the reason for an averting expenditure to a single cause. For example, bottled water can avert exposure to nitrates in groundwater, but it also offers other attributes that may be valued, such as flavor and portability. If so, it would be an exaggeration to ascribe the entire added cost of bottled water to averting nitrate exposure.

Beach and Carlson (1993) found evidence of willingness to pay for worker safety and water quality using a hedonic analysis in which these attributes helped to explain differences among herbicide prices. Their study draws strength from its base in market transactions. But it suffers from weaknesses related to its econometric specification. One weakness is the possible omission of useful explanatory variables describing herbicide attributes (for example, the timing window of herbicide application) that could result in biased parameter estimates. Another weakness is that data measuring health and environmental risks must be publicly available for each compound evaluated. Apart from these reservations, this method provides only a partial measure of herbicide safety benefits, since it cannot measure the consumer surplus of those consumers willing to pay more than the market price.

The contingent valuation method (CVM) is another approach to the measurement of non-market costs or benefits (Carson, 1991; Cropper and Freeman III,

1991). In CVM studies, survey respondents are presented with a hypothetical market for a good or service from which they are invited to "buy" that hypothetical good or service. Their WTP for that good or service is measured by the choices that are made, contingent on the existence of the hypothetical market. While it is widely conceded that CVM offers a theoretically comprehensive approach to the measurement of WTP, it has been much criticized for flaws in its implementation. These flaws are that (1) respondents do not fully understand the hypothetical good or service; (2) they do not take seriously a hypothetical budget constraint; and (3) they may respond strategically (for example, if they believe they may actually have to pay for the good or service in the future, they may underbid their true willingness to pay) (Carson, 1991; Diamond and Hausman, 1994).

Two previous studies have attempted to apply CVM to measure WTP for reduced environmental or health risks due to pesticide use (Higley and Wintersteen, 1992; Mullen et al., 1997). Both suffered from response rates of less than 25 percent, as well as vaguely described hypothetical goods, and potential strategic bias due to open-ended WTP questions.

NEW EVIDENCE ON THE VALUE OF HERBICIDE SAFETY

In order to examine whether and how much corn farmers value safety and environmental quality in herbicides, a more careful CVM survey was designed and mailed to Michigan corn growers in the summer of 1995 (Owens et al., 1997b). The survey focused on the herbicide atrazine. It is widely used for weed control in corn, and has been available for more than thirty years. EPA scientists believe that atrazine poses certain health and environmental risks. The key survey questions elicited WTP for three hypothetical formulations of atrazine: one that did not leach, one that was not at all carcinogenic, and one that was nontoxic to fish. So the atrazine in the CVM survey was identical to conventional atrazine in all respects except the one stated variant for the formulation described. Because the respondents were already familiar with atrazine and because the hypothetical formulations differed from it in a single attribute only, the respondents were expected to understand the good thoroughly. Respondents were offered a specific price for the hypothetical herbicide formulation and were asked whether or not they would buy at that price. If they were willing to buy it at that price, they were asked how many acres they would use it on. This accept/reject market setting approximates the conditions under which a farmer would buy herbicides. It also discourages strategic behavior. By stratifying the sample into 15 different price offer combinations, WTP could be inferred from purchase choices. The questionnaire also elicited information on other topics that include current herbicide practices in corn production, awareness of health and environmental risks from atrazine, information sources on weed management, and background questions about the respondent and their farm. The response rate was 54 percent.

Do Farmers Care about Health and Environment? How Much?

The survey results reveal that many farmers are willing to purchase safer herbicides, but their willingness depends upon price and risk perception. When there is
no price differential for the safer atrazine formulation compared to conventional
atrazine at the prevailing atrazine price of $3.00 per acre, then the percentage of
respondents willing to purchase is 63 percent for the non-leaching formulation, 54
percent for the non-carcinogenic formulation, and 33 percent for the fish-safe
formulation (Owens et al., 1997b). If the safer atrazine formulations cost $5 per
acre more than conventional atrazine, the percentage of respondents willing to
adopt drops to 21 percent, 14 percent and 10 percent, respectively. The range of
elicited adoption rates is presented by price for each of the formulations reviewed
in figure 16.1.

On the average, those farmers willing to purchase the new, safer formulations of
atrazine are willing to pay more than double the current price of atrazine, as
shown in table 16.1. The demand equations from a set of double hurdle econometric models show declining WTP for the safer atrazine formulations as the number
of acres applied increases. Based on these demand equations, an average purchaser would pay $4.40 per acre for the non-leaching attribute on 40 acres, or
$4.92 per acre for the non-carcinogenic attribute above and beyond the cost of or-
dinary atrazine. (The fish-safe formulation was, on average, purchased for fewer
than 40 acres (Owens, 1997, Chapter 4)). Compared with a baseline price of $3.00
per acre for regular atrazine, these figures represent an average WTP of up to 164
percent more to acquire a single safety attribute.

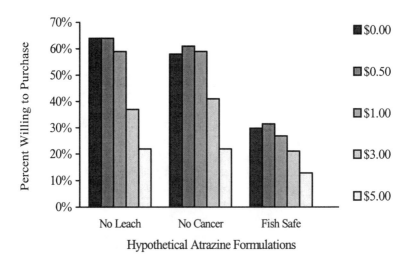

**Figure 16.1 Percent Willingness to Purchase Hypothetically Safer Atrazin-
Formulations by Price Premiums**

Source: Owens et al., 1997b.

TABLE 16.1 Average Respondent's Estimated Total, Mean and Marginal Willingness to Pay for Three Safety Attributes in Atrazine Formulations on 40 Acres of Corn

Formulation	Total WTP on 40 acres	Mean WTP	Marginal WTP
	dollars	dollars per acre	dollars per acre
Non-leaching	175.80	4.40	1.26
Non-carcinogenic	196.78	4.92	1.82
Fish-safe	118.17	2.95	0.00

Source: Authors' estimates.

Farmers Perceptions of Herbicide Health and Environmental Risks

Farmer perceptions of health and environmental risks play a key role in willingness to adopt the non-leaching formulation of atrazine (Owens et al., 1997b). In the Owens survey, the respondents were first asked whether they were aware of scientific assessments of eleven health and environmental risks, which were based on information drawn from the atrazine label and material safety data sheets. Next, the respondents were asked if they agreed with these scientific assessments. Only in the instance of the leachability of atrazine had the majority of respondents heard of the risk cited. Although most respondents had heard of it, only 33 percent agreed with the scientific opinion on atrazine's propensity to leach or felt that it was too conservative. Instead, most respondents believed that scientific opinion overstated actual atrazine leaching. Similar results were found for the other health risks reviewed. The importance of this perception effect is manifest in probit regressions to explain adoption of the new non-leaching formulation, where awareness of the risk significantly enhances the probability of adoption (Owens et al. 1997a).

POLICY IMPLICATIONS

The empirical results from this CVM survey provide clear answers to the first two questions posed in this chapter. "Do farmers care about health and environmental quality?" The answer is yes—a large share of the farmers interviewed would be willing to purchase a safer corn herbicide, even if it cost more than their current one. Of those willing to buy the safer one, the average respondent would pay double to get any one of the three health and environmental attributes on their first 40 acres. The second question was "How do farmers perceive the health and environmental risks related to agricultural water pollution?" Survey results indicate that farmers often are not aware of environmental and health risks, and when they are, they do not agree that risks are as great as believed by the scientists who developed the herbicide labels and material safety data sheets.

This research reveals that untapped opportunities are available to take advantage of many farmers' desires to protect health and environmental quality. Even producers of a low value commodity (such as field corn) are willing to pay extra to insure greater safety. However, many are either unaware or are dubious of the risks posed by certain agricultural chemicals. If they can be persuaded that the risks are real, many farmers would be willing to pay to reduce those risks and WTP could take the form of adopting more costly alternative technologies.

If public policy is to use private risk concerns to advance the public good, knowledge generation and diffusion must play central roles. Knowledge of health and environmental risks is a *sine qua non* for voluntary adoption of safer production practices. This is as true today for environmental and chronic health risks as it ever has been for farm equipment safety risks.

Research into environmental risk assessment is a major part of what the EPA already does. Likewise, the development of reduced-risk technologies is already under way, not only at the federal and state levels, but also in the private sector (Magretta, 1997). Risk communication, however, still appears to be inadequate. Risk information tends to be so complex that most individuals opt for traditional beliefs or seat-of-the-pants decision rules. An area that is ripe for additional research and government intervention is the development of simplified standards of likely risk exposure. Ecolabeling, as proposed by van Ravenswaay and Blend (this volume), demonstrates the potential appeal of "green" products if the benefits can be communicated succinctly and effectively. This potential appeal is equally valid for the agricultural inputs marketed to farmers as it is for the food and fiber products marketed to consumers.

ENDNOTES

1.   This research is based, in part, upon work that was supported by the U.S. Department of Agriculture's Economic Research Service under agreement #43-3AEN-4-80097. Any opinions, findings, conclusions, or recommendations expressed in this publication are those of the authors and do not necessarily reflect the view of the U.S. Department of Agriculture.

REFERENCES

Abdalla, C.W. 1990. "Measuring Economic Losses from Groundwater Contamination: An Investigation of Household Avoidance Costs." *Water Resources Bulletin* 26: 451–463.

_____. 1994. "Groundwater Values from Avoidance Cost Studies: Implications for Policy and Future Research." *American Journal of Agricultural Economics* 76: 1062–1067.

Abdalla, C.W., B.A. Roach, and D.J. Epp. 1992. "Valuing Environmental Quality Changes Using Averting Expenditures: An Application to Groundwater Contamination." *Land Economics* 68: 163–169.

Beach, E.D. and G.A. Carlson. 1993. "Hedonic Analysis of Herbicides: Do User Safety and Water Quality Matter?" *American Journal of Agricultural Economics* 75: 612–623.

Benbrook, C.M., E. Groth III, J.M. Halloran, M.K. Hansen, and S. Marquardt. 1996. *Pest Management at the Crossroads*. Yonkers, NY: Consumers Union.

Carson, R.T. 1991. "Constructed Markets," in J.B. Braden and C.D. Kolstad, eds., *Measuring the Demand for Environmental Quality*, pp. 121–162. Amsterdam, ND: North-Holland.

Cropper, M.L. and A.M. Freeman III. 1991. "Environmental Health Effects," in J.B. Braden and C.D. Kolstad, eds., *Measuring the Demand for Environmental Quality*, pp. 165–211. Amsterdam, ND: North-Holland.

Diamond, P.A. and J.A. Hausman. 1994. "Is Some Number Better than No Number?" *Journal of Economic Perspectives* 8: 45–64.

EPA (U.S. Environmental Protection Agency). 1990. "National Pesticide Survey: Summary of EPA's National Survey of Pesticides in Drinking Water Wells." Follow-up to EPA Report No. 5709-90-003, Office of Water, and Office of Pesticides and Toxic Substances, EPA, Washington, DC.

GAO (U.S. General Accounting Office). 1993. "Pesticides: Issues Concerning Pesticides Used in the Great Lakes Watershed." GAO/RCED Report No. 93-128 to the Chairman of the Subcommittee on Oversight of Government Management, Committee on Governmental Affairs, U.S. Senate, Washington, DC.

Hallberg, G.R. 1989. "Pesticide Pollution of Groundwater in the Humid United States." *Agriculture, Ecosystems and Environment* 26: 299–367.

Higley, L.G. and W.K. Wintersteen. 1992. "A Novel Approach to Environmental Risk Assessment of Pesticides as a Basis for Incorporating Environmental Costs into Economic Injury Level." *American Entomologist* 38: 34–39.

Kellogg, R.L., M.S. Maizel, and D.W. Goss. 1992. *Agricultural Chemical Use and Groundwater Quality: Where are the Potential Problem Areas?* Washington, DC: USDA/SCS.

Magretta, J. 1997. "Growth through Global Sustainability: An Interview with Monsanto's CEO, Robert B. Shapiro." *Harvard Business Review* 75: 78–88.

Mullen, J.D., G.W. Norton, and D.W. Reaves. 1997. "Economic Analysis of Environmental Benefits of Integrated Pest Management." *Journal of Agricultural and Applied Economics* 29(2): 243–253.

NRC (National Research Council). 1993. *Soil and Water Quality: An Agenda for Agriculture.* Washington, DC: National Academy Press.

Nowak, P., S. Wolf, H. Hartley, and R. McCallister. *Final Report: Assessment of 1992 Wisconsin Atrazine Rule (Ag30).* College of Agriculture and Life Sciences, University of Wisconsin, Madison, WI.

Owens, N.N. 1997. "Farmer Willingness to Pay for Herbicide Safety Characteristics." Ph.D. Dissertation, Department of Agricultural Economics, Michigan State University, East Lansing, MI.

Owens, N.N., S.M. Swinton, and E.O. van Ravenswaay. 1995. "Farmer Demand for Safer Pesticides." Department of Agricultural Economics, Staff Paper No. 95-27, Michigan State University, East Lansing, MI.

_____. 1997a. "Double Hurdle Marketing Analysis for Safer Herbicides." Selected paper presented at the annual meeting of the American Agricultural Economics Association in Toronto, Ontario (27–30 July).

_____. 1997b. "Farmer Demand for Safer Corn Herbicides: Survey Methods and Descriptive Results." Michigan Agricultural Experiment Station, Research Report No. 547, Michigan State University, East Lansing, MI.

Padgitt, M. 1997. "Pest Management on Major Field Crops." *AREI Updates 1997: 1 (Updates on Agricultural Resources and Environmental Indicators).* USDA/ERS, Washington, DC.

Porter, M. E. and C. van der Linde. 1995. "Green and Competitive: Ending the Stalemate." *Harvard Business Review* 73: 120–134.

Swinton, S.M. 1998. "Less Is More: Why Agro-Chemical Use Will Decline in Industrialized Countries," in G.A.A. Wossink, G.C. van Kooten, and G.H. Peters, eds., *Economics of Agro-Chemicals.* Aldershot, UK: Ashgate.

Taylor, H. 1996. "1995 Nutrient Use and Practices on Major Field Crops," *AREI Updates 1996: 2 (Updates on Agricultural Resources and Environmental Indicators).* USDA/NRED, Washington, DC.

Zilberman, D. and M. Marra. 1993. "Agricultural Externalities," in G.A. Carlson and J.A. Miranowski, eds., *Agricultural and Environmental Resource Economics*, Chapter 6, pp. 221–267. New York, NY: Oxford University Press.

17 AGRICULTURAL PRODUCTION CONTRACTS TO REDUCE WATER POLLUTION

Scott M. Swinton
Mei-chin Chu and Sandra S. Batie
Michigan State University, East Lansing, MI
Taiwan Institute of Economic Research,
Taipei, Taiwan, Republic of China
Michigan State University, East Lansing, MI

The growing vertical integration of the agricultural sector presents opportunities for flexible incentives to reduce nonpoint-source pollution (NSP) of water. Some agricultural production contracts offer farmers incentives that may lead to environmental degradation due to high rates of agrochemical use. At the same time, some businesses are recognizing the advantages of using environmental responsibility for competitive advantage and are taking an interest in developing a reputation for environmental stewardship. Principal agent theory offers an economic framework with which to evaluate contract designs for acceptability according to both profitability and environmental criteria. A seed corn production contract illustrates the potential to use contracts to reduce nitrate leaching. This form of production contract also illustrates the key elements of acceptable contracts between processing companies and growers.

THE GROWING ROLE OF PRODUCTION UNDER CONTRACT

Although many farmers may be willing to adopt environmental stewardship practices voluntarily (Swinton, Owens and van Ravenswaay, this volume) this willingness only matters for policy design if farmers make their own management decisions. A growing part of agricultural production, however, is carried out under contract. Drabenstott (1994) reported that by 1990, contracts already governed 90 to 100 percent of marketings in poultry, eggs and processing vegetables. The proportion of fresh vegetables and hogs produced under contract had begun to accelerate, each jumping 10–15 percent in the 1980–90 decade. Although the proportion of cereal grains produced under contract was still under 5 percent in 1990, it has been rising (Urban, 1991).[1]

Agricultural production contracts have tended to focus on incentives for the producer to contain costs while meeting quality specifications for the processor. Producer processor contracts in agriculture have not served as a mechanism for the improvement of environmental quality, as evidenced by the near neglect of this topic in the comprehensive overviews of flexible incentives presented elsewhere (Batie and Ervin, this volume; Segerson, this volume). But the traditional focus of agricultural production contracts could change as agricultural and food businesses begin to recognize the potential to be both "green" and "competitive" in the words of Porter and van der Linde (1995). Van Ravenswaay and Blend (this volume) have documented the growing consumer interest in ecolabels to certify invisible environmental attributes of consumer products. Clearly, the interest is growing. What is less clear is whether it is feasible to incorporate "green" production practices into enforceable agricultural production contracts. This issue raises two research questions to be investigated here:

Is it possible to design a mutually acceptable contract between farmer and processor that would enhance environmental quality?

If so, what would be the characteristics of such a contract?

These questions are addressed by analyzing one type of contract between a seed corn conditioning firm and a farmer. The environmental quality issue of interest here is nitrate contamination of groundwater in two southwestern Michigan counties having sandy soils over a shallow aquifer. Many wells sampled in St. Joseph and Cass counties, Michigan, exceed the maximum contaminant level of 10 parts per million of nitrate nitrogen, which was established under the Safe Drinking Water Act. While the cause of the high nitrate levels is unknown, the two counties are home to both intensive livestock production and specialty crop agriculture, notably seed corn, potatoes, and processing vegetables. There has been speculation that these high nitrate levels may be due, in part, to heavy nitrogen fertilization by seed corn farmers who exceed norms recommended by the leading seed corn company (Peterson and Corak, 1994).

The nature of seed corn production in southwestern Michigan gives some basis for believing these speculations. The combination of a shallow water table and of sandy soils has made the area ideally suited to center pivot irrigation. Irrigation stabilizes crop yields, making the area attractive to a variety of specialty crop processors, which include seed corn companies. One dozen seed corn companies compete for area farmers who are skilled at growing seed corn and who control fertile, irrigated land. Farmers, in turn, compete with one another to grow seed corn for the companies that offer the most attractive contractual terms. Foremost among these companies is the one that offers the contractual terms to be described here.

Seed corn production requires special care. Hybrid corn seed is harvested from female plants that are planted at a ratio of three female rows to one male row. In order to insure varietal purity, the female rows must have their tassels removed (detasseling), the male rows must be destroyed after their tassels have shed their pollen, and fields must be scouted for off-type plants (roguing). Because the in-bred corn lines (which are used to produce the hybrid seed) are smaller and less vigorous than hybrid corn, they are more susceptible to weed and insect pests. These factors, combined with the high cost of seed, mean that each acre of seed corn requires costly management. The seed corn companies generally take respon-sibility for pest management, detasseling, roguing, and harvest. Seed corn farmers control irrigation, tillage, and fertilization. The contracts specify the responsibili-ties of each party, the payment formula, and the condition that the seed corn com-pany retains ownership of all seed corn plants growing in the farmer's field.

PRINCIPALS, AGENTS AND SEED CORN

The general principles of contract design are central to understanding seed corn production contracts. Contract design is best viewed in the context of the eco-nomic theory of principal and agent. This theory examines how one individual or institution (the principal) can persuade another (the agent) to do as the principal wishes, even when the principal cannot observe all that the agent is doing. For our purposes, the principal is the agricultural processing and marketing firm, and the agent is the farmer who considers accepting a production contract. In the standard principal agent model, the principal seeks to maximize utility (or profit) subject to two constraints:

The agent must find the contract sufficiently attractive to be acceptable (partici-pation constraint),

The agent must be enticed to do things the way the principal wants them done (incentive compatibility constraint).

These two constraints are both incorporated in the design of the leading form of seed corn production contract in North America (Shaw et al., 1989; Chu et al., 1997a; Chu, 1997). Under that contract, the seed corn company pays the grower a flat fee and a variable payment.[2] The flat fee part of the payment is designed to compensate the grower for the value of a hybrid corn crop that could have been grown on the land where seed corn is being raised. This part of the payment is de-signed to meet the participation constraint. The second component is a variable payment based on how much the grower's seed corn yield deviates from the re-gional average yield among growers who produce the same variety. This compo-nent can be positive or negative, and is geared to magnify the marginal value of seed corn yield. The resultant high value of marginal yield gains encourages growers to strive for maximum yields, thereby meeting the seed corn company's incentive compatibility constraint. High yields allow the seed corn conditioning company to attain its volume goals while minimizing the acreage of seed corn fields that require costly monitoring and inbred seed. Contracts like this one are referred to as tournament contracts, because they encourage competition among agents to meet the goals of the principal.

This tournament contract appears to exaggerate the tendency of many U.S. Corn Belt farmers to fertilize for the best possible growing conditions rather than for average or below average conditions (Babcock and Blackmer 1994). Cognitive focus on the best weather case makes sense, because ordinary nitrogen fertilization rates may fall short of crop needs during very favorable growing conditions. Such a shortfall could cause a grower's yield to slip below the area average that season, causing the variable payment in the contract to be negative. Worse yet, a string of sub-par years can induce the processing firm to cut back on contracted acreage with a given grower, switching contracted acreage over to more productive farmers (Preckel et al. 1998). As a result, many seed corn growers apply nitrogen heavily. That nitrate contamination of groundwater is increasing in two southwest Michigan counties where seed corn production prevails may be a consequence of this behavior.

An Empirical Model of Seed Corn Production under Contract

Modeling the feasibility of alternative contract designs to reduce nitrate leaching in this chapter is accomplished in two steps. First, the seed corn grower's decisions are modeled empirically as a whole farm, mathematical programming problem. This approach is chosen in order to specify the grower's implicit participation constraint (that growing seed corn should be more lucrative than the opportunity cost of foregone earnings from alternative crop enterprises). Second, the net revenue to the seed corn firm is modeled as the residual gross margin after subtracting contract payments from the gross sales revenue. Examination of the changes in grower net income, company gross margin, and projected nitrate leaching under alternative scenarios allows analysis of the incentive compatibility constraint. This sequential modeling approach permits greater farm-level detail than would a true two-level mathematical programming model that encompassed simultaneous optimization by both the principal and the agent (Candler et al. 1981).

The representative farm in our model has 1200 cultivatable acres, with the opportunity for the farmer to rent up to 500 additional acres. The farmer may contract with the seed corn company for up to 500 acres of seed corn production. As a full-time worker with two part-time workers available, the farmer may choose to raise commercial corn, seed corn, or soybean. The farmer may also opt to cash rent land for potato production. The two types of corn may be rotated with any of the other crops, but other crops may not be rotated with one another. Potato rental land must be in a two-year rotation with another crop. Both commercial and seed corn may be fertilized at low, medium, or high levels. Those levels correspond to yields that are close to the long-term physiological maximum as modeled by the Ceres-Maize component of the Decision Support System for Agro-technology Transfer (DSSAT) version 3.0 crop growth model suite (Tsuji et al., 1994). Average simulated yields from 1951 through 1992 are used for the baseline, deterministic farm model. The model is developed using the Purdue Crop/Livestock Linear Program (PC/LP) version 3.2 (Dobbins et al., 1994), as modified for southwestern Michigan. PC/LP is designed to capture time constraints and machinery capacity in detail.

A quadratic risk programming analysis is subsequently developed using a General Algebraic Modeling System (GAMS) version of the same model (Brooke et al., 1988; Dobbins et al., 1996). The risk model applies a mean variance, constant absolute risk aversion (CARA) expected utility function to yield means and variances. Nitrate leaching and yields for seed corn, commercial corn and soybean are simulated with DSSAT from 42 years of weather data, while potato rental rates remain constant. Results from two risk-aversion levels are reported here, including risk-neutral (profit maximizer) and highly risk-averse (CARA exponent = .0001). A chance constrained analysis of probability of exceeding specified levels of nitrate leaching was also conducted, but is not reported here (Chu, 1997).

Four contractual approaches to reducing nitrate leaching are modeled (Table 17.1). They are (a) effluent standards, (b) technology standards restricting agronomic practices, (c) fees on excess nitrogen fertilizer use or leaching, and (d) rearrangement of the fixed and variable payment component of contract payment. A fifth option, the provision of additional information, was not compatible with the mathematical programming model. These contract designs use many of the same incentive mechanisms as public environmental policies do (Khanna et al., this volume; Segerson, this volume), except for the alteration of the contract payment.

TABLE 17.1 Contract Design Alternatives Considered

Effluent Standards

   Restrict nitrate leaching from the whole farm to 35 pounds per acre
   Restrict nitrate leaching from seed corn fields to 35 pounds per acre

---

Technology Standards

   Restrict maximum nitrogen fertilizer use to 107 pounds per acre
   Prohibit rotation with potatoes

Fees on Effluents or Inputs

   $0.30 per-pound fee on nitrate leaching above 30 pounds per acre
   $0.15 per-pound fee on nitrogen fertilizer above 90 pounds per acre

---

Adjust Contract Payment Structure

   Fixed payment of $150 per acre plus a variable payment of $5.28 per bushel of Seed corn[a]
   Fixed payment of $253 per acre plus a variable payment of $3.96 per bushel of Seed corn[b]
   Fixed payment of $230 per acre plus a variable payment of $3.96 per bushel of seed corn[c]

---

[a]Baseline case.

[b]Maintains expected net revenue of farmer equal to baseline case.

[c]Maintains expected net revenue of the seed corn company equal to the baseline case.

The model implicitly assumes that environmental monitoring information on nitrate leaching is available free of cost. Contract enforcement issues are discussed separately.

The specific scenarios are designed to reflect the relative reductions of nitrate leaching or nitrogen fertilizer application. The 42-year crop growth simulations for the 9 seed corn production systems (which are three nitrogen fertilizer levels in seed corn when it is rotated with seed corn, soybeans or potatoes) yield mean annual leaching values for each system. The median fertilizer rate among the 9 seed corn systems is 107 pounds per acre. The median level of resultant nitrate nitrogen leaching is 35 pounds per acre. Routine heavy nitrogen fertilization of potatoes leads this crop to cause greater nitrate leaching than any other. As a result, one way to limit nitrate leaching is to restrict rotation with potatoes, despite the lucrative cash rents offered for potato land. The contract fee levels are specified to be adequate to cause a change in cropping systems. Finally, two alternative contract payment structures were chosen to retain neutrality of mean net revenues either for the farmer or for the seed corn company.

Results from Alternative Contract Specifications

Results from the model runs are compared using two forms of dominance analysis. The first, contract acceptability dominance, focuses on the likelihood that a new contract specification would be mutually acceptable to both the seed corn company (the principal) and the grower (the agent). Contract acceptability dominance evaluates each pair of contracts separately for the grower (based on mean and variance of net income) and for the processor (based on mean gross margin and mean nitrate leaching from seed corn fields). For the processor, contract A dominates contract B if contract A causes less or equal mean nitrate leaching with greater or equal mean gross margin, with one of the inequalities holding strictly. For the grower, a contract is considered acceptable if it causes less than a 1 percent reduction in risk-weighted mean income, which is defined as the certainty equivalent of the CARA expected utility functions (Chu et al., 1997c; Robison and Barry, 1987). Revenues within ±$1 per acre are not considered significantly different for the grower or the seed corn company.

The second form of dominance analysis applied is called cost efficiency dominance. This is designed to measure the cost efficiency of alternative nitrate leaching reduction measures. Under cost efficiency dominance, contract A dominates contract B if contract A either reduces leaching at a lower unit cost for the grower without increasing the unit cost for the processor, or if contract A reduces the unit cost for the processor without increasing the nitrate leaching from the grower's fields. Hence, cost acceptability dominance does not require that both parties benefit. Unit cost is measured by the reduction in the grower's expected utility or the processor's gross margin per pound per acre of leaching reduction. Contracts that are not dominated are considered to be in the efficient set according to these Pareto-style dominance criteria.

Table 17.2 Contracts Not Dominated Under Contract Acceptability and Cost Efficiency Dominance for Risk-neutral and Highly Risk-averse Decision Makers

Risk-aversion Level	*Contract Acceptability*	*Cost Efficiency*
Risk Neutral	Restrict SNL[a] ≤ 35 pounds per acre.	Restrict SNL[a] ≤ 35 pounds per acre.
	Restrict ANL[b] ≤ 35 pounds per acre.	No rotation with potatoes.
	Restrict N[c] ≤ 107 pounds per acre.	$0.10 per-pound fee on SNL[a] > 30 pounds per acre.
	No rotation with potatoes.	Payment: $253 per acre + $3.96 per bushel.
	$0.10 per-pound fee on SNL[a] > 30 pounds per acre.	Payment: $230 per acre + $3.96 per bushel.
Highly Risk Averse	$0.10 per-pound fee on SNL[a] > 30 pounds per acre.	Restrict SNL[a] ≤ 35 pounds per acre.
	$0.15 per-pound fee on N[c] > 90 pounds per acre.	Restrict ANL[b] ≤ 35 pounds per acre.
	Payment: $253 per acre + $3.96 per bushel.	No rotation with potatoes.
	Payment: $230 per acre + $3.96 per bushel.	

[a]SNL = nitrate leaching under seed corn fields.
[b]ANL = average nitrate leaching under all fields.
[c]N = nitrogen fertilizer rate.

Source: Chu, 1997.

The results in table 17.2 confirm that contracts can be designed that are both cost effective in reducing NSP and mutually acceptable to the processor and grower. Depending on the grower's level of risk aversion, different contract designs could be acceptable to both the grower and the seed corn company. The $0.10 per-pound fee on nitrate leaching over 30 pounds per acre meets the contract acceptability criterion for both risk-neutral and risk-averse growers. Under the cost efficiency criterion, both the restriction on nitrate leaching from seed corn fields to no more than 35 pounds per acre and the prohibition of rotation with potatoes are undominated contracts for both levels of risk aversion. No contract design, however, meets both the contract acceptability criterion and the cost efficiency criterion for both levels of risk aversion. The absence of a win-win contract results from the special character of the potato land rental contract, which leads to the most nitrate leaching but also offers the most stable income source in the model. The risk-averse grower suffers a major reduction in expected utility if potatoes are disallowed, which is the case under both the contract that forbids potatoes and the one that restricts nitrate leaching in seed corn fields to under 35 pounds per acre. Hence, no contract is acceptable to the risk-averse grower if it bars potato land rental.

The two most attractive contracts are not dominated under three of the four categories in table 17.2. The $0.10 per-pound fee on nitrate leaching over 30 pounds per acre in seed corn fields meets contract acceptability for both risk-aversion levels and is cost efficient for the risk-neutral grower. However, the fee is not sufficient to induce the risk-averse grower to change production practices, so it fails the cost efficiency criterion in that instance. The ban on rotation with pota-toes is highly cost efficient, since it leads to a reduction in nitrate leaching from 28 to 53 pounds per acre. Such a contract, however, would be unacceptable to the risk-averse grower.

Enforceability also must be a criterion for contract choice. While an effluent fee on nitrate leaching may be a "first-best" approach from a theoretical standpoint, it is clearly impractical to implement in the field. The restriction on crop rotation with potatoes, on the other hand, is easily observed and enforced. This special case, in which nitrate leaching is so heavily affected by one specific rotation crop, the restriction on potato rotations is a practical proxy for unobservable nitrate leaching.

When interpreting these results, two caveats deserve attention. First, the entire analysis presented in this chapter presupposes that the processor firm desires to develop a "green" reputation. Only in that case will the processor firm care about NSP. Second, special circumstances are needed to identify a contract that is mutu-ally acceptable to both parties, cost-effective at reducing NSP, and easily en-forced. The appeal of the contract design based on a technology standard (crop rotation) has promise only in special instances, notably when the contracted com-modity is not the primary cause of NSP but an associated practice is a major pol-luter.

With these caveats, this research offers insights about whether, and how, pro-duction contracts can reduce agricultural water pollution. There appear to be a va-riety of feasible contract designs that can enhance environmental quality, but their relative merits vary according to the contract setting. If the environmentally friendly contract affects the stability of earnings, its appeal will vary according to the risk aversion of the contracting parties. If the contract governs the grower's engagement in an extraneous activity, such as production of a additional crop, this restriction may be quite observable and enforceable. Opportunities to alter the contract payment tend to shift the cost of environmental risk reduction between the processor (the principal) and the grower (the agent), so their feasibility de-pends upon the relative power of each. Where the processor's payments to grow-ers exceed the true level required to meet the participation constraint, the proces-sor may be able to oblige the growers to bear some or all of the economic costs (including opportunity cost) of enhanced environmental quality. If the processor is only just meeting the participation constraint at the margin, however, the proces-sor may have to absorb the full economic cost of being "green" by bribing the growers to change practices.

POLICY IMPLICATIONS

The contract design research results presented in this chapter indicate that private sector contracts to encourage environmentally responsible production practices are not only feasible, but can also be enforceable. The analysis highlights some of the criteria that must be accommodated in the design of a viable "green" production contract. These criteria will vary from one contract setting to another. In particular, contract designs that reduce risks to the environment and to human health require that the contracting parties are aware of those risks and that they have available low cost, practical means of monitoring compliance with contract terms. Both of these elements suggest a role for government policy that hinges on research and knowledge diffusion.

Regarding risk awareness, there is growing evidence that consumers and producers will respond to information about environmental and health risk (van Ravenswaay and Blend, this volume; Swinton, Owens and van Ravenswaay, this volume). North American and European governments are heavily involved in supporting research on environmental risk. The increasing awareness of these risks is spawning reduced risk technologies in both the public and the private sectors (Magretta, 1997). These developments, combined with the growth of production contracting in North American agriculture, are the necessary conditions for "green" contracts to become commercially desirable.

In order for "green" agricultural production contracts to become feasible, low cost and effective means to monitor compliance with environmental performance specifications are required. The research reported here reveals one case in which low cost monitoring could be achieved through the proxy of forbidding certain crop rotations. In the near future, many production contracts are likely to follow the path of monitoring convenient indicator variables. Van Ravenswaay and Blend (this volume) report several cases in which ecolabeling does not actually track environmental impacts, instead it tracks more easily observed measures of pesticides and fertilizer management.

If consumer interest in the reduction of risks to human health and environmental quality continues to grow, tracking rough indicators of risk may cease to be acceptable. Consumer demand to see evidence of actual risk reduction may create a major agenda for research and development of low cost ways to measure risk outcomes. Some risk-measurement devices, like pesticide residue detectors, are already available. Others, for example NSP leaching monitors, will require much more research. For many pollutants, there may never be a low-cost way of tracing the level of contribution from a specific production area. In such cases, indicator variables will continue to be important. It will become imperative, however, to demonstrate how reliably an indicator correlates with measured environmental risk outcomes. Given the importance of environmental risk monitoring to the success of "green" production contracts, inexpensive but reliable risk measures and indicator variables will be an important area for public research in the years ahead.

ENDNOTES

1. The authors wish to acknowledge valued ideas and data assistance from Brian Baer, Craig Dobbins, Otto Doering III, Rodney King, Rebecca Pfeifer, Paul Preckel, and Joe T. Ritchie. This research is based, in part, upon work that has been supported by the Cooperative State Research, Education, and Extension Service, U.S. Department of Agriculture under agreement #94-37102-0839. Any opinions, findings, conclusions or recommendations expressed in this publication are those of the authors and do not necessarily reflect the view of the U.S. Department of Agriculture.

2. This seed corn contract design can be written algebraically as,

$$s(y) = [\alpha(y - y^0) + Q]\beta P, \text{ (Chu et al., 1995; Chu, 1997)}$$

1. in which s denotes a per-acre payment from the seed processor to the contractor-grower; y is grower yield of seed corn per acre; yo is the average yield per acre from the same seed corn variety that is raised by all growers in the region; Q is the baseline commercial corn yield; P is the price of commercial corn per bushel; α is a coefficient that transforms seed corn yield into commercial corn equivalents; and β is a price premium adjustment coefficient. Because the seed corn inbred varieties have lower yields than commercial hybrids, the value of α is always made greater than one. The transaction costs that are associated with contract production are compensated through a price premium adjustment coefficient, β, which also is greater than one. Note that payment s(y) can be written in a linear form that consists of two parts: (1) a fixed payment, and (2) a variable payment, which depends on the observed output level.

REFERENCES

Babcock, B.A. and A.M. Blackmer. 1994. "The Ex Post Relationship between Growing Conditions and Optimal Fertilizer Levels." *Review of Agricultural Economics* 16: 353–362.

Batie, S.S., D.E. Ervin and M.A. Schulz, eds. 1998. *Business-led Initiatives in Environmental Management: The Next Generation of Policy?* Michigan Agricultural Experiment Station, Special Report No. SR-92, East Lansing, MI.

Brooke, A., D. Kendrick and A. Meeraus. 1988. *GAMS: A User's Guide.* San Francisco, CA: Scientific Press.

Candler, W., J. Fortuny-Amat, and B. McCarl. 1981. "The Potential Role of Multilevel Programming in Agricultural Economics." *American Journal of Agricultural Economics* 63: 521–531.

Chu, M. 1997. "Designing Production Contracts to Reduce Agricultural Non-point Source Pollution." Ph.D. Dissertation, Department of Agricultural Economics, Michigan State University, East Lansing, MI.

Chu, M., S.M. Swinton, and S.S. Batie, with C. Dobbins, O. Doering III, and J. T. Ritchie. 1997a. "Designing Contracts to Reduce Agricultural Non-Point Source Pollution." *Taiwanese Agricultural Economics Review* 2: 187–209.

Chu, M., S.M. Swinton, and S.S. Batie. 1997c. "A Risk Programming Approach to Designing Contracts to Reduce Nitrate Leaching." Selected paper presented at the annual meeting of the American Agricultural Economics Association held in Toronto, Ontario, Canada (27–30 July), Department of Agricultural Economics, Staff Paper No. 97-14, Michigan State University, East Lansing, MI..

Dobbins, C.L., Y. Han, P. Preckel, and D.H. Doster. 1994. "Purdue Crop/Livestock Linear Program (PC-LP) User's Manual. Version 3.2," Department of Agricultural Economics and Cooperative Extension Service, Report No. C-EC-6, Purdue University, West Lafayette, IN.

Dobbins, C., M. Etyang, H. Onal, and P.V. Preckel. 1996. "Imitation PC-LP á la GAMS." Unpublished GAMS program code. Department of Agricultural Economics, Purdue University, West Lafayette, IN (April).

Drabenstott, M. 1994. "Industrialization: Steady Current or Tidal Wave?" *Choices* 9: 4–7.

Hamilton, N. D. 1990. *What Farmers Need to Know about Environmental Law: Iowa Edition.* Des Moines, IA: Drake University Agricultural Law Center.

NRC (National Research Council). 1993. *Soil and Water Quality: An Agenda for Agriculture.* Washington, DC: National Academy Press.

Peterson, R. and S. Corak. 1994. "Nitrogen Response in Seed Corn Production," in D. Wilkinson, ed., *Proceedings of the Forty-eighth Annual Corn and Sorghum Industry Research Conference 1993*, pp. 169–189. Washington, DC: American Seed Trade Association, Inc.

Preckel, P.V., T.G. Baker, M. Chu and J. Eide. 1998. "Tournament Contracts: It's Not Easy Being Green." Unpublished manuscript. Department of Agricultural Economics, Purdue University, West Lafayette, IN.

Robison, L.J. and P.J. Barry. 1987. *The Competitive Firm's Response to Risk*, p. 38. New York: Macmillan.

Shaw, D.J., W.H. Howard, and L.J. Martin. 1989. "Costs, Prices, and Contracts in the U.S. and Canadian Seed Corn Industries." Department of Agricultural Economics and Business, Bulletin No. AEB 89-8, University of Guelph, Guelph, Ontario, Canada.

Tsuji, G.Y., G. Uehara, and S. Balas, eds. 1994. *DSSAT version 3: A Decision Support System for Agrotechnology Transfer*. University of Hawaii, Honolulu, HI.

Urban, Thomas N. 1991. "Agricultural Industrialization." *Choices* 5: 4–6.

18 DESIGN VERSUS PERFORMANCE STANDARDS TO REDUCE NITROGEN RUNOFF: CHESAPEAKE BAY WATERSHED DAIRY FARMS

C. Line Carpentier and Darrell J. Bosch
International Food Policy Research Institute
Washington, DC
Virginia Polytechnic Institute and
State University, Blacksburg, VA

States in the Chesapeake Bay drainage area have a goal of reducing nitrogen and phosphorus, which flow into the Bay, by 40 percent by the year 2000. Voluntary cost-share programs and education have helped farmers reduce loadings from agriculture. Nutrient reduction targets, however, are still not being met. Further reductions could be achieved through regulatory design standards.

The potential cost reductions from targeting a 40 percent reduction in nitrogen runoff are estimated for dairy farms in the Lower Susquehanna watershed portion of the Bay. With perfect information on farmers' compliance costs, 50 percent of the sample farms are targeted. Targeting reduces aggregate farm compliance and transaction costs. Total gross margins from farms with the targeted performance standard average $12,381 more than they do under a uniform performance standard. A targeted design standard that requires strip-cropping has lower costs than the targeted performance standard, but it cannot achieve the 40 percent reduction in nitrogen loadings that are desired for the watershed.

Study results imply that policymakers should set clear environmental goals and give farmers maximum flexibility in how they achieve these goals. The assignment of responsibilities to reduce pollution should also be made flexible so that greater reductions can be assigned to farmers with lower pollution reduction costs.

INTRODUCTION

In 1983, Pennsylvania, Maryland, Virginia, the District of Columbia, the Chesapeake Bay Commission and the Environmental Protection Agency signed the Chesapeake Bay Agreement (the Agreement) by which all parties agreed to work together to protect and restore the Chesapeake Bay (the Bay). In 1987, these groups agreed to reduce controllable nitrogen and phosphorus that entered the Bay by 40 percent by the year 2000. An estimated 16 percent bay-wide reduction in total phosphorus was achieved from 1984 to 1992 while nitrogen levels did not change significantly (CBP, 1994). In 1992, the Agreement partners reaffirmed their commitment to the 40 percent reduction goal, but they agreed to extend the time commitment beyond the year 2000. They also adopted tributary strategies that called on states to target their nutrient reduction strategies according to nutrient problems within each river basin (VCBP, 1993).

Control efforts focused on agriculture because agriculture contributes an estimated 39 percent of the nitrogen and 49 percent of the phosphorus that enters the Bay (CBP, 1996). States in the Bay's watershed have emphasized voluntary incentive programs using cost-share and educational programs to stimulate the farmers' adoption of best management practices. Limited public funds for voluntary cost-share programs and slow progress in achieving desired reductions in nitrogen and phosphorus pollution have stimulated interest in other policy tools, such as regulations. For example, Pennsylvania passed the Nutrient Management Act in 1993, which required farms that have two or more animal units per acre (one animal unit equals 1,000 pounds, liveweight) to develop and maintain nutrient management plans (Beegle, 1994). Localities may also have their own requirements. For example, several counties in the Bay's watershed require poultry operations to have nutrient management plans, which document that they have access to enough land to apply poultry manure at or below agronomically recommended rates.

REGULATORY APPROACHES FOR WATER QUALITY

At least five types of regulatory approaches can be used to reduce agricultural pollution (Anderson et al., 1990)

- Design standard—a firm is required to use a given type of pollution reduction technology,
- Performance standard—a limit is placed on the amount of agricultural pollution,
- Quotas and Use Restrictions—limits are placed on the amount of outputs produced or inputs used in production,
- Licensing and Registration—a license is required to sell or to apply chemicals or other substances, and
- Activity Permits and Management Plans—legal permission is required to engage in selected activities. (p. 65)

The evaluation of regulatory alternatives focuses on their relative costs (Bohm and Russell, 1985). In this chapter, the focus is on the first two regulatory approaches because of their general applicability and because they represent very different approaches to regulate agriculture for pollution control. Performance standards can be viewed as types of flexible incentives because farmers are given choices as to how to meet the stated pollution control objective (Batie and Ervin, this volume). Design standards, however, do not give farmers flexibility on how to achieve the pollution control objective.

In this chapter, regulatory costs of achieving water quality goals are assumed to equal the sum of the farmers' compliance costs plus their public transaction costs. Effects of regulations on consumer food costs are not considered. Compliance costs equal the reduction in net farm income resulting from the implementation of required practices, or structures, as well as opportunity costs from having to eliminate or reduce profitable enterprises to comply with the regulation.

Transaction costs are the costs of information, of contracting and of enforcement to achieve the pollution control goal (Krier and Montgomery, 1973). Information costs are the costs of determining the set of actual farm practices, the actual loadings in the watershed and the farmers' adoption practices (or reduced loadings) on farms in the watershed. Contracting costs are the administrative and staffing costs that are involved in contacting targeted farmers; in reaching agreements with farmers about practices that must be adopted (or loading reductions that must be reached); and in writing up a contract to create the legal position that is necessary to implement policies. Enforcement costs are the expenses incurred when determining whether or not pollution-reducing practices have been implemented, whether pollution reductions have been achieved, and whether to impose and/or to extract penalties from noncomplying farms. These costs also include litigation expenses should the required practices or pollution reductions be appealed.

Design Standards

Design standards require farmers to follow a set of prescribed practices in terms of the inputs used, production technologies employed and the types and/or amounts of crops or livestock produced (Bohm and Russell, 1985). For example, one type of design standard could be to limit the nitrogen fertilizer application to the season in which crop uptake occurs, such that it could reduce the quantity of nitrates that leach into the groundwater.

Compliance costs of design standards reduce farm net returns because the adoption of required practices (or the installation of structures) and the farmers opportunity costs (if they prohibit or limit profitable enterprises) can be expensive. Design standards have relatively high compliance costs because they limit the farmers' ability to search for pollution control strategies that are best suited to conditions on their particular farms. Also, they do not allocate pollution control among farmers according to the farmers' individual costs (Batie and Ervin, this volume; Abler and Shortle, 1991).

Transaction costs for uniform design standards are low compared to performance standards because the prescribed practices are usually simple and easy to observe. The information costs of determining where practices need to be applied and of contracting and enforcing these practices with uniform design standards are also lower than they are for other policies. Litigation costs, which are included in enforcement, are likely to be low because the standards are usually simple to implement and precedents have been set for design standards that are imposed on farmers. Low transaction costs make design standards appealing to policymakers.

Performance Standards

Performance standards establish a ceiling on allowable pollution (Bohm and Russell, 1985). Performance standards should have a lower aggregate farm compliance cost than design standards do for comparable levels of pollution control. The reason for this is that they allow farmers to choose how best to meet the standard and they give incentives to the agricultural sector on how to develop new lower cost technologies to reduce pollution (Batie and Ervin, this volume). The compliance and transaction costs of performance standards are functions of the degrees of pollution reduction. Higher levels of pollution reduction may impose more complex practices and higher costs on the farmer. This could mean that more contracting and enforcement costs would be necessary to specify the desired practices and to insure that the desired reduction would be achieved. Performance standards have high transaction costs due to the difficulty of measuring or estimating farmers' pollution, and because a large number of practices may have to be employed to achieve the desired reduction. Agricultural nonpoint-source pollution (NSP) is diffuse and is possibly subject to long time lags between when pollutants leave the farm site and when they reach the surface or groundwater. A simulation model would be needed to estimate pollution from agricultural practices on specific farms (de Coursey, 1985). Examples of such models include the Lake Okeechobee Agricultural Decision Support System (LOADSS) (Negahban et al., 1995) and the Chesapeake Bay Watershed Model (Donigian et al., 1991; Thomann et al., 1994).

If farms have unequal compliance costs, the assignment of greater reductions to farmers with lower compliance costs would be preferable to the assignment of equal reduction to all farmers and could lower aggregate compliance costs. This is called allocative effectiveness (Batie and Ervin, this volume). The minimum control costs to reduce pollution to the specified level would be achieved by assigning pollution reduction so that all farmers, who are required to reduce pollution, have equal marginal compliance costs plus marginal transaction costs. A minimum cost allocation would imply that those farms with relatively high marginal compliance plus marginal transaction costs would not be required to reduce pollution.

While the targeted performance standard has low compliance costs relative to the uniform performance and design standards, the effects of targeting on transaction costs are uncertain. Targeting could increase information costs (required to estimate farm compliance costs and pollution levels) relative to design standards. Information costs for targeted and uniform standards would be similar because information on all farms would have to be collected in either case. Contracting and enforcement costs could decline with targeting because fewer farms would have to be selected and monitored for compliance compared to a uniformly applied standard. Litigation costs could be high with both targeted and uniform performance standards due to the complexity of practices and because farmers may challenge the validity of the models that would be used to estimate pollution. Because targeting would reduce the number of farms on which standards would have to be enforced, enforcement costs could be lower compared to uniform standards. Total transaction costs of targeting could be lower than the costs of uniform performance standards but they could still be higher than the uniform design standard. The conditions for allocating pollution reductions in order to minimize combined transaction and compliance costs are described mathematically in the appendix of this chapter.

We present a case study in this chapter in which the costs of design standards, uniform standards and targeted performance standards are compared for the dairy farms in the Lower Susquehanna Watershed portion of the Bay watershed. This study focuses on nitrogen runoff because of the limited success there has been in controlling this nutrient in the Bay and because phosphorus runoff is already declining due to point-source pollution (PSP) control.

A CASE STUDY OF AGRICULTURAL POLLUTION CONTROL

The Lower Susquehanna Watershed (the Watershed) consists of 5 million acres, of which 1.5 million are agricultural. The Watershed, located mainly in Pennsylvania with a small portion in Maryland, contributes an estimated 130 million pounds of nitrogen and 4 million pounds of phosphorus to the Bay (Hamlett and Epp, 1991). The largest single source of this nutrient loading in the Watershed originates from agriculture (EPA, 1992). Consequently, the Watershed was chosen by the U.S. Department of Agriculture's Economic Research Service, the National Agricultural Statistics Service, and the Natural Resource Conservation Service as the location for a detailed Area Studies' Survey (the Survey) of field- and farm-level agricultural and conservation practices at Natural Resource Inventory (NRI) sites. The NRI, which is conducted every five years, contains numerous physical attributes of randomly selected cropland and pastureland sites.

The Survey includes detailed economic and management data on more than 500 randomly selected NRI sites (weighted for the soil hydrological groups) in the Watershed. This analysis focuses on 237 sites that are operated as dairy farms: Pennsylvania and Maryland have 232 and 14 of these farms, respectively—9 farms were discarded because they had missing information. Of the sample dairy farms, 37 percent have sales between $0 and $99,999, 39 percent have sales between $100,000 and $249,999, 17 percent have sales between $250,000 and $499,999, and 7 percent have sales that exceed $500,000.

Farm Compliance Cost Model

In the farm compliance cost model, the farmer is assumed to maximize total gross margins, which are defined as gross revenues minus variable costs. Revenues are obtained from crop and livestock enterprises, which can be produced with differing technologies that involve different input combinations. Crop output and pollution depend on soil characteristics that affect crop productivity and the potential for runoff. Profits are maximized subject to physical resource constraints—such as land, labor facilities, livestock facilities, and pollution constraints. If a performance standard is imposed, the sum of nitrogen-runoff delivery that results from the crop enterprises has to be less than the farm's allocation of nitrogen-runoff delivery. The allowable nitrogen-runoff delivery depends upon whether a targeted or uniform performance standard is being evaluated. If a design standard is imposed, the nitrogen-delivery constraint would be eliminated and a constraint is added that requires the use of a specified practice.

With the targeted performance standard, a sequential optimization ensures that compliance and transaction costs for the watershed are minimized. Each farmer's objective function is first maximized in a baseline that has no constraints on nitrogen runoff. The objective function is maximized with 20 percent, 40 percent, 60 percent and 80 percent reductions in nitrogen runoff delivery. The farmer's compliance cost for a given runoff reduction is the reduction in farm net revenue relative to the baseline net revenue. A watershed-level optimization model chooses the least costly assignment of runoff reductions to farms considering both compliance and transaction costs.

The maximization problem is solved using a linear programming model, SUSFARM, (Bosch et al., 1995), which is written in the General Algebraic Modeling System (GAMS) (Brooke et al., 1992). Input files, which contain the information specific to each farm from the Survey, are read by GAMS and each farm is solved sequentially.

Livestock Enterprises

Poultry broilers, beef cow-calf enterprises and hog farrow-to-finish enterprises can be produced with a unique ration. Four rations are available for the dairies: (1) alfalfa corn silage; (2) corn silage only; (3) alfalfa hay only; and (4) alfalfa haylage. Milk production per-cow is a function of herd size (Ford, 1992). Feed requirements are, in turn, a function of milk production. Livestock facilities are assumed fixed in the short run and herd size cannot exceed the number of livestock each farmer reports in the Survey. No more than 25 percent of manure production can be spread in any one season—unless the farmer reports having manure storage facilities or having constructed facilities at a fixed cost per-unit of manure capacity. Manure spreading and storage costs are taken from Ritter (1990). Livestock and crop sale prices are Pennsylvania-weighted averaged prices from the 1988–1992 period, livestock variable input costs and crop variable input costs are from Pennsylvania farm enterprise budgets (PCES, 1992). All costs and prices are expressed in 1992 U.S. dollars.

Crop Enterprises

SUSFARM distinguishes 36 rotations that affect potential soluble nitrogen runoff and sediment adsorbed nitrogen runoff. Rotations refer to a sequence of crops (alfalfa, corn grain, corn silage, grass pasture, wheat, soybeans, oats, grass hay and rye cover) and types of tillage (conventional, reduced, no-till and none). Rotations can be produced with or without contour strip-cropping (Camacho, 1992; USDA, 1991). Crop yields are based on the soil type at the sample site, which is assumed to apply to the whole farm (Serotkin, 1993). Each farmer's total land, cropland and pastureland are based on responses from the Survey.

Crop nutrients can be supplied as animal manure and/or as commercial fertilizer. Nitrogen can also be obtained from precipitation, legume fixation, legume carryover and the mineralization of the soil's organic matter. Losses of nitrogen from manure through volatilization (while in storage and after spreading) and from seasonal nitrogen runoff and leaching (between the time of spreading and of crop uptake) are subtracted from the nitrogen availability.

This study is based on the 1990–1995 commodity program requirements prior to the Federal Agricultural Improvement and Reform Act (HR 2854) (USDA, 1996). It is assumed that participating farmers are required to have a total of the programmed crop that is planted (or considered planted) which is (1) at least as large as the base acres minus the set-aside plus flex acres and (2) less than their base acres minus their set-aside acres. Program participants are also allowed to enter the 0–85 program, which allows them to idle any fraction of their base minus set aside and 15 percent flex, and to receive 85 percent of the deficiency payment for the idled acres.

Nitrogen Applications and Delivery

Mass balanced equations in SUSFARM require that the nitrogen—from the soil's organic matter, precipitation, commercial fertilizer, manure, legume fixation and legume carryover—be equal to, or exceed, the crop uptake—after the volatilization, leaching and nitrogen runoff are accounted for. Nitrogen runoff is reduced when the fertilizer is applied nearer to the season (one of the four seasons that is in the model) of plant uptake and by incorporation. Soluble nitrogen runoff depends on (1) the amount of fertilizer that is applied onto the field; (2) the time at which fertilizer is applied; (3) the method of fertilizer application; (4) the crop rotation the fertilizer is applied to; (5) and the hydrological soil group onto which the fertilizer is applied (USDA, 1979 and 1986; Yagow et al., 1990; Novotny and Chesters, 1981). It is assumed that all soluble nitrogen runoff is delivered to the nearest surface body of water.

Sediment adsorbed nitrogen loss is a fraction of the amount of soil erosion that occurs in a rotation, which is based on the Universal Soil Loss Equation (USLE) (Wischmeier and Smith, 1978). The delivery of the sediment adsorbed nitrogen to surface water depends on the distance of the field to water, the land cover, and the slope and nature of the land cover along the flowpath to the receiving body of water (Shanholtz and Zhang, 1988; Heatwole et al., 1987).

Regulatory Scenarios

Activity hours and transaction costs for five regulatory scenarios—manure storage, strip-cropping, uniform performance, targeted performance for targeted farms and targeted performance for non-targeted farms—are estimated by identifying and budgeting the costs of activities that are required to target and to enforce nitrogen-runoff reductions (Carpentier, 1996). These are shown in table 18.1 and are discussed below.

TABLE 18.1 Activity Hours and Transaction Costs per Farm to Implement Regulatory Policies

	Activities			
	Information	*Contracting*	*Enforcement*	*Total*
	hours			
Manure Storage				
Initial (hours)	1.00	0.50	0.00	1.50
Update (hours per year)	0.00	0.00	3.00	3.00
Annualized cost ($)[a]	4.00	1.00	77.00	82.00
Strip-cropping				
Initial (hours)	5.00	0.50	0.00	5.50
Update (hours per year)	0.50	0.00	3.00	3.50
Annualized cost ($)	28.00	1.00	77.00	106.00
Uniform Performance				
Initial (hours)	73.00	0.00	0.00	73.00
Update (hours per year)	12.00	4.00	3.40	19.40
Annualized cost ($)	496.00	109.00	86.00	691.00
Targeted Performance *Non-targeted farms*				
Initial (hours)	77.00	0.00	0.00	77.00
Update (hours per year)	0.00	0.00	0.00	0.00
Annualized cost ($)	233.00	0.00	0.00	233.00
Targeted Performance *Targeted Farms*				
Initial (hours)	80.00	0.00	0.00	80.00
Update (hours per year)	16.00	7.00	3.40	26.40
Annualized cost ($)	610.00	158.00	86.00	854.00

[a] Costs are based on a ten-year horizon and a real interest rate of 5 percent, with an actualization factor of 7.7217. Travel costs include an average of 30 miles at $0.25 per mile. Hourly wages are $23, $25 and $32 for technicians, agronomic experts, and attorneys, respectively.

Source: Carpentier, 1996.

Manure Storage

The manure storage standard requires farmers to have manure storage (equal to six months of manure production) for each livestock type. Farmers must be notified to build manure storage facilities (requiring 0.5 hours per farm). An agent travels to each farm to verify that manure storage is present (1 hour per farm). The annual costs of information and of contracting for manure storage are $4 per farm and $1 per farm, respectively. Enforcement requires a visit by an agent to the farm every year to ensure that the storage is adequate for the number of livestock confinement units. The verification of manure storage and the filing of the proof of its existence require 3 hours per year. The total annualized transaction costs are $82 per farm.

Strip-cropping

In this study, the strip-cropping standard requires farmers on the more erosive soils to strip-crop corn, a practice recommended on steep slopes (Dillaha, 1990). Strips of contour-planted corn are alternated with closely grown crops that add more canopies, such as small grains or hay. For each site in the Survey, erosion is estimated for no-till corn silage using the USLE equation. If the estimated erosion factor exceeds the estimated soil loss tolerance factor (T-factor), the farmer is required to strip-crop his or her corn with small grains or alfalfa. NRI T-factors for each site are taken from the Survey (USDA, undated database). The C and P factors for crop rotation, tillage and conservation practices are based on the Pennsylvania Technical Guide (USDA, 1991).

Upon the T-factor estimation, the regulatory agent travels to each farm, which requires 5 hours of initial time, to determine whether erosion, as estimated by the USLE, exceeds the T-factor on each farm. An additional 0.5 hour is needed per farm each year to determine if field ownership has changed. Annual information costs (including travel costs) are $28. Contracting involves notifying farmers which fields require strip-cropping (0.5 hour per farm). Enforcement includes a visit to the farm every year (three hours per farm) to ensure that the corn, which is grown on soils that are designated for strip-cropping is actually strip-cropped. The annual transaction cost is $106.

Uniform Performance Standard

A uniform 40 percent reduction in nitrogen runoff is applied to all farms. The agency estimates a baseline pollution loading level for each farm and determines the practices that are needed to achieve the 40 percent reduction. Because it is difficult to monitor pollution from individual farms, a simulation model, such as the Erosion Productivity Impact Calculator (EPIC) (Williams et al, 1989), is used to estimate the baseline. An agent must travel to the farms and gather field boundary information, calibrate the simulation model for the area and estimate a delivery ratio for each farm. Delivery of nutrients from the edge of the farmers' field to the surface body of water must also be estimated (Shanholtz and Zhang, 1988).

The practices that are required to achieve a 40 percent nitrogen runoff reduction for each farm are identified for the uniform performance standard. An estimated 73 hours per farm are initially required to gather the necessary information for the uniform performance standard and 12 hours are required each year to update the plan to account for changing economic conditions.

Contracting the uniform performance standard requires that an agent return to each farm, which is estimated to take two hours, to discuss feasible plans with the farmer and to reach an agreement that is satisfactory to both the farmer and the agent. Farmers must be presented with alternative sets of practices for their farms and must be allowed to choose their preferred set of practices among these. An attorney then writes the contract with the agreed upon practices. Four hours are allowed each year at an annual cost of $109.

Enforcement consists of verifying that contracted practices are followed. Because farm adjustments are likely to be complex, this is estimated to require 3.4 hours annually at a cost of $86. Total annualized transaction costs are $691 per farm.

Targeted Performance Standard

The allocation of the burden of control that achieves the 40 percent reduction in nitrogen delivery at least-cost is found using sequential optimization. This procedure first minimizes costs at the farm and then at the watershed level. Shadow prices are estimated for nitrogen-delivery constraints equal to 20 percent, 40 percent, 60 percent and 80 percent. The shadow prices are used to derive each farmer's marginal compliance cost curve. A mixed integer programming model (ALLOCATI) minimizes watershed costs of reducing nitrogen delivery (Carpentier, 1996). Watershed costs equal the sum of compliance costs (approximated by the shadow prices of nitrogen reduction on each farm) plus transaction costs.

The initial costs of visiting the farm and calibrating the crop simulation model are the same as for the uniform performance standard but an additional bioeconomic model is required to simulate shadow prices of the nitrogen-runoff constraint. These shadow prices are subsequently used in the ALLOCATI model to find the social cost-minimizing allocation, which represents the allocation of responsibilities among farmers that minimize compliance plus transaction costs.

In table 18.1, transaction costs are estimated separately for targeted and non-targeted farms. Non-targeted farms are assigned an annualized information cost of $233 per farm for the 77 hours that are required to estimate their marginal compliance costs. Non-targeted farms are assigned zero reduction because of their estimated high compliance costs and, thus, have no contracting and enforcement costs.

Initial information on targeted farms requires 80 hours to gather. More time is needed for targeted farms than for non-targeted farms because farming practices must be specified that achieve the social cost-effective burden of control for each farm. Targeted farms need 16 hours of the agent's time to annually update this information, so the total annualized information costs for these farms are $610.

Contracting costs per farm may be higher for the targeted standard than for the uniform standard because many of the targeted farms are subject to larger reductions than they are under the uniform standard. Farmers may disagree with the imposed reductions or with the practices that are cost minimizing for their farms. Assuming that the allocation is fixed and that farmers can offer alternative practices, the technician is assumed to have to go back to the office and repeat the analysis with the farmer's proposed alternatives. Thus, two visits to each targeted farm are required to evaluate the farmers' proposed alternative practices. Enforcement costs for each targeted farm are assumed to be the same as for the uniform standard. Total transaction costs are $854 per farm.

Results of Farm Compliance Costs

In the baseline (unrestricted) case, average total gross margins are $134,472 per year and nitrogen delivery averages 852 pound per farm (table 18.2). Farmers milk an average of 136 cows and have 275 acres of harvested crops, over a third of which are alfalfa. No-till is the leading form of tillage. This indicates its higher profitability compared to reduced and conventional tillage.

Manure Storage

Manure storage requirements affect 81 farmers who have no manure storage. A total of 1,660 pounds of nitrogen delivery are curtailed in the watershed at a total compliance cost of $45,978 (table 18.3). Storage reduces nitrogen runoff by a relatively small amount because 66 percent of farms already have storage, some winter spreading continues with manure storage because some farms have lower labor costs in winter when excess fixed labor is available and possibly the model understates baseline soluble nitrogen runoff. Once the standard is applied, the 81 farms that did not have manure storage build six months worth of storage—the minimum required by the standard. For dairies, a vertical wall and roof-covered type of storage is assumed (Ritter, 1990).

Once manure storage facilities have been built, it is economically advantageous to spread manure in seasons of crop uptake to better utilize manure nitrogen. In this simulation, winter and fall spreading decline by about 28 percent while spring and summer spreading increase. As a result, soluble runoff declines by 20 pounds per farm on the 81 farms that build storage. This represents an average 7 pounds reduction over all 237 farms (table 18.2).

Strip-cropping

This standard, which is applied to 106 farms, curtails 47,970 pounds of nitrogen delivery and reduces total gross margins by $206,427 in the watershed (table 18.3). Compliance costs of $882 per farm are almost the same as the difference in average total gross margins between the strip-cropping standard and the baseline (table 18.2). Slight differences are due to computing average total gross margins over the 237 farms in the baseline and 234 farms in the strip-cropping standard.

TABLE 18.2 Effects of Two Performance Standards and Two Design Standards on Nitrogen Losses, Output, and Returns on Dairy Farms

Outcomes [b]	Baseline	Design Standards		Performance Standards [a]	
		Manure Storage	Strip Cropping	Uniform	Targeted
Total Gross Margins ($)	134,472	134,278	133,601	118,488	130,869
Nitrogen Delivery (pounds)	852	845	647	511	511
Leaching (pounds)	3,204	3,106	3,380	2,984	3,112
Livestock Enterprises (breeding units)					
Cows	136	136	136	132	135
Hogs	17	17	16	33	21
Poultry [c]	2,014	2,014	2,040	2,023	2,014
Other Cattle	14	14	8	2	11
Manure (tons dry matter)					
Winter Spreading	31	19	30	3	18
Spring Spreading	110	121	107	96	113
Summer Spreading	102	111	107	98	102
Fall Spreading	44	35	44	73	50
Total Manure Production	286	286	285	270	282
Manure Incorporated	39	40	38	118	55
New Manure Storage	0	17	2	20	5
Crop Enterprises (acres)					
Alfalfa	102	102	105	81	98
Corn Grain	57	57	53	33	52
Corn Silage	57	57	56	52	55
Wheat	59	59	59	98	71
Total Harvested Crops	275	275	273	264	276
Rye Cover	2	2	3	7	2
Idle Crop Land	0	0	0	5	0
Grazed Pasture	41	41	41	29	41
Idle Pasture	66	66	64	77	66
Conventional Till	20	20	21	16	20
Reduced Till	70	70	70	127	85
No-till	113	113	108	64	102
Strip-cropped Acres	91	91	148	150	139
Commercial N (tons)	17	17	16	11	16

[a] Standards are for a 40 percent reduction in nitrogen delivery.

[b] All values are averaged over 237 farms (except for strip-cropping which was averaged over 234 farms and uniform performance standard averaged over 236 farms).

[c] Poultry reported as number of birds sold.

Source: Carpentier, 1996.

TABLE 18.3 Reduction in Nitrogen Delivery and Total Control Costs with Two Performance and Two Design Standards

	Design Standards		Performance Standards	
	Storage Manure	*Strip Cropping*	*Uniform*	*Targeted*
	Reduction in Nitrogen Delivery (pounds)			
Total Pounds Reduction	1,660	47,970	80,817	80,817
Pounds Reduced per Farm [a]	7	205	342	341
Number of Farms Targeted	81	106	236	119
Pounds Reduced per Farm Targeted	20	453	342	679
	Compliance Costs ($)			
Total Compliance Cost	45,978	206,427	3,803,848	853,911
Per Farm Compliance Cost	194	882	16,118	3,603
Compliance Cost per Pound	28	4	47	11
	Transaction Costs ($)			
Total Information Cost [b]	324	2,968	117,056	100,084
Info. Cost per Targeted Farms	324	2,968	117,056	72,590
Info. Cost per Non-targeted Farms	0	0	0	27,494
Contracting and Enforcement Costs	6,318	8,268	46,020	29,036
Total Transaction Costs	6,642	11,236	163,076	129,120
Per Farm Transaction Cost	28	48	691	545
Transaction Costs per Pound	4	0[c]	2	2
	Control Costs ($)			
Total Control Costs	52,620	217,663	3,966,924	983,031
Per Farm Control Costs	222	930	16,809	4,148
Control Cost per Pound [d]	32	5	49	12

[a] All averages are computed over 237 farms for the manure storage and targeted performance standard, 236 farms for uniform performance standard, and 234 farms for strip-cropping unless otherwise indicated.

[b] Information for non-targeted farms is needed to decide which farms should not be targeted. Information for targeted farms is needed to decide how much reduction to impose.

[c] Exact cost is $0.23 per pound

[d] Total control costs per pound equal the sum of compliance and transaction costs. Differences are due to rounding.

Source: Carpentier, 1996.

The average area that is strip-cropped per farm increases by 63 percent from 91 acres in the baseline to 148 acres per farm (table 18.2). Hog numbers, cattle numbers and manure production are reduced slightly, but other practices are similar to the baseline levels. The strip-cropping standard does not achieve the 40 percent total nitrogen delivery reduction goal; it would have to be combined with other policies in order to achieve this goal. The strip-cropping standard is the only policy for which a trade-off exists between surface and groundwater quality. Estimated leaching increases by two percent while estimated nitrogen delivery decreases by 24 percent.

Uniform Performance Standard

The uniform performance standard reduces total gross margins in the watershed by $3,803,848 and total nitrogen delivery by 80,817 pounds (table 18.3). The average of total gross margins for 236 farms (one farm was dropped because it could not meet commodity program provisions under the uniform standard) decreases by 12 percent from the baseline to $118,488 (table 18.2). Total nitrogen delivery declines to an average 511 pounds per farm. Total leaching, which was restricted to not exceed the baseline of 3,204 pounds, decreases slightly to 2,984 pounds. The number of dairy cows and other cattle decreases slightly, while hog numbers double to 33 head (breeding units). Alfalfa and corn acreage declines while wheat, which provides winter cover as well as revenue, increases to 98 acres.

Strip-cropping, incorporation of manure and shifting the timing of manure application are major intensive margin adjustments made to achieve the 40 percent reduction performance standard. The construction of an average of 20 tons of manure storage (dry matter basis) allows winter spreading to decrease from 31 tons to three tons. With strip-cropping, manure incorporation triples to 118 tons. Reduced tillage acreage increases from 70 to 127 acres, while no-till acreage is reduced to allow the increase in manure incorporation. Over 50 percent of planted acres are strip-cropped. In addition to these intensive margin adjustments, an average of five acres of cropland and an additional 11 acres of pastureland are idled compared to the baseline.

Targeted Performance Standard

The targeted performance standard reduces nitrogen delivery in the watershed by 80,817 pounds, as does the uniform standard, but it reduces total gross margins by much less than the uniform performance standard, $853,911 (table 18.3). The targeted performance standard is achieved using the same management practices as the uniform standard, but at a much lower intensity. This results in a mean average gross margin of $130,869 per farm (table 18.2). Farms with high marginal compliance and transactions costs are not targeted and they total 118. Seven farms abate 42,487 pounds (80 percent of their baseline), 11 farms abate 17,599 pounds (60 percent of their baseline), and 28 farms abate 11,874 pounds (40 percent of their baseline). These 46 farms together, out of the total 237, contribute 89 percent of the required reduction in nitrogen delivery for the watershed. The remaining 73 farms abate 20 percent of their baseline.

A few farms contribute large reductions because they tend to be on sites with the highest nitrogen delivery potential (such as farms that are on steep slopes and are close to water). These farms also tend not to use any management practices in the baseline to reduce nitrogen runoff, such as manure storage and strip-cropping. The combination—of a large initial delivery and a small number of management practices in use (with many available alternatives to control delivery)—results in very low marginal compliance costs for these farms. For example, the seven farms that reduce 80 percent of their baseline have a mean baseline delivery of 7,587 pounds and mean shadow prices of $1 per pound, $2 per pound, $18 per pound, and $29 per pound for the 20 percent, 40 percent, 60 percent and 80 percent reductions, respectively. Farms that are not targeted have a mean baseline delivery of only 403 pounds and their associated shadow prices are $179, $449, $949 and $1,766 for the 20 percent, 40 percent, 60 percent and 80 percent reduction levels, respectively.

Farm total gross margins with targeting are $12,381 more per farm per year than they are under a uniform performance standard. The types of practices adopted are similar to the uniform standard but they are less widely adopted. Strip-cropped acreage, new manure storage construction and wheat acreage all increase relative to the baseline while winter manure spreading declines. Compliance costs are more unequally distributed under targeting. The 20 targeted farms bearing the highest compliance cost have $662,140 in compliance costs or 78 percent of the total. Under the uniform standard, the 20 farms bearing the highest compliance costs have a total cost of $1,462,980 or 38 percent of the total.

Control Cost Comparison

Targeted Standard versus the Uniform Performance Standard

Compliance costs and transaction costs were shown to be less for the targeted performance standard than they are for the uniform performance standard (table 18.3). Compliance costs are $3,603 per farm for the targeted standard compared to $16,118 per farm for the uniform standard. The lower compliance cost for targeting reflects the lower intensity of overall farm adjustments that are required when farms with low marginal costs of reducing deliveries are selectively targeted for larger reductions in nitrogen delivery.

Transaction costs per farm are $545 for the targeted performance standard compared to $691 for the uniform performance standard. Total information costs are lower for the targeted standard than for the uniform standard. The information on the 236 farms must be updated every year for the uniform standard, while the information on loadings and required reductions has to be updated for only 119 farms with the targeted performance standard (table 18.1). If changes in technology or other economic conditions were to cause non-targeted farms to have lower compliance costs than targeted farms, then estimates of loadings for non-targeted farms would have to be updated as well, and transaction costs for the targeted standard would increase. Contracting and enforcement costs were lower under the targeted performance standard than under the uniform standard because only 119 farms have targeted performance standards whereas 236 farms have uniform standards.

Design Standard versus Performance Standard

When applied to agricultural pollution sources, performance standards are defined as the least-cost combination of practices that achieve the goal of a 40 percent reduction in nitrogen watershed delivery. The private compliance costs of design standards could not be less than the compliance costs of the performance standard (perfectly targeted design standards). The aggregate compliance plus transaction costs of design standards could be lower if the decrease in transaction costs were sufficient to outweigh the increase in compliance costs.

Manure storage curtails 1,660 pounds of nitrogen runoff at a compliance cost of $28 per pound (table 18.3). The strip-cropping standard curtails 47,970 pounds at a compliance cost of $4 per pound. Because manure storage and strip-cropping do not achieve the 40 percent required reduction in nitrogen, their compliance costs are not comparable with those of the performance standards. Strip-cropping and manure storage only abate total watershed nitrogen delivery by 24 percent and 1 percent, respectively.

Strip-cropping is better compared to a targeted rather than to a uniform performance standard because the strip-cropping requirement is targeted at specific soils having a high potential for erosion. When strip-cropping and targeted performance standards are compared for approximately the same amount of control (20 percent nitrogen runoff reduction for targeted performance and 24 percent reduction for strip-cropping), they have nearly the same compliance cost per pound ($4). Strip-cropping has lower transaction costs than the targeted performance standard ($0.23 per pound versus $2) and lower total control costs per pound ($4.23 versus $6.00).

The transaction costs of design standards are much less than those of performance standards. Transaction costs per farm (averaged over all farms in the sample) are only $28 and $48 for the manure and strip-cropping standards, respectively. This is because these practices are easy to observe and to verify. Transaction costs per pound of watershed nitrogen delivery curtailed are less for strip-cropping ($0.23) compared to $2 for the uniform and targeted performance standards and $4 for the manure storage standard.

Compliance costs make up 87 percent to 96 percent of total control costs for all performance and design standards. Consequently, the difference in compliance costs determines the relative ranking of design and performance standards in terms of their total control costs. This result contradicts Kohn's (1991) hypothesis that transaction costs are more important than compliance costs to determine the social cost-effectiveness of policy instruments.

Limitations and Further Research Needs

Only dairy farms are analyzed in this study. The Lower Susquehanna watershed also contains large numbers of other livestock farms (for example, beef cattle, hogs and sheep), cash-crop farms and poultry farms. Large poultry operations often pose environmental problems because not enough land is available for environmentally sound manure management. Including these other farm types with potentially different compliance costs for reducing nitrogen watershed delivery might influence the allocation of control burdens. The inclusion of other farm types would likely have increased the variability of compliance costs and further reduced compliance costs for the targeted performance standard compared to the uniform standard.

Increasing the nitrogen runoff control options to include hauling livestock waste off the farm, installation of buffers and filter strips, and other practices might have reduced control costs on many farms. More research is needed to determine how the inclusion of more control options would affect the benefits of targeting.

This analysis is based on the 1990 through 1995 commodity program provisions. Restrictions on crop acreage that were produced by commodity program participants were relaxed in 1996 with the Federal Agricultural Improvement and Reform Act (FAIR). Further research is needed to determine how the removal of these restrictions affects farmers' baseline nitrogen runoff and costs of reducing runoff.

Farmers make crop management decisions based on soil characteristics at the field level. For example, Green et al. (1996) discovered that irrigation technology adoption depends on field slope and soil permeability. Wu and Segerson (1995) discuss potential biases in estimates of groundwater pollution potential when field variability is ignored. VanDyke et al. (1998) realized that ignoring within-farm soil variability biases the estimates of farmers' costs of reducing nitrogen pollution. Reflecting each farm's heterogeneous soils and slopes might reduce inter-farm variability of compliance costs and reduce the benefits of targeting compared to those estimated in this study. Further work should be done to estimate how soil variability within the farm affects relationships between farm characteristics and costs of reducing pollution.

Litigation costs, which were not considered in this analysis, can significantly increase enforcement costs. Since targeting reduces the number of farms that are affected by regulations, it may also reduce litigation costs compared to uniform standards. Also, targeting farms with a high potential to pollute, could be more defensible in court and could result in less litigation compared to uniform standards where farms with low pollution potential are regulated. Litigation costs would probably be higher for performance standards than for design standards, because there are precedents for design standards in agriculture and these standards usually involve relatively simple measures. Under a performance standard, pollution reductions would probably have to be verified with simulation models that might be challenged in court.

Transaction cost estimates in this study are approximations, because data on transaction costs are lacking. The effects of errors in transaction costs on the total control costs should be relatively small because transaction costs are smaller than compliance costs. Sensitivity analysis revealed that a reduction of marginal transaction costs by 50 percent increased the number of farms that targeted under a performance standard by 7 percent (127 farms), while dividing transaction costs by 4 increased targeted farms by 12 percent (133 farms). Doubling marginal transaction costs decreased the number of farms targeted by 16 percent (100 farms) and multiplying transaction costs by 4 decreased the number of farms targeted by 37 percent (75 farms).

Incentive effects, not considered here, would favor performance standards (Batie and Ervin, this volume). Further research is needed to quantify the incentive benefits of performance standards.

Further analysis of the aggregate impacts of uniform performance, targeted performance standards and design standards is needed. Research should consider the differing effects of regulatory standards on consumers' food costs, net returns to producers and costs to taxpayers.

CONCLUSIONS AND IMPLICATIONS

Given limited public funds, policymakers may rely on regulations to protect water quality. This study compares the costs of design and performance standards. Design standards are expected to have high compliance costs because of their inflexibility, but they are expected to have relatively low transaction costs because they are easy to observe. Performance standards have more flexibility and lower compliance costs than design standards do, but they have higher transaction costs due to their complexity. Targeted performance standards should have lower compliance costs compared to uniform design standards, the information costs to determine which farms to target, however, may be higher. The effects of different regulatory standards on total control costs must be evaluated empirically.

An empirical comparison of two design standards (manure storage and strip-cropping) and two performance standards (uniform and targeted) is carried out for dairy farms in the Lower Susquehanna portion of the Chesapeake Bay Watershed. The uniform performance standard has the highest control cost per pound of nitrogen delivery reduction. The manure storage standard is second—its relatively high cost per pound is due to its low reduction in nitrogen delivery. The targeted performance standard is third in total control cost per pound followed by the strip-cropping standard. However, strip-cropping does not achieve the required 40-percent reduction in nitrogen delivery. When strip-cropping and targeted performance standards are compared for approximately the same reduction in nitrogen delivery, the targeted performance standard has a slightly lower per-pound compliance cost, but has higher total control costs.

Political viability of regulations could vary greatly among regulatory standards. Design standards that are simple and not too costly may be more acceptable than performance standards. Performance standards would likely require crop growth models linked to an economic model (bio-economic model) in order to estimate pollution and costs from different practices. Such models might not be understandable or acceptable to farmers—particularly if they were to impose large compliance costs. Targeted standards would be less acceptable than uniform standards because of perceived unfairness, unless compensation would be provided to farmers. However, targeting of regulatory design standards has won acceptance in some cases in which there has been a widely perceived pollution problem and in which targeting has addressed areas with the highest incidences of pollution. For example, farmers on sandy soils in some parts of Nebraska are required to conduct soil nitrogen tests and to keep records of nitrogen applications. They are not allowed to apply nitrogen during fall or winter (Williamson, 1988).

The results of this study imply that policymakers should focus on setting clear environmental goals and should allow farmers maximum flexibility when achieving these goals (Batie and Ervin, this volume). The assignment of responsibilities for the reduction of pollution in order to meet these environmental goals should also be flexible, so that farmers with lower costs of reducing pollution may be assigned greater reductions. Other policy instruments should also be adopted or adapted to increase the farmers' flexibility in meeting pollution goals. For example, pollution-rights trading (among nonpoint-source polluters, and with point-source polluters) should be promoted as a way of providing incentives to shift pollution reductions to farmers with lower costs of pollution control (EPA, 1992). The exchange of information should be facilitated between farmers who have low costs of reducing pollution and point-source polluters or other nonpoint-source polluters with higher abatement costs who wish to trade for pollution rights. Voluntary cost-share programs should be tailored to assist farmers with the lowest costs of pollution reduction (Ervin and Graffy, 1996; Batie and Ervin, this volume). Results of this study suggest that the increase of farmers' flexibility when meeting pollution reduction objectives would have high net benefits in terms of reducing overall farm compliance costs.

REFERENCES

Abler, D.G. and J.S. Shortle. 1991. "The Political Economy of Water Quality Protection from Agricultural Chemicals." *Northeastern Journal of Agricultural and Resource Economics* 20: 53–60.

Anderson, G.D., A.E. De Bossu, and P.J. Kuch. 1990. "Control of Agricultural Pollution by Regulation," in J.B. Braden and S.B. Lovejoy, eds., *Agriculture and Water Quality: International Perspectives.* Boulder, CO: Lynne Rienner Publishers.

Beegle, D. Undated. *Nutrient Management Legislation in Pennsylvania.* Agronomy Extension Fact Sheet No.40, Pennsylvania Cooperative Extension Service, University Park, PA.

Bohm, P. and C.S. Russell. 1990. "Comparative Analysis of Alternative Policy Instruments," in A.V. Kneese and J.L. Sweeney, eds., *Handbook of Natural Resource and Energy Economics.* New York, NY: North-Holland.

Bosch, D.J., C.L. Carpentier, and R. Heimlich. 1995. "A Farm Model for Evaluating Nonpoint Source Pollution Abatement Programs." Department of Agricultural and Applied Economics, SP-94-03, Virginia Polytechnic Institute and State University, Blacksburg, VA (February).

Brooke, A., D. Kendrick, and A. Meeraus. 1992. *GAMS: A User's Guide, Release 2.25.* Washington, DC: The World Bank.

Camacho, R. 1992. *Chesapeake Bay Program Nutrient Reduction Strategy Reevaluation Report No. 8: Financial Cost Effectiveness of Point and Nonpoint Source Nutrient Reduction Technologies in the Chesapeake Bay Basin.* Interstate Commission on the Potomac River Basin, ICPRB Report No. 94-4, Washington, DC (December).

Carpentier, C.L. 1996. "Value of Information for Targeting Agro-pollution Control: A Case Study of the Lower Susquehanna Watershed." Ph.D. dissertation, Department of Agricultural and Applied Economics, Virginia Polytechnic Institute and State University, Blacksburg, VA.

CBP (Chesapeake Bay Program). 1994. *Trends in Phosphorus, Nitrogen, Secchi Depth, and Dissolved Oxygen in Chesapeake Bay, 1984 to 1992.* U.S. Environmental Protection Agency, CBP/TRS 115-94, Washington, DC (August).

_____. 1996. *State of the Chesapeake Bay, 1995.* Annapolis, MD: Chesapeake Bay Program Office.

deCoursey, D.G. 1985. "Mathematical Models for Nonpoint Water Pollution Control." *Journal of Soil and Water Conservation* 40: 508–513.

Dillaha, T.A. 1990. "Role of Best Management Practices in Restoring the Health of the Chesapeake Bay: Assessments of Effectiveness," in M. Haire and E.C. Krome, eds., *Perspectives on the Chesapeake Bay, 1990: Advances in Estuarine Sciences CBP/TRS41-90.* Gloucester Point, VA: Chesapeake Research Consortium (April).

Donigian, A.S., B.R. Bicknell, A.S. Patwardhan, L.C. Linker, D.Y. Alegre, C.H. Chang, and R. Reynolds. 1991. *Watershed Model Application to Calculate Bay Nutrient Loadings: Final Findings and Recommendations.* Study prepared by Aqua-terra Consultants, Computer Sciences Corporation for the U.S. Environmental Protection Agency Chesapeake Bay Program, Washington, DC.

EPA (U.S. Environmental Protection Agency). 1992. *Incentive Analysis for Clean Water Act Reauthorization: Point-source/Nonpoint-source Trading for Nutrient Discharge Reductions.* Office of Water and Policy, Planning and Evaluation, U.S. Environmental Protection Agency, Washington, DC.

Ervin, D.E. and E.A. Graffy. 1996. "Leaner Environmental Policies for Agriculture." *Choices* 11: 27–33.

Ford, S.A. 1992. *1991 Pennsylvania Dairy Farm Business Analysis.* Extension Circular. The Pennsylvania State University, Cooperative Extension, University Park, PA.

Green, G., D. Sunding, D. Zilberman, and D. Parker. 1996. "Explaining Irrigation Technology Choices: A Microparameter Approach." *American Journal of Agricultural Economics* 78: 1064–1072.

Hamlett, J.M. and D.J. Epp. 1991. *Evaluation of the Impacts on Water Quality of Conservation Practices in the Chesapeake Bay Program.* Environmental Resources Research Institute Report No. 9107, Pennsylvania State University, University Park, PA (August).

Heatwole, C.D., V.O. Shanholtz, E.R. Yagow, and E.R. Collins, Jr. 1987. *Targeting Animal Waste Pollution with Virginia GIS.* American Society of Agricultural Engineers, Paper No. 87-4049, Saint Joseph, MI.

Kohn, R.E. 1991. "Transactions Costs and the Optimal Instruments and Intensity of Air Pollution Control." *Policy Sciences* 24: 315–332.

Krier, J.E and W.D. Montgomery. 1973. "Resource Allocation, Information Cost and the Form of Government Interventions." *Natural Resources Journal* 13: 89–105.

Lee, W.F., M.D. Boehlje, A.G. Nelson, and W.G. Murry. 1980. *Agricultural Finance.* Ames, IA: The Iowa State University Press.

Negahban, B., C. Fonyo, W.G. Boggess, J.W. Jones, K.L. Campbell, G. Kiker, E. Flaig, and H. Lal. 1995. "LOADSS: A GIS-based Decision Support System for Regional Environmental Planning." *Ecological Engineering* 5: 391–404.

Novotny, V. and G. Chesters. 1981. *Handbook of Nonpoint-source Pollution Sources and Management.* New York, NY: Van Nostrand Reinhold Environmental Engineering Series.

Pennsylvania Cooperative Extension Service. 1992. *Pennsylvania State Farm Enterprise Budgets.* Department of Agricultural Economics and Rural Sociology, The Pennsylvania State University, University Park, PA (April).

Ritter, W.F. 1990. *Manual for Economic and Pollution Evaluation of Livestock Manure Management Systems.* Pennsylvania Department of Environmental Resources, Bureau of Soil and Water, Harrisburg, PA (March).

Serotkin, N., ed. 1993. *Pennsylvania State Agronomy Guide.* Pennsylvania State Cooperative Extension Service, University Park, PA.

Shanholtz, V.O. and N. Zhang. 1988. *Supplementary Information of Virginia Geographic Information System.* Department of Agricultural Engineering, Virginia Polytechnic Institute and State University, Special Report No. VirGIS 88-21. Blacksburg, VA: VirGIS Laboratory.

Thomann, R.V., J.R. Collier, A. Butt, E. Casman, and L.C. Linker. 1994. *Response of the Chesapeake Bay Water Quality Model to Loading Scenarios, CBP/TRS 101-94.* Study prepared by the Modeling Subcommittee of the Chesapeake Bay Program for the U.S. Environmental Protection Agency, Washington, DC: (April).

USDA (U.S. Department of Agriculture). Undated. *National Resources Inventory Database.* USDA/SCS, U.S. Department of Agriculture, Washington, DC.

_____. 1979. *Animal Waste Utilization on Cropland and Pastureland: A Manual for Evaluating Agronomic and Environmental Effects.* Science and Education Administration, USDA Utilization Report No. 6, Washington, DC (October).

_____. 1986. *Urban Hydrology for Small Watersheds.* Soil Conservation Service, U.S. Department of Agriculture, Washington, DC.

_____. 1991. *Pennsylvania Technical Guide.* Soil Conservation Service, U.S. Department of Agriculture, Harrisburg, PA.

_____. 1996. *Agricultural Outlook Supplement.* Economic Research Service, U.S. Department of Agriculture, Washington, DC (April).

VanDyke, L.S., D.J. Bosch, and J.W. Pease. 1998. *Impacts of Within-farm Soil Variability on Nitrogen Pollution Control Costs.* Unpublished paper. Department of Agricultural and Applied Economics, Virginia Polytechnic Institute and State University, Blacksburg, Virginia (March).

VCPB (Virginia Chesapeake Bay Program). 1993. *Virginia's Tributary Strategies.* Richmond, VA: Virginia Department of Environmental Quality (April).

Williams, J.R., C.A. Jones, J.R. Kiniry, and D.A. Spanel. 1989. "The EPIC Crop Growth Model." *Transactions of the ASAE* 32: 497–511.

Williamson, D. 1988. "Implementation of the Nebraska Nitrate Control Legislation," in American Water Resources Association's *Nonpoint Pollution: 1988—Policy, Economy, Management and Appropriate Technology.* Minneapolis, MN.

Wischmeier, W.H. and D.D. Smith. 1978. "Predicting Rainfall Losses—a Guide to Conservation Planning." *Agriculture Handbook No. 537.* Washington, DC: U.S. Department of Agriculture.

Wu, J. and K. Segerson. 1995. "On the Use of Aggregate Data to Evaluate Groundwater Protection Policies." *Water Resources Research* 31: 1773–1780.

Yagow, E.R., V.O. Shanholtz, J.W. Kleene, S. Mostaghimi, and J.M. Flagg. 1990. *Annual Estimation of Nitrogen in Agricultural Runoff.* American Society of Agricultural Engineers (ASAE), Paper No. 90-2054. St. Joseph, MI.

APPENDIX: COST MINIMIZING ALLOCATION OF POLLUTION CONTROL AMONG FARMS

Total control costs of uniformly applied performance standards (CC_{PU}) are

(1) $$CC_{PU} = \sum_i [C_{iPU}(r_i) + TC_{iPU}(r_i)].$$

C_{iPU}, the ith farm's compliance cost of achieving the standard, is an increasing function of r_i, the level of pollution reduction on the ith farm. TC_{iPU}, the transaction costs of achieving the standard, is also an increasing function of pollution reduction.

Assigning greater reductions to farmers with lower compliance costs could lower aggregate compliance costs of uniformly applied performance standards. Minimum total control costs of the targeted performance standard (CC_{PT}) are

(2) $$Min\ CC_{PT} = Min \sum_i [C_{iPT}(r_i) + TC_{iPT}(r_i)]$$

subject to,

(3) $0 \le r_i \le e_i$, and

(4) $$\sum_i [e_i - r_i] \le Z.$$

Pollution reduction on any farm, r_I, must be less than the farm's unconstrained pollution level, e_i. The standard requires that total pollution over i farms not exceed a designated level Z in the watershed.

Minimization of total costs occurs under the following conditions

(5) $$\frac{\partial C_i}{\partial r_i} + \frac{\partial TC_i}{\partial r_i} - \lambda \ge 0,$$

(6) $$r_i \left[\frac{\partial C_i}{\partial r_i} + \frac{\partial TC_i}{\partial r_i} - \lambda \right] = 0,$$

(7) $$\lambda \left[\sum_i [e_i - r_i] - Z \right] = 0,$$ and

(8) $\lambda \ge 0.$

Each farm's marginal compliance costs plus transaction costs must equal or exceed λ, the shadow price for the pollution constraint, which is the incremental cost of a one-unit reduction in Z (the allowable pollution in the watershed). Farms for whom the marginal increase in transaction plus compliance costs exceeds λ are not required to reduce pollution ($r_i = 0$). Farms required to reduce pollution have marginal compliance plus transaction costs equal to λ. If the sum of pollution were less than Z, the shadow price of the pollution constraint would be zero ($\lambda = 0$). The shadow price of the pollution constraint cannot be negative.

19 REGULATORY TAKINGS ISSUES: THE CASE OF GRASS FIELD-BURNING RESTRICTIONS IN EASTERN WASHINGTON STATE

Ray Huffaker and Stina Levin
Washington State University, Pullman, WA

This chapter develops a framework for evaluating the constitutional feasibility and social efficiency of regulations that employ flexible incentives. A preliminary framework summarizes the principles that the U.S. Supreme Court applies to decide whether a regulation satisfies the constitutional protections of private property. The Court's approach is assessed for its ability to promote socially efficient regulation and resource use. A modified decision framework is developed to better promote these objectives, which are applied to determine the constitutionality and economic efficiency of grass field-burning restrictions in eastern Washington State.

INTRODUCTION

Regulations that employ flexible incentives set performance standards on firm behavior and/or on environmental quality. These standards can be implemented with subsidies, taxes and/or economic penalties for noncompliance (Batie and Ervin, this volume; Segerson, this volume). Such regulations are designed to modify traditional private land uses, and, to the extent that they impose economic costs on landowners, they could incite a constitutional challenge as an infringement of landowner rights under the Takings Clause of Amendment V of the U.S. Constitution. This chapter develops a framework for analyzing whether regulations that employ flexible incentives would survive such a challenge and would remain as a valid regulatory exercise of governmental power in particular circumstances. The particular circumstances that are analyzed in this chapter are the regulations concerning grass field burning in eastern Washington State.

REGULATORY TAKINGS

The U.S. Constitution authorizes the federal[1] and state[2] governments to restrict personal freedom and property rights for the protection of public safety, health, morals and general prosperity. This authority is referred to as constitutional power when it is exercised by the federal government, and as police power when it is exercised by state governments (Nowak et al., 1983). The federal government's exercise of its constitutional power is expressly restricted by the Due Process[3] and Takings[4] clauses of Amendment V of the U.S. Constitution. The individual state's exercise of police power is expressly restricted by the Due Process[5] and Equal Protection[6] clauses of Amendment XIV of the U.S. Constitution. U.S. Supreme Court (Court) decisions have made the Equal Protection Clause applicable to the federal government (through its interpretation of the Due Process Clause of Amendment V)[7], and have made the Takings Clause applicable to the state governments (through its interpretation of the Due Process Clause of Amendment XIV).[8]

For more than two hundred years, the Court has attempted to determine when governmental interferences that concern private property constitute compensable takings. The only clear result of these determinations is that the compulsory physical encroachment of land by either type of government is considered to be a compensable taking. As governmental interference with an owner's use of land becomes less physical, there is increasing uncertainty as to whether compensable takings have occurred. This constitutes the regulatory takings issue that is currently receiving so much attention by federal and state judiciaries and by legislatures, environmentalists, property rights advocates and scholars, which include economists.

The Court's failure to provide a simple, consistent and concise resolution of the regulatory takings issue has encouraged many of the above interested parties to attempt to provide one. At one end of the spectrum, federal and state legislatures are in various stages of considering takings statutes that have been largely initiated by property rights advocates. These would greatly expand the legal circumstances under which police power regulations constitute compensable takings (for example, the State of Washington's Private Property Regulatory Fairness Act that was enacted by the legislature in 1995 and was later repealed by voters). At the other end of the spectrum, some economists contend that it is socially inefficient to consider any regulations as compensable takings (Bromley, 1993). Other economists allow for the existence of regulatory takings, and recommend compensation rules that encourage efficient levels of governmental regulation and private investment under various land-use scenarios (Innes, 1995; Micelli and Segerson, 1994).

The entire spectrum of legislative and disciplinary attempts to rationalize regulatory takings policy offers a wide, and often conflicting, variety of perspectives and analytical frameworks. Selecting among these frameworks, or crafting one's own, requires an evaluation of their comparative advantages relative to the Court's regulatory takings jurisprudence. This chapter provides such an evaluation and formulates a regulatory takings framework. This framework is useful for analyzing the constitutionality of regulations that rely on flexible incentives as a means to encourage the adoption of environmental technologies in agriculture.

The chapter follows a two-step approach to formulate a regulatory takings framework, which is used in the grass field-burning case study. The first step is to construct a preliminary framework (presented as a flowchart) that summarizes the tests and principles that the Court applies to determine whether a regulation constitutes compensable takings under the Takings Clause of the U.S.Constitution. This preliminary framework is based on the majority opinions of the past three major court cases that dealt with regulatory takings. In these cases, the Court wrestled with striking a balance between the public good generated by regulations and the exactions placed on the regulated parties. This preliminary framework also accounts for the associated minority opinions, which are based on earlier Court cases that were never overruled. These minority opinions claim that police power regulations can never constitute compensable takings. The second step is to determine how this preliminary framework can be modified to encourage socially efficient levels of governmental regulation and private investment under various land-use scenarios. This chapter also reviews past scholarly work that has recommended modifications be made to the regulatory takings framework. It then evaluates the effectiveness of these recommendations for the promotion of regulatory efficiency and adjusts the Court's takings framework accordingly.

U.S. SUPREME COURT: DUE PROCESS, EQUAL PROTECTION AND TAKINGS JURISPRUDENCE

How the Court implements the constitutional safeguards of private property, which are given by the Due Process, Equal Protection and Takings Clauses of the U.S. Constitution is summarized in Figure 19.1.

DUE PROCESS AND EQUAL PROTECTION CHALLENGES

The first discussion regards the due process and equal protection challenges to government regulation. In the *Lawton v. Steele* [9] (*Lawton*) case, the Court formulated a two-part test that a regulation must survive in a due process challenge. Generally, a regulation must fulfill a legitimate public purpose and not be unduly oppressive on the regulated party. A regulation that fails the *Lawton* test is invalidated. The application of this test has been limited largely to regulations that restrict personal liberties, since the Court decided in *United States v. Carolene Products Company* [10] (*Carolene*) that regulations that restrict economic interests would receive much less scrutiny under due process challenges (and thus, be harder to invalidate). The jagged line on both sides of the reference to *Carolene* in figure 19.1 signifies that the Court can use the *United States v Carolene* case to short circuit the due process consideration of economic regulation.

The Court also uses the *Lawton* two-prong test to determine whether a regulation that classifies persons can survive an equal protection challenge. [11] This is denoted by the arrow that runs from the Equal Protection box to the Due Process box in figure 19.1.

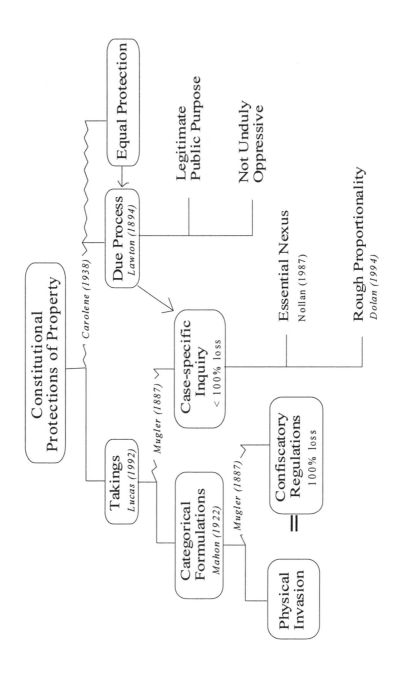

FIGURE 19.1 Constitutional Protections of Private Property as Implemented by the U.S. Supreme Court.

TAKINGS CHALLENGE

The Court's takings jurisprudence can be traced to Justice Holmes' majority opinion and Justice Brandeis' dissent in *Pennsylvania Coal Co. v. Mahon (Mahon)*,[12] in which the Pennsylvania Coal Company challenged Pennsylvania's Kohler Act. This act prohibited coal mining, which resulted in ground subsidence under residential dwellings, as a regulatory taking under Amendment V of the U.S. Constitution. Holmes' majority opinion stated that the Kohler Act constituted a taking because it went too far in destroying the company's right to extract its coal. Holmes reasoned that the police and eminent domain powers exist in a continuum. When a regulation goes too far, it constitutes a taking. Curiously, the remedy to the problem was an invalidation of the Kohler Act—the same remedy as in due process challenges—even though the Constitution requires just compensation. The Court did not settle upon a compensation remedy until 1987. It decided that the remedy was a temporary compensation from the time the regulation became takings until its trial (if the government repealed the regulation), or that the remedy was a permanent compensation (if the government elected to rule that the remedy stand, which is an inverse condemnation).[13]

In Mahon, Brandeis argued that the Court should apply its doctrine from an earlier case, *Mugler v. Kansas (Mugler)*.[14] *Mugler* challenged a Kansas statute that prohibited the brewing of beer, which had forced him to close his brewery. The *Mugler* Court ruled that exercises of police power that validly preserve the public health, safety or welfare never constitute takings. Since regulations that fail to preserve these purposes violate the Due Process Clause of the U.S. Constitution, *Mugler* effectively stands for the proposition that police power regulations cannot be takings.

Although *Mugler's* proposition—in which it was ruled that police power regulations could not be takings—obviously conflicts with *Mahon's* too-far doctrine, neither has been overruled. The too-far test continues to be a staple of the Court's analysis, and *Mugler* continues to be favorably cited by all sides of the Court and of the lower courts. Figure 19.1 illustrates how these competing doctrines have been incorporated in the past three major regulatory takings cases that the Court has considered.

In *Lucas v. South Carolina Coastal Council (Lucas)*,[15] the Court set out the basic outline of its modern regulatory takings jurisprudence. David Lucas purchased two residential lots on a South Carolina barrier island for the purpose of building single family homes. Subsequently, the state passed the Beachfront Management Act, which barred such construction. Lucas argued that the Management Act constituted a taking, since it deprived him of all viable economic use of his property. In deciding this case, the Court determined that a regulation could constitute a compensable taking by falling under a categorical formulation.[16] One categorical formulation is the physical invasion of private property by the government (that is, eminent domain), which is traditionally an exercise of police power that requires compensation. Another categorical formulation that requires compensation is called confiscatory regulations.[17] Confiscatory regulations satisfied *Mahon's* too-far doctrine. The *Lucas* Court, determined that too far meant total deprivation of beneficial use since "from the landowner's point of view, [it is] the equivalent of a physical occupation."[18] It also determined that a regulation, which deprives an owner of all beneficial use of his or her property, is not confiscatory if the state can elucidate that the owner's use, which is restricted by the regulation, is already restricted by the owner's title or by federal or state statutes.

In following *Mahon*, the *Lucas* Court reversed a lower court ruling, which had relied on the *Mugler* doctrine that "no compensation is owing under the Takings Clause regardless of the regulation's effect on the property's value."[19] The dissenting opinions in the *Lucas* case continued to rely on the *Mugler* doctrine.[20] Since the Court refused to overrule the *Mugler* doctrine, the *Mugler* doctrine may be used as authority to prevent a future decision from using the confiscatory regulation categorization to justify compensation. This scenario is represented in figure 19.1 by the jagged line that runs from both sides of the *Mugler* cite to the Confiscatory Regulations box.

Although an owner who suffers less than a 100 percent loss cannot claim the confiscatory regulation categorical formulation, the *Lucas* Court held that he or she may still be entitled to compensation after case-specific inquiry into the public interest that is advanced in support of the restraint.[21] If a future majority decision were to re-embrace the *Mugler* doctrine—in which it was decided that police power regulations never constitute compensable takings—then case-specific inquiry would be aborted. This is accounted for by the jagged line that runs from both sides of the *Mugler* cite to the Case-Specific Inquiry box (figure 19.1).

The Court's case-specific inquiry is a two-pronged test that a regulation must pass to avoid being classified as a compensable taking. The Court formulated this test using two recent decisions. In *Nollan v. California Coastal Commission (Nollan)*,[22] the Court held that an essential nexus[23] must exist between the restriction, or the exaction, imposed on the owner by the regulation and by a legitimate state interest.[24] For example, the *Nollan* Court considered the constitutionality of a building permit condition—which was imposed by the California Coastal Commission (CCC) on Nollan's plan to replace a single-story structure that would block the public view of the Pacific Ocean from California's coastal highway with a two-story structure. The CCC required Nollan to mitigate the public's blocked view by providing public access across his property (that is, parallel to the beach and to the highway). The Court held that this permit condition would not improve the public view that would be impaired by the Nollan development, and therefore it lacked the essential nexus necessary to pass muster under the Takings Clause of the U.S. Constitution.

In *Dolan v. City of Tigard (Dolan)*,[25] the Court added the second prong to its case-specific inquiry. If the Court finds that an essential nexus exists, then it must determine whether a rough proportionality[26] exists between the private costs that are exacted on the regulated owner and the external social costs of his or her development. If the regulation is out of proportion in this sense, then it constitutes compensable takings. For example, Dolan sought the City of Tigard's permission to construct a paved parking area and to increase the size of her store. Tigard approved Dolan's permit subject to the condition that she dedicate the portion of her lot, which is within the ten-year floodplain of a bordering creek, to the city and that she dedicate an additional 15-foot strip of land adjacent to the floodplain as a pedestrian or bicycle path. The Court found that an essential nexus existed between the two dedication conditions and the legitimate state's interests in flood control and traffic management. However, the Court decided that these conditions failed the rough proportionality test because "[t]he city has never said why a public greenway, as opposed to a private one, was required in the interest of flood control,"[27] or demonstrated " . . . that the additional number of vehicle and bicycle trips generated by the petitioner's development reasonably relate to the city's requirement for a dedication of the pedestrian or bicycle pathway easement".[28]

In summary, the *Lucas, Nollan* and *Dolan* decisions indicate that only very extreme regulations are considered takings on constitutional grounds. Regulations that deprive the owner of the total economic use of property are compensable takings so long as the restricted behavior is not already proscribed by the title to the property, by common law and/or by statute. Regulations that deprive the owner of less than total economic use constitute takings only if they are poorly designed. (That is, they are neither reasonably related to a legitimate social purpose nor do they place a burden on the regulated owner out of proportion to any social costs he or she may perpetrate.) The decisions also indicate that the Court's case-specific inquiry overlaps significantly with its inquiry under the Due Process and Equal Protection Clauses of the U.S. Constitution. The essential nexus test requires that the regulation satisfy the Due Process legitimate public purpose test and the rough proportionality test, which is similar to the Due Process unduly oppressive test. This is the significance of the arrow running between the Case-Specific Inquiry box and the Due Process box in figure 19.1.

SOCIO-ECONOMIC EFFICIENCY OF TAKINGS JURISPRUDENCE

Increasingly, economists have turned their attention toward evaluating the Court's takings jurisprudence on economic efficiency grounds from society's viewpoint, and have suggested various modifications. Innes (1995) offers the following definition of economic efficiency from a social viewpoint:

Economic efficiency requires that net economic benefits, the difference between benefits and costs across all people in an economy, be at the highest possible level. Net benefits include benefits to the general public from government provided goods, such as national defense and park land; costs of activities to persons other than those engaging in the activities, such as the costs of chemical plants' toxic waste to neighboring residents; and all private benefits of activities and uses.[29] (p. 6)

This section reviews past work that investigated the efficiency of compensating property owners for regulatory takings. This information is used to assess the economic efficiency of the Court's takings jurisprudence, which is set out in figure 19.1, and is used to suggest efficiency-enhancing modifications.

PAST WORK

The efficiency implications of compensation depend on the context of private economic development. Some authors assume that a private owner has a parcel of land that can be developed in several ways (Blume et al., 1984; Miceli and Segerson, 1994). Typically, the most valuable private use generates negative spillover effects that the government may choose to regulate in the future. In the event of regulation, the owner loses his or her initial investment. The two tasks for an efficient compensation policy are to elicit efficient investment levels from the landowner and to elicit efficient regulatory choices from the government. These tasks, unfortunately, exist in a trade off. The payment of compensation acts as a disincentive for excessive regulation by the government. It also creates a moral hazard problem because the landowner, who expects compensation in the event of regulation, is encouraged to over-invest in the use that creates the negative spillover effect.

Miceli and Segerson (1994) analytically derive two compensation rules that (1) achieve an optimal balance between efficient regulation and moral hazard, and (2) provide a standard to determine when a regulation has gone too far (as in the *Mahon* case). The *ex post* rule grants compensation when the social benefits of the regulation are less than the costs that are exerted on the regulated property owner, and is designed to encourage efficient regulation. The *ex ante* rule is designed to encourage efficient land use and awards compensation to a landowner who is engaged in an efficient use of his land before a negative spillover effect is regulated (for example, a feedlot that is established prior to a residential development).

There are two major implications of the Miceli and Segerson (1994) paper that apply when modifying the Court's takings analysis to promote social efficiency. First, there is no reason to expect that the Miceli and Segerson compensation rules would set the socially efficient too-far standard at the Court's level of a 100 percent economic loss, and thus that the Court's too-far analysis can guarantee social efficiency under general circumstances. Second, since the broad application of the *ex post* rule to past Court decisions proves to be very similar to the balance between public and private interests that are associated with the rough proportionality test, Miceli and Segerson's socially efficient implementation of the too-far rule appears to be redundant to the Court's case-specific inquiry. Both implications support severing the too-far analysis from the Court's takings jurisprudence to promote economic efficiency from a social viewpoint. The Court's current implementation of the too-far test cannot guarantee efficiency, and an implementation that guarantees efficiency would duplicate other portions of the Court's analysis.

In many situations, Innes (1995) contends that development is better measured along the extensive margin (that is, the number of land parcels developed) rather than along the intensive margin (that is, the level of investment on a single parcel), as studied previously. Focusing on the extensive margin of development raises a different set of efficiency concerns, since some landowners develop their parcels earlier than others do. A consequence of this approach is that the government encounters heterogeneously developed properties when deciding whether and how to regulate takings. According to Innes (1995), efficiency dictates that the government takes the undeveloped parcels first—other things being equal. Consequently, in the absence of compensation, landowners have an incentive to develop their parcels earlier to reduce the risk of losing value because of regulation. This rush to develop results in parcels that are developed at an inefficiently rapid rate.

Although compensation is sufficient to remove the incentive to rush development, Innes (1995) argues that it is not necessary to restore efficiency. The necessary condition, he claims, is that the government allocate the private costs of regulation to the owners of developed and undeveloped parcels equally (that is, to provide equal protection to landowners). In this way, undeveloped parcels do not bear a disproportionately large share of regulatory losses compared to developed parcels, and the rush to develop is alleviated. Compensation is one means of spreading this cost equally, but Innes (1995) favors other schemes, such as taxing landowners whose developed property is not taken and using the proceeds to compensate landowners whose undeveloped property is taken. He argues that compensation discourages government regulation, and results in governmental bodies that allow parcels to be developed too quickly. He further argues that, contrary to common belief, compensation is not needed to discourage governmental over-regulation. The government's taxation powers also allow compensation to benefit from private development. This provides an adequate disincentive to over-regulate development. The broad application of Innes' (1995) equal protection compensation rule is that compensation should be awarded only if the Court determines that the landowner was not ". . . afforded equal protection and, hence, the same treatment and opportunities as were owners of similarly situated properties."[30] He asserts that the application of the Court's case-specific inquiry ". . . may promote inefficient government behavior and, as a result, may not be in the public interest."[31]

An informative way to evaluate the equal protection compensation rule is to consider the appropriateness of applying it to the *Nollan* and *Dolan* cases. This rule would award compensation in these cases if there were ". . . other property owners who had obtained development rights akin to those requested by [Dolan/Nollan] and who were not confronted with a similar price for these rights."[32] This type of rule may be misapplied to the *Dolan* and *Nollan* cases for two major reasons.

First, the rationale that requires the equal protection compensation rule (that is, to prevent a race to develop) is not relevant to the facts of either case. Dolan applied to the city of Tigard for a permit to redevelop her hardware store under existing state and local regulatory regimes. Nollan applied to the California Coastal Commission for a permit to redevelop his coastal residence, also under an existing state regulatory scheme. Consequently, neither had any incentive to engage in a race to develop before the impending regulation was implemented. Application of an equal protection compensation rule would focus all the attention on combating a source of economic inefficiency—which does not appear in the facts and which is to the neglect of the more relevant factors discussed below. The facts of both *Nollan* and *Dolan* appear more consistent with the development context posed by Miceli and Segerson (1994)—that is, a given parcel with a number of potential uses. This implies that applying their compensation rules to these cases would enhance economic efficiency. As shown above, the broad application of their *ex post* compensation rule is very similar to the Court's rough proportionality test.

Second, applying a compensation rule that focuses on equal protection does not reach the main issue of overreaching regulation that is relevant in both cases—that is, whether or not the regulatory agency's permit conditions imposed economic burdens on the landowners that are out of proportion with the social costs of their developments. The Court refused to adopt a purely equal protection compensation rule for this reason in the *Lucas* case because,

> . . . a regulation specifically directed to land use no more acquires immunity by plundering landowners generally than does a law specifically directed at religious practice acquire immunity by prohibiting all religions. [This] approach renders the Takings Clause little more than a particularized restatement of the Equal Protection Clause.[33] (p. 1027)

As explained above, Innes (1995) contends that compensation is not necessary to prevent regulation from plundering landowners generally because the government's taxation powers allow it to share in the benefits of private development. Governmental taxation powers, however, did not prevent the over-regulation that was found by the Court in either the *Dolan* or *Nollan* case. In the *Dolan* case, the reason could be that Tigard's over-burdensome permit conditions were to the city's benefit. Tigard would have expanded its greenway property along Fanno Creek at Dolan's expense and would have collected increased property taxes from her redevelopment. In *Nollan*, the regulatory agency that set the permit conditions (that is, the California Coastal Commission) is a state agency that has no taxing authority.[34] Thus, an equal protection compensation rule relies on a check to government over-regulation that was inoperable in these cases. In the absence of such a check, compensation is effective in providing landowners with constitutional protection against poorly drafted and over-burdensome regulations, as envisioned by the Court's case-specific inquiry.

An interesting final issue in this matter is whether Innes' rule would be well applied to the specialized development context that is consistent with a race to develop, such as that found in the *Lucas* case. Lucas purchased two residential lots under a regulatory regime that allowed him to construct single-family homes, but a subsequent change in regulations barred such development. Application of an equal protection compensation rule under these circumstances would remove the incentive for similarly situated landowners to develop parcels at an inefficient rate. However, would the rule protect Lucas and similarly situated landowners from over-regulation? Innes (1995) contends that the equal protection compensation rule would not be required to discourage the government from over-regulating because the government would forego tax revenues if it were to pass regulations that decrease land values. The problem with this contention is that, similar to the *Nollan* case, the regulatory agency (that is, the South Carolina Coastal Council) was a state agency without taxing authority. Although the equal protection compensation rule would have resolved the inefficiency that would result from a race to develop, it would have left landowners without operative protection against in-efficiencies that would have been caused by government over-regulation.

EFFICIENCY ENHANCING MODIFICATIONS TO THE COURT'S TAKINGS JURISPRUDENCE

The confiscatory regulation component of the Court's categorical formulations should be severed. The Court's implementation of this component promotes so-cially inefficient regulation and land use, and an efficient implementation dupli-cates the Court's case-specific inquiry.

The Court's case-specific inquiry should be retained. Its operation is consistent with the Miceli and Segerson (1994) compensation rules when attention is focused on the intensive margin of investment. When the extensive margin is of interest, case-specific inquiry can be adapted to discourage a race to develop by increasing the importance of equal protection considerations in determining an essential nexus between the regulation and a legitimate state purpose. As explained above, the essential nexus test is very similar to the legitimate public purpose test for due process, which is used to resolve any equal protection challenges to regulation. The modified version of the Court's takings jurisprudence to promote socio-economic efficiency is depicted in figure 19.2. The confiscatory regulations cate-gorical formulation is removed, and the Court's case-specific inquiry applies, no matter what the economic loss to the regulated party.

THE CASE OF GRASS FIELD BURNING

The Court's takings jurisprudence, modified according to figure 19.2, is well equipped to evaluate the constitutional feasibility and social efficiency of flexible incentive regulations. These regulations assume many forms and are applied to a myriad of circumstances (Batie and Ervin, this volume; Segerson, this volume). Thus, consistent with the Court's case-by-case approach, their legal and economic performance must be evaluated for particular circumstances. The regulation of seeded grass field burning in eastern Washington State provides a timely and rele-vant set of circumstances for illustrative purposes.

Bluegrass seed production, which comprises about 60,000 planted acres, is a long-standing component of agriculture in the eastern portion of Washington State. Bluegrass is a perennial and is harvested for its seed each season. After harvest, the grass stubble is burned as an effective means of enhancing next season's seed crop. The smoke from grass field burning reduces air quality, which generates a number of negative spillover effects. These effects include decreased visibility that results in increased traffic accidents, decreased recreational opportuni ties, less aesthetically pleasing views and the potential toward severe health problems for people who suffer from chronic cardiopulmonary conditions.

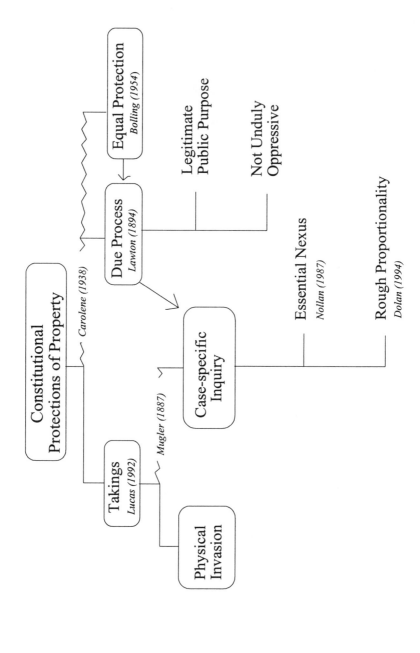

FIGURE 19.2 Takings Analysis Modified to Promote the Social Efficiency of Regulation and Resource Use.

In 1996, the State Department of Ecology (Ecology) reacted to increasing complaints over this grass burning by issuing an emergency ruling that called for a 33 percent reduction in the number of acres that could be burned in the state. A permanent rule that required an additional 33 percent reduction was adopted in 1997. A recent study commissioned by the Washington State Department of Ecology (DOE) estimates the probable social benefits of the burned acreage reduction to be 8.4 million dollars, and the costs to the industry to be 5.6 million dollars (Holland et al., 1996).

Ecology's restrictions on grass field burning are types of flexible incentive for the adoption of environmental technologies in agriculture. They give grass growers the flexibility to respond to usage limits in a non-prescribed manner, for example, by changing their production technology or location. The restrictions represent an exercise of the state's police power to protect public health and safety under Amendment X of the U.S. Constitution. The issue of interest here is whether this exercise of police power should be found to constitute a takings under the U.S. Constitution's Amendment V.

Following the modified takings analysis shown in figure 19.2, these restrictions do not involve a physical invasion of grass fields by the government, and thus do not qualify as a categorical formulation that requires compensation. Consequently, the regulations must qualify as takings under the Court's case-specific inquiry. There appears to be no rush to burn since all grass growers are under the burned acreage restriction, thus, there is no reason to emphasize equal protection compensation in the case-specific inquiry.

The first test under the case-specific inquiry is whether there is an essential nexus between the restrictions and a legitimate state interest. An essential nexus likely exists because the state's acreage restrictions on burning are rationally related to a legitimate interest in controlling smoke for the protection of public health and safety. The second test is whether a rough proportionality exists between the costs of the restrictions to the grass producers and the external costs of unrestricted grass field burning to the public. The rough proportionality test requires no precise mathematical calculation,[35] thus the cost-benefit figures cited in DOE's study offer more refined evidence than the Court usually has available. To the extent that these figures are accurate, they offer compelling evidence that the cost of the burning restriction to grass growers is proportionate with the cost of smoke-related negative spillover effects to the public. Thus, Ecology's burning restrictions appear to pass the Court's case-specific inquiry, and social efficiency does not require that compensation be awarded to regulated Bluegrass growers.

Perhaps the growers' best opportunity for compensation is to relax the condition of social efficiency and to apply the too-far rule that defines confiscatory regulations. Interestingly, the early jockeying for political position between grass growers and clean air advocates mirrored the type of analysis underlying the too-far rule. A spokesperson for the grass growers declared in a local newspaper that the public has no constitutional right to breathe clean air. The unstated implications are that growers have the right to use air as a smoke depository, and that any interference with this right constitutes an economic loss to growers for which they should be compensated. If a grower could prove that the economic loss would be complete (that is, 100 percent), the Court might find the burning restrictions to be confiscatory, and would award compensation. For example, the grower could demonstrate that the available technological or locational adjustments required to accommodate the burning restrictions would completely rob bluegrass seed fields of their profitability.

CONCLUSION

Batie and Ervin (this volume) consider the technical difficulties of applying flexible incentives to agro-environmental problems. This chapter considers some legal difficulties and implications for economic efficiency of doing so. Regulations, including those that employ flexible incentives for the adoption of environmental technologies in agriculture, must pass constitutional muster under the Takings, Due Process and Equal Protection Clauses of the U.S. Constitution if they are to stand up to judicial scrutiny. The U.S. Supreme Court has formulated a framework for scrutinizing the constitutionality of regulations, and for determining whether compensation is due to private citizens for their economic losses because of regulation. Traditionally, the Court awards compensation if the government physically invades private property or if a regulation goes too far in imposing a 100 percent economic loss on regulated parties. The alternative is for the Court to engage in a case-specific inquiry and to award compensation (1) if the regulation lacks an essential nexus with a legitimate state purpose, or (2) if it imposes costs, which are not roughly proportional with any social damage that is caused by their unregulated activities on the regulated parties

ENDNOTES

1. "The Congress shall have Power . . . To make all Laws which shall be necessary and proper for carrying into Execution the foregoing Powers, and all other Powers vested by this Constitution in the Government of the United States . . ." United States Constitution Article I, §8, Clause 18.
2. "The powers not delegated to the United States by the Constitution, nor prohibited by it to the States, are reserved to the States respectively, or to the people." United States Constitution Amendment X.
3. " . . . nor shall any person be deprived of life, liberty, or property, without due process of law . . "
4. United States Constitution Amendment V.
5. ". . . nor shall private property be taken for public use, without just compensation." United States Constitution Amendment V.
6. ". . . nor shall any State deprive any person of life, liberty, or property without due process of law . . ." United States Constitution Amendment XIV.
7. "No State shall . . . deny to any person within its jurisdiction the equal protection of the laws." United States Constitution Amendment XIV.
8. *Bolling v. Sharpe*, 347 U.S. 497 (1954).
9. *Chicago, Burlington & Quincy R. R. Co. v. City of Chicago*, 166 U.S. 226 (1897).
10. *Lawton v. Steele*, 152 U.S. 133 (1894).
11. *United States v. Carolene Products Co.*, 304 U.S. 144 (1938).
12. *Bolling v. Sharpe*, supra note 7.
13. *Pennsylvania Coal Co. v. Mahon*, 260 U.S. 393 (1922).
14. *First English Evangelical Lutheran Church v. County of Los Angeles*, 482 U.S. 304 (1987).
15. *Mugler v. Kansas*, 123 U.S. 623 (1887).
16. *Lucas v. South Carolina Coastal Council*, 505 U.S. 1,003 (1992).
17. Ibid. at 1,015.
18. Ibid.

19. Ibid. at 1,017.
20. Ibid. at 1,010.
21. See Justice Blackmun's dissent at 1047, and Justice Steven's dissent at 1064.
22. Ibid. at 1,015.
23. *Nollan v. California Coastal Commission*, 483 U.S. 825 (1987).
24. Ibid. at 837.
25. Ibid. at 837.
26. *Dolan v. City of Tigard*, 512 U.S. 374 (1994).
27. Ibid. 386 and 391.
28. Ibid. 393.
29. Ibid. 395.
30. Innes at 6.
31. Ibid. at 43.
32. Ibid. at 6.
33. Ibid. at 43.
34. Lucas at footnote 14, p. 1,027.
35. The California Coastal Commission is composed of 16 members, some of whom may work for local government agencies with taxing authority. However, the enacting statute requires that members be geographically dispersed over the coastal regions of the state, thereby decreasing the chances that the commission's decisions would be swayed according to the desires of any single local area. [California Coastal Act: California Code, Division 20, §30301].
36. Dolan at 391.

REFERENCES

Blume L., D. Rubinfeld, P. Shapiro. 1984. "The Taking of Land: When Should Compensation be Paid?" *Quarterly Journal of Economics* 100: 71–92.

Bromley D. 1993. "Regulatory Takings: Coherent Concept or Logical Contradiction?" *Vermont Law Review* 17: 647–682.

Holland D., K. Painter, D. Scott, P. Wandschneider, and D. Willis. 1996. *Estimates of the Benefits and Costs from Reductions in Grass Seed Field Burning.* Report submitted to Washington Department of Ecology under the terms of Inter-agency agreement No. C9600164, Washington, DC.

Innes R. 1995. "An Essay on Takings: Concepts and Issues." *Choices* First Quarter: 4–44.

Micelli T and K. Segerson. 1994. "Regulatory Takings: When Should Compensation be Paid?" *Journal of Legal. Studies* 23: 749–776.

Nowak J., R. Rotunda, and J. Young. 1983. *Constitutional Law (Second Edition)*, p. 480. St. Paul, MN: West Publishing Co.

20 PROVIDING FOR THE COMMON GOOD IN AN ERA OF RESURGENT INDIVIDUALISM

Alan Randall
The Ohio State University, Columbus, OH

Faith in the legitimacy of big government and its ability to solve the problems that concern ordinary people has declined precipitously, and ideological individualism is on the rise. Privatization is all the rage but leaves us in isolation paradoxes of various kinds. The answer lies in institutional innovations based on the lessons of game theory: seek problem-scale solutions and replace existing conflict with win-win incentives for sustainable cooperation. People are inventing such institutions and making them work while mainstream economists, bogged down in their market failure paradigm, have barely noticed what is going on.

Agriculture generates more than its share of isolation paradoxes. With its emphasis on restructuring incentives so as to convert conflict into opportunity for mutual gains, isolation paradox thinking is hospitable to many of the policy tools currently being developed and promoted under the heading of flexible incentives. This chapter concludes with applications to agricultural pollution from nonpoint sources and biodiversity/habitat protection. These are problems for which traditional solutions have failed conspicuously.

INTRODUCTION

Harking back to the progressive faith in scientific government at the beginning of this century, and the mid-century confidence in government's power to correct the failures of the market, it becomes obvious that we are now in an era of resurgent individualism and concomitant skepticism about public institutions. This shift in thinking was surely boosted by the events of 1989 when the massive experiment in Soviet style collectivism was exposed as bankrupt, but it began much earlier. Intellectual roots can be found in the academic writings of Arrow (1951), Tiebout (1956) and Coase (1960), as well as the popular writings of Ayn Rand (1944) during the same period. The Goldwater nomination in 1964 provided an early in-dication that individualism was starting to catch on with the public.

The mid-century notion of market failure—elaborated on by economists in a taxonomy that includes public goods, common property resources, externality and natural monopoly—has come under scathing attack from individualists who ask what policy implications could possibly arise from market failure when the fail-ures of government are even more pervasive. Nevertheless, an essential reality remains. There exist many situations, called isolation paradoxes, in which indi-vidual actions fail, but a cost allocation can be found such that everyone would be better off with coordinated action than with no action at all. Insistence on individ-ual action, or none at all, leaves everyone isolated and ineffective, but the search for arrangements that make cooperative action beneficial to all concerned may be rewarding. Rather than a simplistic dichotomy of market or government, the iso-lation paradox concept suggests openness to solutions that invoke a variety of in-stitutional forms—private enterprises, voluntary associations and govern-ment—from the most local level to the national scale and beyond. Building on a combination of abstract theory (from game theory, political science and econom-ics) and on emerging experience, it is possible to identify some of the characteris-tics of policies and processes that are effective in breaking the isolation paradox.

Isolation paradoxes abound in agriculture. Examples include pollution (in which the difficulty of monitoring nonpoint-source pollution (NSP) has precluded the public from enjoying the benefits of adequate controls and farmers from profiting from gainful permit trades and biodiversity) and habitat protection (in which the fragmentation of land into private parcels and failure to devise incentives for co-operation among landowners have denied the public adequate provision for biodi-versity and denied farmers the opportunity to profit from the potential value of their land as habitat).

The isolation paradox concept suggests an acceptance of a diversity of institu-tional forms. With its emphasis on restructuring incentives so as to convert con-flict into opportunity for mutual gains, it also seems to incorporate many of the policy tools that are currently being developed and promoted under the heading of flexible incentives (Batie and Ervin, and Segerson, this volume). While flexible incentives take on many forms—especially if one's definition is as inclusive as that of Batie and Ervin (this volume) and of Segerson (this volume)—the concept surely includes performance regulation as opposed to technology regulation, miti-gation, restoration banking, trading in pollution control credits and trading in habitat protection credits. Each of these possibilities is discussed below.

In this chapter, I will elaborate on the argument outlined above, and will con-clude with illustrations using the exemplars of NSP control and biodiversity pres-ervation.

THE PROGRESSIVE DREAM OF SCIENTIFIC GOVERNMENT

For the first half of the twentieth century, the progressive movement dominated thinking about the role of government in society. A sharp distinction was made between questions of basic policy and of social value that were the proper role of politics, and questions of an administrative or instrumental nature that were best handled in a scientific fashion by neutral technical experts. In accordance with the progressive belief that scientific government was not only possible but desirable, the role of politics was to be diminished as far as possible and the role of scientific administration was to be expanded. Administration was taken to be largely a mechanical or engineering problem, so it is not surprising that efficiency emerged as the central objective of progressive government (Nelson, 1987).

Economists, of course, had well-developed notions of efficiency and of the theory that efficiency tended to be promoted by markets. By the 1930s, progressive notions that interacted with the extreme economic dislocations of the time were pressed into service to correct the obvious failings of the market. Scientific government was needed for macro-stabilization (Keynes, 1936). Faced with the dilemma that markets might promote efficiency but they seem to have obvious failings in the area of distribution, the social welfare function (Bergson, 1938) was offered as the ultimate progressive device for dealing scientifically with what might otherwise have been thought to be a political question. In case a scientific approach to deal simultaneously with efficiency and distributional concerns seemed implausibly ambitious, economists also offered a benefit-cost test (Kaldor, 1939; Hicks, 1939) for the efficiency of public undertakings. To address a particular set of allocative issues thought to be handled poorly by markets, Pigou (1932), Samuelson (1954) and Bator (1958) elaborated on their theories of market failure. When markets failed to deliver efficiency, it was the duty of government to take corrective action.

Market Failure, Government Fix

The market failure paradigm identifies four kinds of circumstances in which even a fundamentally competitive economy would experience market failure. These are (1) externalities, (2) public goods, (3) common property resources and natural monopoly. For three of these phenomena, the conventional solutions call unambiguously for government action. These solutions are to tax or regulate externalities, to raise revenue for public projects that provide for the public good and to regulate the pricing policies of natural monopolies. For common property resources, the range of endorsed solutions is broader, and reflects the uneasy coexistence of progressive and free market thinking in the conventional economic minds of the day. Regulation, taxation and direct government provision may be suggested, but it is also frequently suggested that the government specify private property rights and then stand aside as emerging markets restore efficiency.

While the market failure paradigm is no longer uncontested, it remains influential among some economists and many technical experts in the executive agencies. It is easy for economists, who mistake the trees for the forest, to perceive the battle lines as drawn—for example, between command-and-control regulations and the more flexible incentive of pollution taxes—when there is a much larger battle raging between progressive government and a resurgent individualism.

THE RESURGENCE OF INDIVIDUALISM

It is now obvious that we are in an era of resurgent individualism and concomitant skepticism about public institutions. The 1995 Congress provided dramatic evidence that individualistic notions were, by then, thoroughly mainstream. The intellectual roots of the revival of individualism, however, can be found much earlier.

Arrow (1951) challenged the foundations of the social welfare function, in effect denying the basis for the scientific solution of distributional questions. Tiebout's (1956) *voting with the feet* model, while ostensibly a modest contribution to the theory of local public finance, related the household to the government in an innovative and, in a sense, subversive way. This relationship is not that citizens work with others in their jurisdiction to make things better, but that they are potential migrants shopping among the competing governments to find the best deal. Coase's (1960) challenge to the progressive theory of market failure was so successful that it became the most cited publication by a living economist.

All of this needs to be placed in the context of broader philosophical trends. Just as optimistic notions of progressive government were supported by optimistic notions of modernism in the philosophy of science, increasing post-modernist skepticism in science has helped feed the growing skepticism that government could solve problems of society. A considerable audience read the popular mid-century individualistic writings of Ayn Rand (1944): the presidential nomination of Barry Goldwater in 1964 provided an early indication that individualism was catching on with the public.

The Attack on Market Failure

Since Coase's 1960 classic paper, the market failure government fix paradigm has been in retreat. Coase (1960), Cheung (1970) and Dahlman (1979) established that externality has little analytical content, in that inefficient externality can not persist unless there are additional impediments to trade among involved parties. Various authors have established that the public goods problem of conventional analyses is really two distinct problems that may occur separately or together. These are nonexclusiveness (that is, the inability to exclude those who benefit and do not contribute or those who impose costs but do not pay) and non-rivalry (that is, no additional costs for providing the good to additional users once a particular amount of the good has already been provided. In the case of ordinary (that is, rival) goods, establishing and enforcing transferable rights—provided that all of that can be done at reasonable cost—will restore efficiency. In the case of non-rival goods, however, efficient arrangements are necessarily more complicated.[1] Randall (1983) argues that whatever valid content there is in the market failure concepts of externality, common property resources and public goods can be captured more simply with the concepts of nonexclusiveness and non-rivalry. These amendments to the market failure paradigm are now widely accepted among economists.

The Property Rights Approach

In 1954, Gordon demonstrated that nonexclusiveness is not conducive to efficiency or to conservation. He conflated the distinction between nonexclusiveness and common property,[2] and argued implicitly for private property solutions. Gordon's analysis predicts the total collapse of the nonexclusive economy unless the government steps in to enforce private property rights or, less plausibly, to impose some unwieldy set of optimal taxes and subsidies. The Coasian analysis of externality also focused on non-attenuated property rights as a sufficient condition for efficiency.

This emerging focus on property rights undermined not only the analytics of the market failure paradigm, but also its progressive government activism. The Coasian analysis drew attention to the possibility of market-like behaviors in many domains of human interaction that were beyond conventional markets. At first glance what might appear to be market failure, in fact, may be an efficient market solution. Thus, the burden of proof was switched to those who claimed market failure in any particular case. As the Coasian tradition developed, it was argued with increasing generality that attenuation of rights was endemic to the public sector itself. The fact that government failure may be an even more pervasive problem than market failure took the argument one rather large step further. A sustained posture of government activism to rectify market failure was not merely unnecessary—it was undesirable.

Thus, the property rights approach asserted that the so-called market failures were caused mostly by attenuated property rights. In this respect, government failure could be even more pervasive than market failure, and privatization was the appropriate policy response to diagnosed allocation inefficiencies.

Can Community Survive the Resurgence of Individualism?

Privatization, deregulation and the incorporation of flexible incentives into government regulations that remain have produced many benefits. Efficiency has been enhanced, waste has been reduced and, to a non-trivial extent, individuals have enjoyed greater freedoms. It seems, however, that there has also been a cost. The loss of progressive faith has undermined the legitimacy of public institutions. Increasing frustration with national institutions is being seen. The proliferation of gated communities and private schools suggests that many households no longer trust even the most local of public institutions to provide basic services like security and education. Shopping around for neighborhoods and *voting with the feet* results in a series of one-shot transactions devoid of any commitment to stay and help work out any problems that cannot be resolved easily. If the essence of *community* is that its members are committed to its continuity (that is, to resolve a wide variety of problems and to provide a wide variety of services within the community structure), then these developments are threats to the community.

In a true community, a repeated game structure encourages the sequential solution of problems. Single issue lobbying, shopping around for public sector services, and the club style provision of local public services without concern for those left out of the club, seem to be inferior substitutes for a true community. The motivations for these substitutes are often less than generous—taking care of oneself rather than joining with others to take care of everyone. The results are sufficiently unsatisfying that a fairly broad based group of public philosophers and opinion leaders are at work trying to develop a new communitarism.

THE ISOLATION PARADOX

While the property rights movement has succeeded in discrediting the progressive paradigm of market failure, it also has its Achilles' heel—the persistent notion that there are some things worth doing that require coordinated action. This common sense intuition has proven sound. First, the property rights school has, self-servingly, paid much closer attention to nonexclusiveness (which may be more amenable to privatization solutions) than to non-rivalry (which is much less so). Second, even when dealing with nonexclusiveness, the property rights school has been fixated on the false dichotomy of non-attenuated property rights or economic collapse, and has become insensitive to the diversity of institutional forms that have promise.

In the past two decades, several novel, but related, approaches have emerged to shed new light on the possibilities for collective action. These approaches include game theory formulations of the non-rivalry and nonexclusiveness problems (Sen, 1967; Runge, 1981); resource allocation mechanisms (Hurwicz, 1973); the theory of teams (Marshak and Radner, 1971); incentive compatible mechanisms (Groves and Ledyard, 1980); and principal agent models (Arrow, 1986).

An early and influential game theoretic formulation was the prisoners' dilemma, a game in which individuals who were unable to communicate with each other had to choose either a cooperative or non-cooperative strategy in order to play. Payoffs were set up so that each player was better off if he defected while the others cooperated, but every player preferred the *all cooperate* outcome to the *all defect* outcome. Nevertheless, in a one-shot prisoners' dilemma, it is the *all defect* outcome that emerges.

By the 1960s, it was widely held that the Samuelson-Gordon analyses of market failure could be reformulated as single period *n*-person prisoners' dilemmas. Such reformulation would, of course, reconfirm Samuelson's and Gordon's prediction of total collapse in the non-rival and/or nonexclusive sectors.

The single period prisoners' dilemma was, however, only the beginning. It was soon realized that the prisoners' dilemma was not necessarily the proper specification for non-rivalry and nonexclusiveness problems (Sen, 1967; Dasgupta and Heal, 1977). As Shubik (1981) observed, games of pure opposition have many uses in military tactics but relatively few applications in economics. In many economic contexts, cooperative behavior is the individually preferred alternative. What is required for stable cooperative solutions is the credible assurance that other players will not defect.

The intuition that coordinated action is an essential, and likely, stable solution for an important set of economic problems is hardly new. Adam Smith discussed the case of one hundred farmers in the upper end of a valley, beyond the reach of the existing barge canal. While all would benefit from extending the canal, none could bear the cost alone. Yet every single one of them would enjoy benefits greater than one one-hundredth of the cost. Acting alone, each could do nothing, but everyone could enjoy a net benefit from coordinated action. The general name given to problems of this kind is *isolation paradox*. An isolation paradox is present whenever individual action fails, but there exists a cost allocation (not necessarily an equal sharing of costs, as in Smith's example) such that all parties would be better off with coordinated action than with no action at all.[3] The essential idea is that where an isolation paradox exists, there is (in principle) the possibility of converting a conflict situation into a sustainable cooperative solution, and we may benefit from exploring that possibility.

The non-rivalry and/or nonexclusiveness problems are reformulated correctly as isolation paradoxes. The prospects for stable cooperative solutions for one-shot isolation paradoxes are much greater than for one-shot prisoners' dilemmas. Correct specification of the game helps. In addition, the prospects for stable cooperation are enhanced when any of the following holds: (1) the game is repeated (preferably, stochastically several times); (2) group contributions are observable to all players; and (3) individual contributions are observable.

This kind of thinking is useful in amending both the *market failure government fix* and the *property rights* approaches. Game theory no longer confirms the Samuelson-Gordon collapse thesis for the non-rival and nonexclusive economics. Stable cooperative solutions are, at least, a possibility in a variety of circumstances. Some insights have been developed that are concerned with the factors that work in favor of stable cooperation. These results take us some distance beyond the idea that individual actions lead to market failures, which only exogenous government fixes can cure. Similarly, they tend to deny the *property rights libertarian* dichotomy that damns all institutional arrangements except private property rights.

The demonstration that, for several relevant classes of games, coordinated strategies permit stable Pareto-efficient cooperative solutions is not entirely comforting. Coordination is likely to be a costly activity, and complete coordination (especially if it requires consultation among all participants) may be prohibitively costly. Private (that is, rival and exclusive) goods markets work well because prices convey sufficient information and incentives to accomplish coordination. Neither centralized management nor direct consultation among all market participants is necessary. The working hypothesis that motivates research on principal agent models is that signaling devices can be developed for adequate and cost-effective coordination, so that cooperative arrangements in large organizations that deal with non-rival and nonexclusive goods are reasonably stable and efficient.

For principal agent models, the following issues are typical. Can total costs of loss and damage be reduced if insured parties have some incentives for loss-avoiding behavior? If the work effort of individual agents cannot be monitored directly, what incentives can the manager devise to encourage agent efficiency without incurring the excessive turnover of agents? If the effluents from individual polluters cannot be monitored fully, can the pollution control authority devise incentives for reasonably efficient pollution control?

Each of these problems is characterized by hidden action (the agent can take some action that is unobserved by the principal) or hidden information (the agent knows something the principal does not). An interesting variant is the problem of a single principal and many agents, in which the principal can observe the combined output of all agents but he can not observe the individual output of any one of the agents. The relevance of this kind of thinking to nonexclusiveness and nonrivalry problems is obvious.

The literature on principal agent problems is substantial and often highly mathematical. No attempt at careful review and evaluation is offered here, but some impressions can be conveyed. Considerable progress has been made in modeling information requirements and group performance—given various combinations of problems and incentives. Results about information requirements provide indirect evidence on the transaction costs that are associated with various arrangements. While principal agent models reconfirm the efficiency of price signals in a neoclassical competitive economy, they offer no support for the *private property or total collapse* thesis of the libertarians. A wide variety of workable arrangements (with outcomes that approach Pareto efficiency in some cases and outcomes that avoid collapse in many others) can be identified for diverse problems that exhibit aspects of non-rivalry and/or nonexclusiveness.

Institutions to Solve the Isolation Paradox

Solutions that break the isolation paradox do not have to involve the government or (even worse, in today's political environment) big government. Individuals may act together to form and maintain clubs in order to get the job done. Many entities that call themselves clubs (for example, local health and fitness clubs) are actually private for-profit enterprises. Today, one can readily imagine a private entity resolving the canal extension problem profitably, an option that did not occur to Adam Smith, just as city water delivered to my home by an investor-owned corporation was not considered in times past.

The isolation paradox concept suggests openness to solutions that invoke a variety of institutional forms (such as private enterprise, voluntary associations and government) from the most local level to the national level and beyond. Given the centrality of information and coordination, the array of feasible institutions is continually shifting as information, communication and exclusion technologies develop. For particular problems, the appropriate institutions will be consistent with the dimensions and scale of the problem itself, and with the prevailing technologies and political realities. For example, to protect biodiversity one can conceive of profitable private genetic reserves (including nature reserves that are operated by corporations, voluntary associations or governments); clubs that are supported by members and donors who operate in markets to enhance both private and government conservation efforts; and governments that operate as facilitators of consensual agreements among stakeholders, legislators, regulators and resource managers. Flexibility is the key in both institutional forms and in the incentives that those institutions transmit.

Building on a combination of abstract theory (for example, game theory, political science and economics) and experience, it is possible to identify some of the characteristics of policies and policy processes that are effective in breaking the isolation paradox. These are elaborated here.

Seek Problem-scale Solutions

National solutions proposed as panaceas to local and regional problems are currently out of fashion. There are some good reasons for this. National scale solutions often do not make sense in the local context. Whether they make sense or not, solutions that are imposed from distant capitals seldom enjoy the local commitment necessary for their success. Indeed, it makes sense to seek solutions that are scaled to the problem at hand, and to a considerable degree, to seek solutions that are fashioned by those involved most directly. Nevertheless, a framework of national laws and policies remains necessary to provide parameters within which local solutions can be negotiated.

A major element in the national framework is property rights. It behooves us to remember that property rights are the creation of the government that defines and secures them. They evolve over time in response to changing circumstances. The current property rights movement is not really about promoting the efficiency advantages of non-attenuated property rights in general, nor is it about protecting existing property rights. Instead, the main concern of the property rights movement is to extend property rights in ways quite inconsistent with recent political history: broadening the conditions under which property owners may demand compensation for private losses because of regulation in the public interest, and reversing the 25-year old principle that the *polluter pays*.

More generally, there is an inherent tension between the advantages of problem-scale solutions and the need for national policy. Industries operating on a global scale, for example, have proven more than willing to use the current enthusiasm for state and local institutions to create prisoners' dilemmas for their own benefit. We observe this when states and localities find themselves in destructive competition to attract firms with tax abatements and/or relaxed enforcement of environmental controls.[4] An effective policy process encourages problem-scale solutions within a framework of national policy. It does not simply set states and localities adrift and wish them well.

Establish a Long-term Process that involves all of the Legitimate Interests

Since the 1970s, public participation has been an important part of the process for resolving resource management issues. Since the 1980s, involvement of all significant stakeholders has been considered essential. What is relatively new is the notion of committing the participants to a long-term continuing process that is supported by the theory of repeated games and by practical experience. Rather than merely commenting on a solution proposed by professional managers (a typical way of implementing public participation), participants over time actually work out solutions to the problems of concern. A long-term continuing process has obvious advantages: it allows time for participants to develop an understanding of each others' interests and objectives; to gather and interpret essential information; and to develop solutions that will be broadly acceptable. An advantage is that after a few rounds, individuals tend to become committed to bringing the process itself to a successful conclusion. If the default outcome is recognized broadly as unsatisfactory and participants come to see the failure of the process as bad in and of itself, conditions are favorable for a successful process.

Establish a Shared Vision

The process starts by defining goals at the community level and the values that underlie those goals. The objective is to develop and articulate a shared vision—a statement of what it is that the community values and seeks to become. During this process the stakeholders, whose most immediate interests would seem to be in conflict, frequently discover that their basic values and vision of the future are, in fact, quite compatible. At this stage, it helps to define the problem set broadly. What does this community seek to become, and how can it get there?

Use all of the Tools for Achieving Consensus: Deliberation, Persuasion and Negotiation

Structured discourse and deliberation can often undermine conflict, and careful consideration of information can erode firmly held prior possibilities and open up new ones. It would be a mistake—one than an economist might easily make—to underestimate the value of deliberative processes. Nevertheless, negotiations, real trades and win-win solutions are often essential to break impasses. Flexible incentives are often important elements of the package, in that they tend to reduce the costs of meeting environmental policy targets. They encourage win-win solutions and ease the pain of compliance in cases where win-win solutions prove impossible. Depending on particular circumstances, purchases of land or easements, land swaps, mitigation banking and resources-for-resources compensation can be both efficacious and fair. They help move things toward real solutions that benefit all parties who are directly concerned. A broad definition of the problem set is helpful at this stage as well, because it increases the scope of potential trades and win-win solutions. As with all negotiations, however, it pays to proceed cautiously. It is not uncommon for parties to proclaim a secure *status quo* position that may, in fact, be quite shaky or to exaggerate the costs and adverse employment impacts of proposed environmental policies.

APPLICATIONS

Nonpoint-source Pollution Control

NSP is now a large and highly visible component of the total water pollution problem. This is largely because effective controls have been established for PSP while NSP controls have been much more problematic. This is usually explained by the relative simplicity of monitoring effluents from PSP compared to that of NSP. Without the capacity to monitor effluents at the firm level, it seems that major classes of pollution control instruments have been ruled out—for example, performance standards and pollution taxes. It is easy to assume that pollution control authorities enjoy few options beyond specifying control technologies. They tend to specify best management practices and embark on educational campaigns that encourage farmers to adopt such practices.

Game theory and principal agent models, however, suggest several promising possibilities. While monitoring individual firm contributions remains difficult, it is relatively easy for the agency to monitor total pollution loads at the sub-catchment level. This situation can be analyzed as a standard problem in game theory: the principal cannot monitor individual contributions, but it can readily monitor group output. An interesting twist is that the agents know more about each other than the principal knows about any of them. Farmers in a sub-catchment know a lot about their neighbors' farming practices and their relative contribution to off-site pollution loads. In these circumstances, a promising approach is for the principal to monitor performance and establish incentives for compliance at the group level and then leave farmers in the sub-catchment to police each other. What do we mean by incentives at the group level? The *scapegoat* contract calls for severe punishment of a single randomly chosen agent when group performance is unsatisfactory. Alternatively, unsatisfactory group performance could be managed by penalizing all members of the group. If the *carrot* were preferred to the *stick*, all members of the group could enjoy a reward when group performance surpasses pre-announced targets.

These ideas are no longer only the speculations of academics. The U.S. Environmental Protection Agency lists 13 existing point/nonpoint pollution trading programs. A similar number are under development or consideration. PSP that are unable to meet pollution control performance standards can purchase credits from groups of nonpoint polluters who succeed in reducing their effluents. Programs are set up at the catchment or sub-catchment level. Several levels of government are involved as are point-source polluters (who may be private or public organizations) and nonpoint-source polluters. Presently, it is typical for these trading programs to specify the price for NSP credits and to conservatively calculate the number of credits earned on the basis of computer modeling of what would be accomplished by the adoption of best management practices on agricultural land.

These point-source/nonpoint-source pollution trading programs simultaneously introduced a number of innovations in pollution control policy: (1) point sources of pollution were switched from command-and-control technology standards to performance standards; (2) economic incentives were introduced via permit trading opportunities; and (3) nonpoint polluters earned tradable credits as a group rather than as individuals. Of these innovations, the first two are classic flexible incentives, while the third solves an isolation paradox by providing benefits (income from the sale of credits) to all members of the group whenever a group target is achieved.

Initial results from these point-source/nonpoint-source pollution trading programs have paralleled the experience with point-source pollution trading: The number of actual trades has been surprisingly few (Burtraw, 1996). Nevertheless, considerable savings in pollution control costs have been accomplished, largely through the switch from regulating technology to performance regulation. Point-source polluters have been remarkably successful at finding relatively inexpensive ways to reduce effluents. It is reasonable to expect that, as cost reductions that result from switching to performance standards become exhausted, the frequency of trading will increase.

Discussions of point-source/nonpoint-source pollution trading programs have tended to emphasize the opportunities they provide to introduce economic incentives into NSP control policy. These programs introduce an additional and interesting aspect: Cooperation among landholders in a (sub-)catchment in order to earn credits as a group. A logical next step would be to switch from technology regulation to performance regulation monitored at the group level.[5] This promises to take policy beyond the first stage, in which NSP control was largely ignored. It goes beyond its second stage, which was limited to educational programs to encourage the adoption of best management practices, and onto a third stage, in which the isolation paradox is solved by monitoring and rewarding performance at the group level. Thus, the process takes advantage of the knowledge that nonpoint polluters in the same sub-catchment have about each other's practices. A fourth stage now seems technically feasible, which is to apply the *polluter pays* principle to NSP. Agriculture has resisted *polluter pays* by maintaining the impossibility of monitoring individual farmer contributions to in-stream pollution loads. However, if nonpoint polluters can organize to maximize the number of credits they can sell collectively, there seems to be no reason why they cannot organize to reduce group effluents in order to minimize penalties that are imposed upon the group.

Protecting Biodiversity

While it is possible to think of protecting individual species and varieties in seed banks and various high technology variants thereof, protection of biodiversity typically involves the protection of habitats from human encroachment. If it is granted that habitat protection is a valuable and potentially welfare increasing land use, it remains true that many competing and valuable land uses involve encroachment. Existing endangered species legislation, which has emphasized the protection of critical habitats, has aroused several celebrated controversies when it has been applied to public projects and public lands. The more recent application of habitat protection to private lands has generated controversy of a more widespread and persistent kind.

In this context, at least two problems arise that are amenable to game theoretic or principal agent analyses. The first problem is that private landowners are likely to have more information on whether their land might harbor an endangered species and/or an important section of critical habitat than public agencies would. Yet, when laws to protect biodiversity are structured so as to impose significant costs on landowners, they are discouraged from providing information to relevant public agencies. Polasky and Doremus (1998) identify incentive packages that would encourage landowners to efficiently provide the necessary information by incorporating elements of both the *carrot* and the *stick*.

The second problem is a classic isolation paradox. For many kinds of ecosystems, protection of biodiversity requires large areas of contiguous habitats. This is feasible only if considerable numbers of independent landowners can be encouraged to cooperate with each other, and often, to cooperate with public land agencies. It is revealing that nine out of ten invited representatives of private interests at a recent workshop on ecological policy wanted only to tell horror stories about private property owners being terrorized by the eco-police. Only one representative discussed processes by which property owners, environmentalists, government agencies and other interested parties might work together to devise mutually acceptable solutions. There have been a number of recent successes of this kind (Interagency Ecosystem Management Task Force, 1995). Exactly how these successes were accomplished cannot be explained simply—the process itself is inherently complicated. The various interest groups came to see the legitimacy of each other's positions. Framed in broad terms, goals for the community were not so different across the various interests, so a shared vision was established successfully. But the phenomenon was not entirely sociological. The threat of regulatory action may have motivated property owners to accept environmental targets that required some real sacrifices (Segerson, this volume). The prospect of a long and uncertain regulatory process, if the negotiation process were to break down, could have motivated environmentalists to accept a solution that would require some actual compromise. In other words, real trading may have occurred.

Flexible incentives of a more standard kind may also have a role in habitat protection for biodiversity. Wetland mitigation banking is now an established tool in the regulatory arsenal. It works well in Ohio, where wetland restoration begins with removal of existing drains from land that was wetlands originally. Recent trends in policy to implement Superfund and the Oil Pollution Act have emphasized compensatory restoration, where responsible parties provide restoration projects to compensate the public for environmental damage. From these beginnings, it is only a modest stretch to habitat protection credit trading, as sketched by Hodge and Falconer (1998). The agency would delineate land that is desirable as habitat and would specify the amount to be protected. Habitat protection credits could be specified and traded among property owners subject to various restrictions provided, for example, that migration corridors be provided so that the least-productive land in alternative uses would be assigned to habitat protection. As with all permit and credit trading schemes, initial distribution would be an issue. The arguments, however, would be familiar, despite the unfamiliar context of habitat protection. The analogy with wetland regulation suggests landowners have conservation obligations that could be satisfied on their land, or elsewhere (within limits that are defined by the regulator) if they could persuade another landowner to do the job for them. The analogy with the Conservation Reserve Program suggests that the government might offer landowners the opportunity to profit, rather than to reduce compliance costs, by preserving habitat.

CONCLUSION

Flexible incentives offer new opportunities to engage farmers in environmental protection and in meeting environmental targets for agriculture. These objectives can be achieved through reduced compliance costs and/or new profit opportunities. As such, flexible incentives provide a potential avenue out of the historical impasse regarding environmental regulation, which agriculture often has been able to avoid in the past because of the difficulty of monitoring performance.

This chapter attempted to place flexible incentives in a broader context. Command-and-control regulation is giving way to flexible incentives programs not just because flexible incentives are potentially more efficient, but also because faith in the legitimacy and efficacy of scientific government has declined precipitously. Ideological individualism is on the rise. Big government finds few vocal defenders, while privatization is all the rage but leaves us in isolation paradoxes of various kinds. The answer seems to lie in institutional innovations that are based on some of the lessons of game theory and communitarian political theory—innovations that seek problem-scale solutions and replace existing conflict with win-win incentives for sustainable cooperation. Flexible incentives will find an important place in this larger scheme of things.

ENDNOTES

1. This emphasis on exclusion may lead the reader to assume that solutions to nonexclusiveness and non-rivalry naturally involve the extension of the domain of private property rights. Privatization is not always a feasible cure for nonexclusiveness; if exclusion were costly, other kinds of arrangements could work better. Privatization could fail entirely to provide some kind of non-rival goods, yet that alone is insufficient reason for society to go without.
2. Ciriacy-Wantrup and Bishop (1975) pointed out that the common property resources analysis is really applicable only to pure nonexclusiveness. The tragedy of this analysis is misleading if applied to the myriad of common property institutions that have been developed to handle resource management problems in various traditional and modern societies.
3. My term, isolation paradox, and its definition are in the spirit of Adam Smith's discussion and example. I point this out to minimize confusion that might otherwise arise, given Sen's (1967) idiosyncratic usage, in which isolation paradox is synonymous with n-person prisoners' dilemma.
4. While this problem must be taken seriously, we should not make too much of it. The race to the bottom has its limits. Assume the public likes a clean environment and a considerable array of services that are provided by state and local governments and low taxes. Then, a jurisdiction will find that a strategy of tax and environmental subsidies to attract business is undercut, to some degree, when mobile workers demand higher wages to compensate for the less attractive environment, poorer services and/or higher taxes on households that will inevitably result from such a strategy.
5. A switch to performance regulation of agricultural pollution is likely to encourage innovation in abatement technology and resultant cost savings in agriculture just as it has in other sectors. Khanna, et al. (this volume) provides examples of innovations in cost reducing and environmentally benign agricultural technologies that could be encouraged with appropriate incentives.

REFERENCES

Arrow, K.J. 1951. *Social Choice and Individual Values*. New York, NY: Wiley.
_____. 1986. "Agency and the Market," in K.J. Arrow and M.D. Intrilligator, eds., *Handbook of Mathematical Economics*. Amsterdam, Netherlands: North-Holland.
Bator, F.M. 1958. "The Anatomy of Market Failure." *Quarterly Journal of Economics* 72: 351–379.

Bergson, A. 1938. "A Reformulation of Certain Aspects of Welfare Economics." *Quarterly Journal of Economics* 52: 311–334.

Burtraw, D. 1996. "The SO_2 Emissions Trading Program: Cost Savings without Allowance Trades." *Contemporary Economic Policy* 14(7): 9–94.

Cheung, S.N.S. 1970. "The Structure of a Contract and the Theory of a Nonexclusive Resource." *Journal of Law and Economics* 13: 49–70.

Ciriacy-Wantrup, S.V., and R.C. Bishop. 1975. "Common Property as a Concept in Natural Resources Policy." *Natural Resources Journal* 15: 713–729.

Coase, R.H. 1960. "The Problem of Social Cost." *Journal of Law and Economics* 3: 1–44.

Dahlman, C. 1979. "The Problem of Externality." *Journal of Law and Economics* 22: 141–162.

Dasgupta, P.S. and G.M. Heal. 1977. *Economic Theory and Exhaustible Resources*. Cambridge, UK: Cambridge University Press.

Gordon, H.S. 1954. "The Economic Theory of a Common Property Resource: The Fishery." *Journal of Political Economy* 62: 124–112.

Groves, T. and J. Ledyard. 1980. "The Existence of Efficient and Incentive Compatible Equilibria with Public Goods." *Econometrica* 48: 1487–1506.

Hicks, J.R. 1939. "The Foundations of Welfare Economics." *Economic Journal* 49: 696–712.

Hodge, I. and K. Falconer. 1998. "Institutional Incentives for Sustainable Arable Agricultural Systems." Working Paper, University of Cambridge, Cambridge, England, UK.

Hurwicz, L. 1973. "The Design of Mechanisms for Resource Allocation." *American Economic Review* 63: 1–30.

Interagency Ecosystem Management Task Force. 1995. *"The Ecosystem Approach: Healthy Ecosystems and Sustainable Communities."* Federal Report NTIS # PB95-265609, Springfield, VA.

Kaldor, N. 1939. "Welfare Propositions in Economics." *Economic Journal* 49: 549–552.

Keynes, J.M. 1936. *"The General Theory of Employment, Interest and Money."* London, UK: Macmillan.

Marshak, J. and R. Radner. 1971. *The Economic Theory of Teams*. New Haven, CT: Yale University Press.

Nelson, R.H. 1987. "The Economics Profession and the Making of Public Policy." *Journal of Economic Literature* 25: 49–91.

Pigou, A.C. 1932. *The Economics of Welfare*. New York, NY: Macmillan.

Polasky, S. and H. Doremus. 1998. "When Truth Hurts: Endangered Species Policy on Private Land with Imperfect Information." *Journal of Environmental Economics and Management* 35: 22–47.

Rand, A. 1944. *The Fountainhead*. Indianapolis, NY: Bobbs-Merrill.

Randall, A. 1983. "The Problem of Market Failure." *Natural Resources Journal* 23: 131–148.

Runge, C.F. 1981. "Common Property Externalities in Traditional Grazing." *American Journal of Agricultural Economics* 63: 595–606.

Samuelson, P.A. 1954. "The Pure Theory of Public Expenditure." *Review of Economics and Statistics* 36: 387–389.

Sen, A. K. 1967. " Isolation, Assurance and the Social Rate of Discount." *Quarterly Journal of Economics* 81: 112–124.

Shubik, M. 1981. "Game Theory Models and Methods in Political Economy," in K.J. Arrow and M.D. Intrilligator, eds. *Handbook of Mathematical Economics*. Amsterdam, Netherlands: North-Holland.

Tiebout, C.M. 1956. "A Pure Theory of Local Expenditures." *Journal of Political Economy* 64: 416–424.

21 POLITICAL FEASIBILITY: INSTITUTIONAL LIMITS ON ENVIRONMENTAL REGULATION

William P. Browne
Central Michigan University,
MT PLEASANT, MI

This chapter looks at specific problems and proposed solutions that have been advanced by several policy professionals in their case studies. It then analyzes the likelihood of successful adoptions of the proposals (for example, their political feasibility). In this sense, the chapter is an applied agricultural case that uses the tools of political economics—especially political transactions and the costs of making them.

INTRODUCTION

Some articles are written about making good public policy. This analysis is about structuring acceptable United States public policies, particularly those regulating environmental uses that are politically feasible. The purpose of this chapter is to explain political feasibility; identify institutional factors that create political obstacles to policy change; summarize the conditions that have come to benefit environmental issues and policies; and emphasize things to avoid when proposing environmental policy change. The underlying premise of this analysis is that scientifically good policy often makes for bad programs and politics (Browne, 1998). In the battle between science and politics, if things come to that, politics will win.

POLITICAL FEASIBILITY

Public choice economists understand that American politics are extraordinarily transactional. Dewey and Bentley (1949) summarized the meaning of transactional. Politics are neither about self-action (that is, about doing it yourself) nor about interaction. Rather, what happens in transactions occurs because a diversity of variables come together that influence each other collectively, and seldom do so in a way that attributes effect to any single one of them. Thus, it is as difficult to specify models in public policymaking as it is in science.

This is a frightening thought for those who want to involve themselves in policymaking. The content of American politics cannot be laid out in a nice matrix, nor can it be summarized in a generalized and systematic flow chart. Only the steps in a policy process can be assembled that way. These steps represent little more than a checklist that reveals only where one is and what is left to do. Nothing is suggested about how and why the process works the way it does.

What is methodologically so scary about analyzing a transactional situation (Webber, 1986). Most who propose and advocate regulatory reforms think of themselves as policy experts. Experts, as positivists, provide information and databased guidance on substantive problems. They do the technical and economic analysis that considers environmental consequences and then they propose solutions (Snare, 1995). Often, as Randall (this volume) so aptly illustrates, they can also broker settlements and resolve disputes by negotiating with competing experts before those conflicts go to their respective legislative bodies. Those cooperative efforts tend to be tied well to analysis.

Neither analytical techniques nor expert consultation are sufficient for doing the work that Snare (1995) calls policy troubleshooting. Policy troubleshooters are neither experts nor advocates within the political process. Troubleshooters, instead, are fixers. Their focus is on the transactional nature of the policymaking process—to bring disparate variables together to make policy ideas workable and to pass them into law. While troubleshooters do most of their work with legislatures and administrative agencies, they also frequently help bring the competing public interests and their conflicting views together. Again, as Snare (1995) states, the troubleshooter's philosophy is that less than optimal decisions that pass are far better than the truly optimal ones that slow down and die. Unlike the tools of expert analysis, policy sciences say little about manipulating political variables (Mead, 1983). Consequently, there is no detailed manual for the role of troubleshooting.

Troubleshooters do a rather free-style kind of political feasibility analysis of transactional situations, even when they bring together competing interests and experts. They focus on identifying the players who are likely to affect processes; the motivations and responses of those players; the events and conditions that shape the playing field of politics; and the institutional structures of the specific contexts in which the decisions are being made. Troubleshooters study the limits of what is possible. Forget, they say, about what is best, really nice or quantitatively elegant. Instead, they say, just look at what is likely to be feasible.

As a consequence of this free-style approach, there is precious little written about examining political feasibility. There is a vague sense that it should be done, but generally by someone else. Academics (and former academics) dominating the ranks of policy experts and advisors generally scorn their own free-style involvement in favor of competing through their classic styles. It is like a cross-country ski race in which the young free-style ski and the older traditionalists (who finish much later) just ski on the lanes and pathways.

The traditional expert recommendations of the academically trained only lay out various scenarios and options (Webber, 1986). Then they pass the responsibility of negotiation and resolution on to others. As gatekeepers, public officials determine the constraints of resources, the distributional factors of who wins and who loses, and the previous institutionalization within organizations along with the existing policy base (Majone, 1975). Policymakers are the stakeholders and can best judge subjective reality as to what is probable (Meltsner, 1972)—that is their job.

This traditional expert approach, of course, does not bring about socially optimal policy. Policymakers just do not know nor do they understand the theory or technical details (Weiss, 1977). That is why they initially seek the advice of experts. They only ask that they be advised in terms of their own individual political problems, and their job-specific perspectives. Of course, they must be advised in their own language (Browne and Schweikhardt, 1995).

Specifically what do troubleshooters, who study political feasibility, do? As Dror (1968) noted, the substantive emphasis must be on identifying how to gain support, how to accommodate contradictory policy goals and positions, and how to identify and reconcile diverse values. Also, with great frequency, troubleshooters must specify which policy proposals to avoid and explain why avoidance is best. The starting place is to analyze the institutional conditions of the policy process in general and then to evaluate the contextual landscape of, in this instance, the policy area of environmental regulation.

This is not neglected territory in this book. Batie and Ervin look at several types of flexible incentives and call for the designs of policy approaches that are unobtrusive and low cost for producers. They make it clear that some types of incentives are more feasible than others are. Khanna et al. similarly lay out options to ponder. In a very useful chapter, Segerson outlines a conceptual framework to superimpose on policy analysis for flexible incentives. Much of her commentary is helpful when examining feasibility, and when looking at the tools and techniques of the policy sciences. Academics, therefore, are making considerable progress in thinking politically.

INSTITUTIONAL FACTORS

The entire decision-making arena of American government is institutionally delineated by rules. Nearly every embittered policy player seems to decry the organized special interests (Berry, 1997). These interests are criticized for their proliferation of skewed, biased or just plain erroneous information. Critics especially dislike these organized interests for spreading fear and causing issues to be seen by people from the parochial position of factions rather than as parts of a common good.

Well then, why do special interests continue to exist? Rules are why. In this case, these are basic constitutional rules that are extraordinarily hard to change. Freedoms of speech, assembly and the use of private resources to petition the government make it improbable that anything will be done to severely limit special interests, no matter how much policymakers and the media may resent them.

Thus, institutional rules favor letting multiple advocacy positions proliferate. Their voices are at least as legitimate in American policymaking as those of the experts or of the pure of heart.

What occurs with the institutionalization of rules is that some obvious policy options are made inoperable, or at least, incredibly difficult to pass into law. These bad issues simply go too much against the encompassing political grain for immediate sentiments or even a crisis to change them. Within American government, some of the institutionalizations that matter (for example, free speech) are constitutional. Other rights and powers have been added to the institutional structure because constitutional conditions have made it possible to satisfy multiple factions by gradually adding their wants to the base of public policy (Shepsle, 1986). Equilibrium institutions produce primarily equilibrium policies. Public policies have grown topsy-turvy. With institutional growth, future policy options are further restrained but, at the same time, they are enhanced. An unfortunate thing about many of the chapters in this volume is that the authors still want to change too much in a government that is ruled by equilibrium and by competing and fragmented interests.

There are three additional sets of institutional conditions that are important for understanding what is feasible with environmental regulation. These are the decentralized and fragmented rule making processes, the reliance on citizen incentives to minimize public discontent, and the resulting processes to lower the transaction costs of public decisions (Browne, 1998). Policy players want every opportunity to be involved. They like to bestow favors to their constituency and prefer to do it with minimal difficulty. The evolution of American institutional rules means that policy players can get all three opportunities. Randall (this volume) very much understands this by suggesting that pre-legislative negotiations can pay off.

A few things need to be understood about these institutional conditions. First, compared to other nations, American governments are comparatively weak. They are democratic in processes, diverse in representation and extremely open to public participation. American governments also are separated in their powers, are able to exercise internal checks and balances to ensure that nobody is in absolute authority, and are subdivided into semi-sovereign states and municipalities. Heinz et al. (1993) said it best:

> American government has a hollow core without central authority, and all the decisions that are made are at the core's peripheries. One issue is at least temporarily resolved at a time, each on its own merits but each building on the past. (p. 308)

There is more to this hollow core, however, than just a lack of strong central authority. Especially since the 1960s, dizzying arrays of institutional structures have evolved that share in the decision-making process. Multiple agencies each have a piece of environmental regulatory control. The White House may or may not intervene. At the federal level, Congress decides the limits of that administrative control. Within Congress, authority has been divided between leaders, committees and subcommittees, and rank-and-file members (Shepsle, 1989). Each of these political players is driven by his or her specific legislative interest. Accordingly, none cooperate very well with the others (Bonnen, et al., 1996). The fifty state governments are not much different.

The implication is that nearly anyone in government can be a policy player, when he or she wants to be and on any issue he or she wants. On environmental concerns, anyone can play on behalf of, or can represent, whomever they wish (Cohen, 1992). Of course, this further encourages the proliferation of factional interests. There is almost always some official who will champion a particular faction's views. Everybody, therefore, may as well take a minimal risk and organize. This point is sadly lacking in this book. Too many of its chapters imply that policymaking is orderly and rational.

Secondly, from these conditions come the incentives. North (1990) explained that only incentives bring about changes in the economic performance of nations. People need reasons to participate—they want to pursue an expanding interest. The same is true of government and politics when citizens are shopping for representation in a highly competitive and decentralized public policy marketplace.

Factionalizing interests do not want things to be taken away from them. Rather, they want government to give them very specific things. In order to avoid an unhappy public and an unstable government, public officials respond and offer incentives wherever they can. In any single decision, hopefully they can offer multiple incentives—each one tailored for multiple, but unique, beneficiaries. If public officials do not provide these multiple benefits, the public will always be shopping for someone new. Worse yet, the public may end up following charismatic crackpots. The result is that incentives from government have become the generally accepted standard by which private interests judge public officials' performance.

This brings us to the third and final condition: With so many policy providers, so many interests making policy demands and so many policy issues, American politics cannot be easy. The transactional context is extraordinarily complex and risky. If public policymakers were not constantly attempting to mute competition and to keep the costs of transacting public business as low as possible, processes would strangle to a halt. The public would then erupt, and no officials would want that kind of chaos.

That is where the tinkering tendency of U.S. policymaking comes forward. On any single decision, officials prefer to bargain with the fewest players regarding the least significant and least financially costly items. They want to make the fewest possible changes they can in the existing policy base—perhaps they want to make no changes at all. Thus, they are inclined to tinker and not to make radical policy departures. Many policies are created, but generally it is quite conventional. Institutions have been put in place to reduce the uncertainty of everyday life (North, 1990). They provide opportunities to keep adjusting marginally who gets what. Within this institutional setting, public policy is like a giant erector set. Batie and Ervin understand this, yet they still generally articulate the view that policy experts will put most of the set together. That simply is not the case.

From an institutional standpoint, what does environmental regulation face? It faces a great many who would meddle with it. Far greater, and perhaps insurmountable, it faces public problems with proposals to limit what people see as their basic property rights. It also faces governments that lack enthusiasm for doing things that are controversial or risky in nature. Academics must face this reality or be left out of any policy relevance.

THE POLITICAL STRENGTHS OF ENVIRONMENTAL POLICY

Things are far from bleak. The institutional conditions of American government that were discussed above are but a disadvantageous setting for what otherwise is a favored issue. Some environmental policy proposals do make good policy issues (Browne, 1998). That explains why they have not been pushed aside, indeed, that is why they have spread.

A review of eleven different institutional conditions that are advantageous to good environmental policy proposals is presented below. These are discussed in the order they have emerged over time. Their cumulative effect is transactional and not causally determinable (Dewey and Bentley, 1949). This fact can be seen in a careful reading of this volume, even if some authors miss the point and look for a single cause of policy resolution. All of the authors, however, should better understand the reasons why Americans support environmental policy.

First, the United States and its populace have a long-term and deeply embedded stewardship ethos. This is, of course, linked to its agrarian heritage and values. The public truly believes in protecting the land and the living species (Browne et al., 1992). Through good policy marketing, environmental activists have done a great job of linking their interests and demands to the stewardship ethos. Yet this public belief system, by itself, is far too abstract to be meaningful.

Second, respected and well-supported conservation interests have been active in American politics since the early twentieth century. Numerous public policies have been put in place that established federal parks, forests, game and fish programs, and soil conservation. All this has provided considerable institutionalization in law and in public management. These laws were used to make the judicial system a critical component in the advancement of environmental policy (O'Leary, 1993).

Third, the public has been ardently and enthusiastically using these conservation programs. A raft of local sportsmen's clubs grew up around these activities, occasionally becoming active and mobilizing supporters in politics. The ongoing political activism of these interest groups has helped create a broad popular support base.

Fourth, conservation programs subsequently have suffered some setbacks due to this intensity of public use and have been subject to criticism. Poor fishing and park/forest degradation have been two obvious problems.

Fifth, conservation lobbyists have become convinced that new public policy strategies are needed: The existing policy base has been insufficient and the lobbyists have assumed a greater role of advocacy.

Sixth, environmentalists have had ready proposals and believable explanations (Bosso, 1991). They initially spoke as extensions of the early conservation rhetoric. Yet, they once kept intellectual statements that doted on ecology and on closing circles to a minimum. Environmentalists have learned to broaden their message when they addressed the mainstream political arena. They were careful in their initial demands to neither threaten natural resource use nor to refuse to negotiate.

Seventh, environmental interests have slowly been able to convert that publicly held stewardship ethos into more specific and precise public perceptions (Pursell, 1973). Especially important were their early efforts with teachers, school children and education. By 1995, only 29 percent of the public felt that environmental restrictions were too harsh. Yet, only 1 percent saw the environment as an important policy area (Bosso, 1997).

Eighth, environmentalists have been identified as outsiders (Bosso, 1991). This has helped them win recognition. They have labeled any opponents as powerful conspirators and the public has liked the underdog rhetoric.

Ninth, environmentalists have proceeded slowly. They rarely, at first, challenged multiple established interests at any given time. Instead they proceeded from local government to business and then to farming. They waited until a generally favorable public opinion emerged before they collectively opposed all their adversaries at once.

Tenth, those public officials, who had not been supportive initially—representative as they were—had to eventually fall in line. At least enough of them did. Being publicly popular was like catching a wave while surfing. Environmental policy has provided a good ride.

Eleventh, many interest groups have been organized because there was such a multitude of issues to the environmental movement. This gave public officials numerous causes to champion. These private and public interests have played coalition politics superbly, by working together on common cooperative projects around the country.

In summary, transactional circumstances have worked well to make clean air, clean water, wildlife, soil protection and habitat protection good political issues. Some policy successes have been achieved through superior lobbying, but most have been accomplished as a result of the nature and evolution of public institutions. This is a common thread in many chapters of this volume, although it is never specifically articulated.

Without a doubt, these advantages never made environmental politics easy (Cohen, 1992; Bosso, 1997). Conflicts have been numerous, opponents have been plentiful within every circle of government and progress has been slow. Thus, environmentalists have remained frustrated. The advantages discussed above did make politics possible at the most basic level for the environmentally concerned who have been playing the hard and unpopular game of high transaction costs public policy. This volume suggests that now is the time to give things away and to lower these transaction costs.

DEVELOPING FLEXIBLE REGULATORY INCENTIVES

By contrasting the generalities of institutionalization with the advantages of environmental policy, it is observable that all policy proposals for the environment are not equally feasible in the political arena. The proposals and findings presented in this volume are more feasible than most because they are, as the collective title suggests, laced with incentives and are flexible in their delivery and use. Americans want to see environmental sensibility and policy sense (Yandle, 1997). Policy proposals that do not achieve both will lose. Witness the defeat of the environmentalists on the North American Free Trade Agreement and the 1992–93 Democratic Congress' unwillingness to rewrite and make harsher pollution standards.

What has transpired since that time is, in a related way, more limiting. The property rights argument has struck a very responsive chord among the public and the environmental users. Its logic was well planned, as was the logic of mixed public land use. A politically active and long-time liberal proponent of public policies, who also is a large-scale farmer, related the following observation to me in a personal interview:

> By G . . . d, I've changed my mind and told that to my representatives. I believe in protecting our environment. But those d . . . d prairie potholes belong to my family and me. If we see a reason to fill them, we should be able to do it without reluctance or hesitation. H . . . l, we'll help the ducks in some other way.

He went on to decry that the public land use that he saw as too narrow. Americans and elected representatives believe in property rights and in their independence of judgement. Policymakers who threaten either will have a long, tough ride—and will usually lose.

What is important to observe about the earlier chapters? As Ribaudo and Caswell observe in the overview, the history of agro-environmental regulation is one of direct assistance in winning voluntary compliance. Clients have gained assistance and have been free to accept or decline the accompanying requirements. Breaking from that mold will be very difficult. As Ogg suggests, policymakers will want to bribe farmers. Public officials will not want to accept policy solutions that take opportunities away from producers without compensating them. Khanna et al. hit upon a very feasible approach. They suggest finding areas of sustainable production techniques that will not limit farm production and income. They also suggest designing financial incentives to help farmers adopt them. Batie and Ervin call these suggestions low transaction cost approaches. These have obvious policymaker appeal as long as the stigma of backwardness (sometimes associated with sustainability rhetoric and academic jargon) is avoided.

In the Batie and Ervin chapter, the promotion of self-regulation in accordance with local and regional needs is politically poignant. Other aspects of their typology are not. Increasing user awareness produces uneven results. Public officials are usually skeptical of trying to change public attitudes in any significant way. Citizens already have views and values, and are apprehensive of government interference in their lives. Economic disincentives, which make it costly to continue certain practices on one's own property, would only create nasty rebellions. Governments that have done so, given the existing public attitudes, have experienced decision-making turmoil. Thus, the transaction costs of decision making are too high. Disincentives are likely to be viewed by producers as nothing but punishment, even if the costs are designed to be as low as possible.

The case study proposals provide a second set of important ideas to ponder when considering the feasibility of public policy change. Only a few main themes from among the many interesting chapters need be pointed out to demonstrate that experts can do a little troubleshooting and formulate policy proposals that have a chance.

In looking at wildlife habitat needs, Roka and Main were able to elicit a means to provide many incentives through mutually rewarding public/private partnerships. Getting both collective and selective policy benefits is always good politics. In an interesting addition, the authors ask just how much saving of animals can be afforded. That is a politically critical question that others should have asked. Being feasible does not equate to being politically correct.

By looking at previous attempts to regulate pollution, the Casey and Lynne chapter compares different approaches to agro-environmental regulation. They point out a number of issues that need to be addressed.

These chapters reveal how good science, by being broadly acceptable, incentive-based and not too controversial, can clarify and lead to environmental approaches that are politically feasible. Public policymakers are aware that governments do not need to be punitive in their legislation. Punitiveness might be optimal and desirable to experts, but it is not politically wise. As seen in a few of these chapters, the judicial system may be the environmentalists' favorite official institution. Judges, however, are very reluctant to limit property rights and environmentalists often lose in such cases despite their advantages. There needs be a compelling logic, beyond that of a simple public good, in order to win (O'Leary, 1993).

A whole different approach is presented in other chapters of this volume. As noted in this chapter, public officials do not want to pry too much in people's lives, or be intrusive in their policy education. Two chapters, however, describe how public education can be carried out in positive non-threatening ways so that it is perceived as highly usable information. Van Ravenswaay and Blend illustrate how ecolabeling may be employed and be valuable to both manufacturers and consumers. The ideas on health risk-information benefits and likely popularity are only a bit less specific. Swinton, Owens and Chu strongly suggest that such information can be converted into a usable, if not exactly cherished, policy product. This is especially true if economists follow Randall's vision of cooperative effort, in which they act in brokerage fashion.

Both the ecolabeling and conversion analyses, however, need to better address the question of whether or not these proposals, if put into law, would pay any specific dividends. Public officials would certainly want to know this. Lovejoy's study of soil conservation would provide enough of a reason to interest policymakers and to bring about that question, were it not already a part of conventional political scrutiny. His is a good common sense analysis, which is what policymakers want. The chapter by Huffaker and Levine tends also to be straightforward and usable by public officials, especially when addressing the matter of policy takings.

What exactly are policy takings? That is a neglected question which, if left unde-fined, would lead to inexact policy discussion.

Of course, it would be easy to find appropriate and useful ideas in each of the case study chapters. These academics, as policy experts, convincingly demonstrate that this community of scholars can do much better than has been done in the re-cent past. These talented people can address issues that public officials will actu-ally champion—proposed policies that give rather than take away from producers or agribusinesses and can be made into law with minimal transaction costs. They can do that by building on the traditional advantages of environmental policy.

Still, all is not right with the multitude. Too often, there exists a pronounced tendency to think that the community of policy experts has all the right answers. This still pervades academic thought and remains a scholarly community flaw that is hard to overcome. As a consequence of that flaw, too few of the chapters ad-dress exactly why public policymakers should find these to be compatible propos-als. They also do not spell out what benefits their proposals bring to the political decision-making process. The authors still seem to want to leave those conclu-sions to somebody else. But, quite clearly, academics and analysts are also learn-ing to successfully work the policy process.

CONCLUSION

What has been said in this review? It can be summarized briefly as follows: Ad-dress what is politically possible, understand your proposal's institutional weak-nesses and assess how the policy objectives are advantaged by the evolutionary conditions of political environmental institutions. From there, as a policy analyst, play the game and play it by the rules of politics. If you go by the rules of the aca-demic cloister or by the scientific method, you will simply face political irrele-vancy. The contents of this book illustrate that policy experts are now thinking about how to avoid that deep dark pit of methodological despair. As a conse-quence, their policy impact should once again grow.

REFERENCES

Berry, J.M. 1997. *The Interest Group Society, Third Edition*. New York, NY: Longman.

Bonnen, J.T., W.P. Browne, and D.B. Schweikhardt. 1996. "Further Observations on the Changing Nature on National Agricultural Policy Decision Processes, 1946–1995." *Agricultural History* 70: 130–152.

Bosso, C.J. 1991. "Adaptation and Change in the Environmental Movement," in A.J. Cigler and B.A. Loomis, eds., *Interest Group Politics*, third edition. Washington, DC: Congressional Quarterly Press.

_____. 1997. "Seizing Back the Day: The Challenge to Environmental Activism in the 1990s," in N.J. Vig and M.E. Kraft, eds., *Environmental Policy in the 1990s*, third edition. Washington, DC: Congressional Quarterly Press: 54–58.

Browne, W.P. 1998. *Groups, Interests, and U.S. Public Policy*. Washington, DC: Georgetown Univer-sity Press.

Browne, W.P., J.R. Skees, L.E. Swanson, P.B. Thompson, and L.J. Unnevehr. 1992. *Sacred Cows and Hot Potatoes: Agrarian Myths in Agricultural Policy*. Boulder, CO: Westview Press.

Browne, W.P., and D.B. Schweikhardt. 1995. "Demosclerosis: Implications for Agricultural Policy." *American Journal of Agricultural Economics* 77: 1128–1140.

Cohen, R.E. 1992. *Washington at Work: Back Rooms and Clean Air*. New York, NY: Macmillan.

Dewey, J. and A.F. Bentley. 1949. *Knowing and the Known*, p. 108. Boston, MA: Beacon Press.

Dror, Y. 1968. *Public Policymaking Re-examined*. San Francisco, CA: Chandler.

Heinz, J.P., E.O. Laumann, R.L. Nelson, and R.H. Salisbury. 1993. *The Hollow Core: Private Interests in National Policymaking*. Cambridge, MA: Harvard University Press.

Majone, G. 1975. "On the Notion of Political Feasibility." *European Journal of Political Research* 3: 259–272.

Mead, L.M. 1983. "Policy Science Today." *Public Interest* 73: 165–170.

Meltsner, A.J. 1972. "Political Feasibility and Policy Analysis." *Public Administration Review* 32: 859–867.

North, D.C. 1990. *Institutions, Institutional Change and Economic Performance*, p. 3–4. Cambridge, MA: Cambridge University Press.

O'Leary, R. 1993. *Environmental Change: Federal Courts and the EPA*. Philadelphia, PA: Temple University Press.

Pursell, C., ed. 1973. *From Conservation to Ecology: The Development of Environmental Concern*. New York, NY: Crowell.

Rosenthal, A. 1993. *The Third House: Lobbyists and Lobbying in the States*. Washington, DC: Congressional Quarterly Press.

Shepsle, K.A. 1986. "Institutional Equilibrium and Equilibrium Institutions," in H. Weisberg, ed., *Political Science: The Science of Politics*. New York, NY: Agathon.

_____. 1989. "The Changing Textbook Congress," in J.E. Chubb and P.E. Peterson, eds., *Can the Government Govern?* Washington, DC: Brookings.

Snare, C.E. 1995. "Windows of Opportunity: When and How can the Policy Analyst Influence the Policymaker during the Policy Process." *Policy Studies Review* 14: 407–430.

Webber, D.J. 1986. "Analyzing Political Feasibility: Political Scientists' Unique Contribution to Policy Analysis." *Policy Studies Journal* 14: 545–553.

Weiss, C.H., ed. 1977. *Using Social Science Research in Public Policymaking*. Lexington, MA: D.C. Heath.

Yandle, B. 1997. "Environmental Regulation: Lessons from the Past and Future Prospects," in T.L. Anderson, ed., *Breaking the Environmental Policy Gridlock*. Stanford, CA: Hoover Institution Press.

22 From Adoption to Innovation of Environmental Technologies

Scott M. Swinton and Frank Casey
Michigan State University, East Lansing, MI
Northwest Economic Associates, Vancouver, WA

Agriculture is experiencing an explosion of experimentation with flexible incentives to induce better environmental stewardship. The foregoing chapters present a range of incentives that span both public and private domains with a spectrum from wholly voluntary to largely coercive measures. The unifying element is the freedom of the agriculturalist to choose whether, and how, to adopt environmental technologies.

The case studies in this volume reveal the practical strengths and limitations of the incentive approaches whose conceptual attributes are outlined by Segerson and by Batie and Ervin.[1] The incentive approaches presented are conceived with the objective of inducing the adoption of existing environmental technologies. If continual improvement in agricultural productivity and environmental quality is the goal, however, then the adoption of today's technologies is not enough. An institutional setting must be created in which continual innovation is induced to generate technologies that are both agriculturally productive and environmentally sound (Khanna et al.). This chapter begins by recapping what has been learned about the design of flexible incentives for environmental technologies in agriculture. The chapter closes with observations on how to shape an institutional environment that will continually induce innovations in environmental management.

MATCHING THE INCENTIVE TO THE PROBLEM

The incentives presented here are tailored to specific sets of producer preferences, technology alternatives and institutional settings. At one extreme are cases in which producers need little inducement to adopt environmental technologies—either because those technologies dominate alternatives in profitability or because the producer cares about environmental quality and will willingly trade off some marginal gain in profit for a marginal reduction in health or in environmental risk. At the other extreme are cases in which producers must be forced, or be given, inducements to adopt an otherwise unattractive technology.

One way to illustrate the range of combinations of technology and producer preferences is with a trade-off frontier, as in figure 22.1. Points A, B, C, D and E represent five different technologies that produce specific levels of marketable product (P) and associated levels of environmental quality (EQ). The dashed line connecting points B, D and E with the two axes represents an implicit production possibilities frontier possible through linear combinations of technologies. Each of the indifference curves in the family U_p, U'_p represents the set of points that leaves the producer household indifferent to alternative combinations of product and environmental quality. Indifference curve U_s represents a society's preferences at the scale of an individual producer.

Private Sector-led Approaches

When a new environmental technology offers greater profitability than existing alternatives do, the only incentive necessary for adoption is diffusion of information. This typically entails publicity about its availability, and often training and education in its use. If farmers care about both profitability and environmental quality, then the win-win technologies that dominate on both accounts will be preferred. In figure 22.1, this would correspond to a move from point A to point B or from point C to point D, which would increase both marketable production and environmental quality. Excelling on both of these accounts is characteristic of the broad class of precision technologies that are discussed by Khanna et al. Precision technologies are particularly *a propos* for cases in which wasted inputs can become pollutants. Soil erosion is illustrative: It wastes a productive agricultural resource while it pollutes surface water and reduces the capacity of irrigation canals. The polyacrylamide polymer technology, which binds soil particles, can be viewed as a precision technology that reduces pollution by reducing waste (Parker and Caswell). Drip irrigation is another precision technology that reduces the amount of water that is wasted because it is not taken up by crop plants. Casey and Lynne find that, for some producers, drip irrigation offers the same win-win appeal that triggers ready adoption.

Few technologies strictly dominate the alternatives in *both* profitability and environmental quality. A larger group allows some increase in one alternative at the expense of the other. How willing are farmers to make those trade-offs voluntarily? Growing evidence reveals that most agricultural producers care about environmental quality. Swinton, Owens and van Ravenswaay estimate corn growers' willingness to pay for herbicide safety characteristics. Their research suggests that many farmers would voluntarily make a move analogous to going from point B to point D (figure 22.1). The shift from U_p to the higher indifference curve U'_p leaves the producer household more satisfied and reveals a willingness to pay the opportunity cost of foregone production, P_b - P_d, in exchange for environmental quality gain, E_d - E_b. In order for those trade-offs to be acceptable, however, farmers

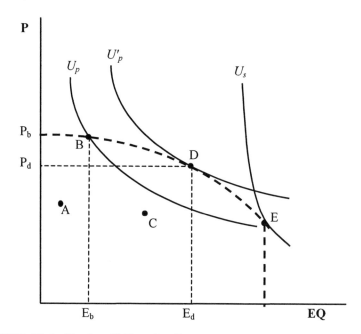

FIGURE 22.1   **Trade-off Frontier Between Marketable Product (P) and Environmental Quality (EQ)**

would require more innovative environmental technologies and better information about the environmental and health risks they would face.

Society's preferences may sometimes demand more environmental quality than farmers would choose because some agricultural practices cause negative environmental externalities that are not borne by the farm household. Such a case might be illustrated by the indifference curve, U_s, in figure 22.1. When consumers demand more environmental quality than farmers would choose, this demand can be transmitted to producers through the price mechanism to induce better environmental stewardship. The pricing of environmental quality can allow intermediation between the preferences of consumers for more environmental quality and the preferences of producers for more income from marketable production. Van Ravenswaay and Blend discuss ecolabeling initiatives that certify to consumers the environmental stewardship attributes that are invisible in the marketed product. Such certification can allow higher prices to capture the consumers' willingness to pay for environmental quality.

In figure 22.2, the valuation of environmental quality induces farmers to produce at point E, with socially desired environmental quality, E_e, while enjoying income equivalent to producing at level P'_e. The Swinton, Chu and Batie chapter emerges from a similar conceptual framework that regards an agricultural processing company's motivation to be viewed as a "green" firm; implicit is the expectation that greenness will be rewarded by customer loyalty or by higher prices. Of course, the private sector is not the only economic actor that is empowered to influence farmer earnings from agricultural goods whose production affects envi-

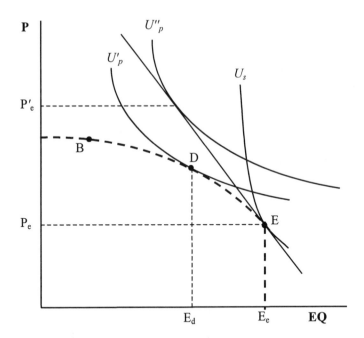

FIGURE 22.2   Trade-off Frontier Between Marketable Product (P) and Environmental Quality (EQ) When EQ Is Priced

ronmental quality. Governments in the United States and elsewhere have been extensively involved in influencing farming practices to favor the environment for more than half a century.

Public Sector-led Approaches

The earliest environmental interventions in agricultural policy were educational efforts to stem soil erosion (Lovejoy). Like virtually all agricultural education and technical assistance efforts (Ribaudo and Caswell), soil conservation education was motivated by the belief that if farmers understood the environmental consequences of their actions, they would change their behavior. Since soil erosion could clearly undermine land productivity, conservation education sought to induce the kind of win-win changes that are illustrated in figure 22.1 by moves from point A to B or from point C to D.

A wide range of financial incentive tools has been devised to permit governments to induce behavioral changes for the sake of reducing environmental risk. Public sector incentives that promote the adoption of environmental technologies can be classified according to who holds what property rights. One broad class of incentives emerges from the assumption that the farmer holds the right to behave as he or she wishes, and this includes the right to pollute. The resulting incentives follow a *carrot* approach, which entice the farmer to behave in the public's best interest. The other broad class of incentives emerges from the assumption that society holds the right to a clean environment. These incentives tend to follow a *stick* approach, which penalizes the farmer if his or her farming practices undermine environmental quality.

Most agricultural environmental policies have evolved from the assumption that the farmer holds the property rights. Under its cost-share programs, the U.S. government has subsidized the cost of agricultural production innovations that have led to reduced soil erosion or to improved water quality (Ribaudo and Caswell; Ogg). Under the 1996 Freedom to Farm Act, the Environmental Quality Incentives Program opened cost sharing to an even wider range of environmentally beneficial agricultural activities as defined by local committees. Federally subsidized land rental, under the Conservation Reserve Program, represents an attempt to reduce soil erosion by reducing agricultural output and/or by switching production to less vulnerable land (Ogg; Ribaudo and Caswell). Government research and extension services have been devoting increasing effort to the research, development and education of alternative production practices and of technologies that are more environmentally benign. The public financing of these services represents another form of subsidy intended to redirect farmer behavior.

Marketable pollution permits have seen little use in agriculture so far. However, they represent a promising avenue for inducing the adoption of agricultural practices that reduce nonpoint-source water pollution by creating and distributing new property rights (Randall).

Although most government interventions have aimed to reduce pollution, Roka and Main report an example in which the policy objective is to protect wildlife habitat. Subsidizing private landowners to protect the Florida panther habitat represents an unusual case in which the government buys an environmental service from private landowners.

The agricultural environmental policies that emerged from the assumption that society holds the right to a clean environment are fewer in number but are likely to become more common. They include design standards, performance standards and taxes (Segerson; Batie and Ervin). Generally, these have emerged in response to strongly held public concerns. Food safety and farm worker safety concerns motivated the laws that authorize bans on dangerous classes of pesticides (Ogg). International pacts to slow global warming encouraged the ban on methyl bromide as a soil sterilant (Deepak et al.). The high water quality risk posed by point-source water pollution instigated the Clean Water Act regulations on concentrated animal feeding operations of more than 1,000 head (Norris and Thurow; Ribaudo and Caswell).

Since the private sector bears the costs of reducing environmental risk under this set of policies, how these costs are allocated and whether they are justified are hotly contested topics. Deepak et al. argue for phased implementation of the methyl bromide ban, and recommend a quota on methyl bromide use. They further suggest that localized quotas would cause less displacement of production (and of farmer incomes) than would national quotas.

In certain instances, targeted *stick* policies have the potential to ameliorate environmental quality more efficiently than do policies that are implemented across the board. The phosphorus emissions tax that targets the Everglades Agricultural

Area is designed to reduce agricultural runoff in an area with an especially vulnerable environment. By allowing tradable emission credits, this provision is designed to encourage economic efficiency via permit trading that reduces phosphorus use (Lee and Milon). Carpentier and Bosch illustrate how geographically targeted agricultural runoff performance standards in the Chesapeake Bay watershed could accomplish a major reduction in water pollution. The high transaction costs of monitoring, however, would only be bearable if a change in property rights were to permit targeted monitoring of farms that are shown to be major polluters.

The role of property rights in the determination of who bears the costs of environmental policy has led to legal challenges of public policies under the "due process" and "takings" clauses of the Fifth Amendment to the U.S. Constitution (Huffaker and Levin). Such challenges are likely to continue, and Huffaker and Levin suggest that the courts may be well placed to make case-specific determinations of liability—given the wide range of environmental, technological and institutional settings in which environmental policy is applied.

INNOVATION FOR FUTURE IMPROVEMENTS

Current experimentation in the design of flexible incentives to encourage adoption of environmental technologies is testing a broad set of policy tools to determine which technologies will work best in what context. The next challenge is to move from policy tools that induce the adoption of *existing* technologies to policy tools that induce the innovation of *new* environmental technologies for agriculture.

Most of the key elements needed for policies that foster innovations in environmental technologies can be found by adapting Hayami and Ruttan's (1985) induced innovation hypothesis to environmental characteristics. This is most easily illustrated for the case of a polluting agricultural input. Hayami and Ruttan posit that a given level of agricultural output results from a combination of inputs under a given technology. That technology can be illustrated by an isoquant that represents combinations of inputs that produce the same level of output. Their special insight is to recognize that, at a given moment in time, the existing level of scientific knowledge makes possible the development of many technologies that involve different input combinations. The specific technologies that get developed are induced by relative input prices in the context of contemporary scientific knowledge (Hayami and Ruttan, 1985). Porter and van der Linde (1995) complement the Hayami-Ruttan model by arguing that carefully designed government regulations can induce environmental innovation that enhance competitiveness.

Technological innovation to reduce agricultural pollution is illustrated in figure 22.3. The figure shows combinations of two inputs that can produce an agricultural product. In initial period 0, production takes place at point A on isoquant T_0, where the cost of producing T_0 of output is at a minimum. Isoquant T_0 represents a technology that was developed under initial scientific innovation possibilities curve I_0 and relative input prices $p_e/p_p^{\,0}$. A change in relative prices makes polluting input X_p much more expensive, so that the slope of the input-price ratio shifts from $p_e/p_p^{\,0}$ to $p_e/p_p^{\,1}$. This, in turn, induces the innovation of new technology, T_1, which yields production at point B. Importantly, the new technology conserves the

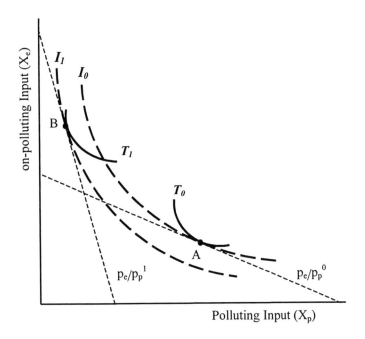

FIGURE 22.3 Induced Environmental Innovation in Agriculture

use of polluting input X_p relative to other technologies that could have been developed along innovation possibilities curve I_1, and especially those that would have been developed under the original input price ratio p_e/p_p^0. (Note that all new technologies developed along I_1 would use inputs more efficiently than they would under the old innovation possibilities curve I_0; this is the essence of the Porter and van der Linde argument that technological improvements may offset the cost of compliance with environmental regulations that induced those improvements.)

Many, but not all, of the environmental policies that are discussed in this volume can induce the kind of environmental technology innovation that is illustrated in figure 22.3. The illustration is driven by a change in relative prices, and several mechanisms can induce such changes. The most obvious ones are those that act directly on input prices, such as environmental cost shares (which reduce p_e) or taxes on polluting inputs (which increase p_p). Environmental liability risk, pollution emissions taxes and performance standards can be viewed as indirect increases in the price of polluting inputs. Conversely, marketable pollution permits can be viewed as a joint product in the agricultural production process, one whose value is realized by producing less than the permitted level of pollution, thereby generating an added marketable product. Whereas the emissions tax increases the implicit cost of the polluting input, the marketable permit reduces the implicit price of the non-polluting input.

An alternative approach to induce environmental technological change is via targeted research into environmental innovations. Environmentally oriented research would shift the innovation possibilities curve so that, unlike the unbiased innovation possibilities curve, it would shift from I_0 to I_1 in figure 22.3, the new innovation possibilities curve (I_1') could save more on polluting input X_p than I_1 would at any point along the curve.

In order to foster environmental technology innovation, the economic environment must change, and innovators must expect the changes to last. Short-term policies will not accomplish this. Hence, cost-sharing inducements to adopt conservation practices are unlikely to foster innovation if embedded in farm bills of only five years duration. Current technology standards also fail to engender environmental innovations. They may dampen innovation by disallowing promising avenues that were not contemplated when the standards were put in place.

INSTITUTIONAL DESIGN FOR ENVIRONMENTAL INNOVATION

Identifying changes in economic incentives that would foster environmental innovation is perhaps the easier task. The harder task is to conceive of an institutional setting that would give rise to these incentives. The policy experimentation discussed in this book testifies to the fact that we are already part way there.

The assignment and structuring of property rights is key. Most environmental property rights are currently structured around the concept that "the polluter pays". As an alternative, Porter and van der Linde (1995) discuss how outcome-oriented environmental policies can facilitate environmental innovation in which private firms prosper while they attain societal objectives for environmental quality. Apart from policies directly targeted at environmental outcomes, environmental benefits can also be induced through the certification of grades and standards for invisible environmental attributes. Standards such as the ISO 14000 series for environmental production (van Ravenswaay and Blend) may be maintained through private or public sector entities. Their existence and certification by an impartial entity is a necessary condition for the development of markets in products that are produced using environmentally benign practices (Swinton, Chu and Batie).

Although outcome-oriented environmental-quality parameters are necessary to create a climate for voluntary environmental innovation, by themselves, they are not sufficient. Where markets for environmental innovations are small, private firms may be chary of investing in technology development. This reluctance to innovate is more likely to occur when innovation costs are high. The development of new pesticides for small acreage crops in the United States is illustrative. Even with well established pesticides, when a federal mandate to renew registration raises the prospect of costly toxicology tests, chemical manufacturers sometimes opt to drop pesticide uses on small acreage crops that are unlikely to produce enough future revenue to compensate the toxicology testing costs. The same pattern—while not documented—surely underlies investment decisions on experimental pesticide compounds. So there are limits to the role that the assignment of property rights can play as an inducement or constraint to shaping desirable social choices in the presence of flexible incentives.

How do we invite innovation where it may be unprofitable? One way to do this is to change the cost structure. For example, the U.S. Food Quality Protection Act of 1996 reduced the financial and bureaucratic transaction costs of biological pest control measures relative to chemical pesticides (Ogg). An alternative way to induce otherwise unprofitable innovation is to raise the costs of failing to meet socially defined environmental norms. Legal liability for environmental contamination can add potential costs for environmentally irresponsible behavior in a flexible way that is open to judicial interpretation and tailoring to the specific setting (Huffaker and Levin; Schmitz and Polopolus).

The direct approach of public research into environmental innovations may work best where private incentives for innovation are difficult to develop. Indeed, this is the historic rationale for the publicly supported land-grant colleges to conduct agricultural research and education in the United States. Recently, public research investment has also been one regulatory response for developing safe pest management practices for small acreage crops. Khanna et al. see an important role for government in the direct support of basic research, development and education, as well as in the facilitation of entrepreneurship for the innovation of improved environmental technology. An ongoing challenge is to ensure that the level of public investment matches societal need when economic markets are absent and when there exists only the political marketplace for ideas.

The political arena is the one in which most institutional innovations must be made. Browne stresses that it is not enough for institutional innovations to be good in theory—they must also be robust enough to survive the political process. This creates a dynamic favoring win-win policies, as well as those without clear losers (like better labeling or product standards). Proposed policies that threaten well-organized interest groups, which include agricultural commodity groups, are unlikely to become law. Yet the same democratic openness that seems to drag radically innovative ideas toward centrist compromise also guards the promise of institutional innovation. An open, participatory political system, however cumbersome, offers the flexibility to give birth to new institutions.

Many complicated environmental problems demand cooperative solutions that come only with difficulty after deliberation, persuasion and negotiation (Randall). Of particular importance are public involvement programs. These and other new institutions continue to emerge from open political systems. As public appreciation grows for environmental sustainability, there is cause for hope that ongoing institutional innovation will foster the development of new environmental technologies for agriculture that are based on flexible incentives.

ENDNOTE

1. All authors who are mentioned and undated are authors of chapters in this volume.

REFERENCES

Hayami, Y. and V.W. Ruttan. 1985. *Agricultural Development: An International Perspective, Second Edition*, pp. 90–91. Baltimore, MD: Johns Hopkins University Press.
Porter, M.E. and C. van der Linde. 1995. "Green and Competitive: Ending the Stalemate." *Harvard Business Review* 73: 120–134.

INDEX